Statistical Tu

for

Johnson and Kuby's

Elementary Statistics

Ninth Edition

Statistical Tutor
for
Johnson and Kuby's
Elementary Statistics
Ninth Edition

Patricia Kuby
Monroe Community College

Australia • Canada • Mexico • Singapore • Spain • United Kingdom • United States

Printed in the United States of America
3 4 5 6 7 07 06 05 04

Printer: Phoenix Color Corp

ISBN: 0-534-39916-9

For more information about our products, contact us at:
Thomson Learning Academic Resource Center
1-800-423-0563

For permission to use material from this text, contact us by:
Phone: 1-800-730-2214
Fax: 1-800-730-2215
Web: http://www.thomsonrights.com

Brooks/Cole—Thomson Learning
10 Davis Drive
Belmont, CA 94002-3098
USA

Asia
Thomson Learning
5 Shenton Way #01-01
UIC Building
Singapore 068808

Australia/New Zealand
Thomson Learning
102 Dodds Street
Southbank, Victoria 3006
Australia

Canada
Nelson
1120 Birchmount Road
Toronto, Ontario M1K 5G4
Canada

Europe/Middle East/South Africa
Thomson Learning
High Holborn House
50/51 Bedford Row
London WC1R 4LR
United Kingdom

Latin America
Thomson Learning
Seneca, 53
Colonia Polanco
11560 Mexico D.F.
Mexico

Spain/Portugal
Paraninfo
Calle/Magallanes, 25
28015 Madrid, Spain

CONTENTS

PREFACE

This Statistical Tutor contains solutions for the odd number exercises in Elementary Statistics, 9th edition, as well as information and assistance. Included at the end of the manual are sections covering Introductory Concepts and Review Lessons on various algebraic and/or basic statistical concepts.
For each chapter, the following are provided:

1) Chapter Preview
2) Chapter Solutions
3) Student Annotations.

Student annotations are printed inside a border and placed before related exercises.
There are several ways for a student to use this manual. One possibility that has proved to be beneficial contains the following steps. Begin by reading the exercise under consideration in the text and attempting to answer the question. Locate the exercise number in the manual. If the exercise was answered, compare solutions. If solutions are not similar or difficulty has been encountered, read the boxed annotations before the exercise in detail.
Try the exercise again. If difficulty remains at this point, review the boxed annotations and the solution. Ask the instructor for aid or verification to get needed additional information.
In several exercises, a concluding question is asked to determine if the overall concept is comprehended. These questions are marked with an asterisk *. Locations of the corresponding answers are given and marked with an * and the exercise number.

SOLUTIONS AND STUDENT ANNOTATIONS

CHAPTER 1 ∇ STATISTICS

Chapter Preview

The purpose of Chapter 1 is to present:

1. an initial image of statistics that includes both the key role statistics has in the technical aspects of life as well as its everyday applicability,
2. its basic vocabulary and definitions,
3. basic ideas and concerns about the processes used to obtain sample data.

Articles reported in USA Today that are published by the U.S. Census Bureau in the Statistical Abstract of the United States, are presented in this chapter's case study.

CHAPTER 1 CASE STUDY

1.1 a. Americans

b. There are 6 Snapshots therefore 6 possible sets of answers.
Penny - each individual's preference to eliminate the penny: eliminate it, keep it, undecided
Excuses - each person's excuse to officer for committing traffic offense.
Tools - kind of the 'tool' each person used during a meeting.
Movie mania - each person's opinion about the state of movies: getting better, getting worse, undecided, and their age.
Road rage - the conditions surrounding each perpetrator's commission of road rage - time, day, light conditions, season, volume of traffic, type of road
Masters - ranking after each round, round score

c. There are many different possible answers. Here are a few of them.

Penny - 65% of those people sampled said they believed the penny should be kept in circulation.

Penny - 32% of those people sampled said they believed the penny should be eliminated from daily use.

Excuses - 59% of those people sampled claimed "they didn't see the sign."

Excuses - 25% of those people sampled begged the officer not to give them the ticket.

Excuses - 23% of those people sampled claimed "they were lost."

Tools - 53% of those people sampled said they used "pens, pencils and paper" during meetings.

Tools - 35% of those people sampled said they used an "erasable marker board" during meetings.

Road rage - "Prime time" means that more road rage offenses occurred between 4 and 6 PM on Friday than at any other time on any other day. The same is true for the other descriptive terms, more occurred during sunny times then non-sunny times and so on.

Jockeying - "Average ranking of the last 20 winners after 3rd round" - 3rd, the average rank position of the last 20 winners at conclusion of 3rd round.

Jockeying - "Average score each round" 4th round - 68.9, the average score recorded by all players during 4th round

d. Answer will vary according to each person's beliefs.

SECTION 1.1 ANSWER NOW EXERCISES

The article in Application 1.1 gives information about the sample (the number of people surveyed). Be watchful of articles that do not give any of this information. Sometimes not knowing something about the sample or survey size causes a question of credibility.

1.2 a. University students
b. 1,746
c. 82% like to snack at home
d. multiple answers

1.3 a. Employed American adults
b. U.S. Labor Department
c. 3,262,120 is the number of cashiers surveyed in 1996, $5.75 is the median(an average) hourly pay rate for the cashiers surveyed.

1.4 Answers will vary.

1.5 a. American travelers who take trips of more than 100 miles.
b. Typical long distance traveler is a 38 year old male whose income is $50,000 or more. He travels by car between 100 and 300 miles, usually to visit friends or relatives and stays at their homes.
c. Answers will vary; different gender, age, income, etc..

1.6 a. American adults
b. Within ± 4%
c. The actual percentage who rate the economic conditions as good could range from 50% to 58%.

1.7 Answers will vary.

SECTION 1.1 EXERCISES

Descriptive Statistics - refers to the techniques and methods for organizing and summarizing the information obtained from the sample.

Inferential Statistics - refers to the techniques of interpreting and generalizing about the population based on the information obtained from the sample.

1.9 a. descriptive b. inferential

1.11 a. USA drivers
b. 837
c. Has driver ever had to suddenly swerve, inadvertently speeded up, know someone who had crash while talking on cellphone.
d. 41% of those surveyed said they had inadvertently speeded up.
e. 0.41(837) = 343

1.13 a. 45%
b. The percentages are from different groups.

SECTION 1.2 ANSWER NOW EXERCISES

1.15 a. Took samples from various spots in the bay
b. New pollution laws enacted.

1.16 The population is all US adults. A sample is the 1200 randomly selected adults. The variable is "allergy status" for each adult. One data is "yes" or the actual allergy "dust" from one respondent. The data is the set of yes/no responses or the set of actual allergies/no allergy. The experiment is the method used to select the adults and wording of the allergy question. The parameter is the percent of all US adults who have an allergy, in this case, 36%. The statistic is the 33.2% based on the sampled adults.

1.17 Parameters give the information for the entire population and has one specific value. Statistics come from samples which can vary in size and method of data collection, therefore giving different measurements for each different sample.

1.18 a. Answers will vary.
b. Answers will vary.
c. Sample averages vary less for samples of size 10 - the larger sample size.

1.19 Possibilities: marital status, ZIP code, gender, highest level of education

1.20 Possibilities: annual income, age, distance to store, amount spent

1.21 Possibilities: marital status, gender, ZIP code

1.22 Possibilities: highest level of education, ranking of department preferences, rating for first impression of store

1.23 a. Score can only be whole numbers (scores are counted).
b. Number of minutes (a length of time) can be any value (time is measured); its accuracy depends on the precision with which it is measured.

1.24 a. Gender, nominal
b. Height, continuous

1.25 a. Residents of Lee County, Florida
b. Households of Lee County, residents, registered adults
c. Income, age, political affiliation
d. Income, continuous; age, continuous; political affiliation, nominal
e. The collected data was counted by categories. The counts were divided by total count to find percentages reported on the Household Income and Age circle graphs. The counts were used to create the bargraph for Political Party.

SECTION 1.2 EXERCISES

A variable is the characteristic of interest (ex. height), where data is a value for the variable (ex. 5'5"). A variable varies (heights vary), that is, heights can take on different values. Data (singular) such as 5'5" (one person's height) is constant; it does not change in value for a specific subject.

An attribute variable can take on any qualitative or "numerical" qualitative information (ex. kinds of fruit, types of music, religious preference, model year - most answers are in words, although model year would have "numerical" answers such as "2003"). An attribute variable can be nominal (description or name) or ordinal (ordered position or rank; first, second,…).

A numerical variable can take on any quantitative information. This includes any count-type and measurable-type data (ex. number of children in a family, amount of time, age, height, area, volume, miles per gallon). A numerical variable can be discrete or continuous. The domain of a discrete variable has gaps between the possible values; there are numerical values that cannot occur. Theoretically, the domain of a continuous variable has no gaps since all numerical values are possible. Do not be confused by data that has been rounded due to scale being used or for convenience reasons.

1.27 a. Severity of side-effects from a particular medicine for a patient
b. attribute(ordinal)

Population - the collection of all individuals, objects, or scores whose properties are under consideration.

Parameter - a number calculated from the population of values.

Sample - that part of the population from which the data values or information is obtained.

Statistic - a number calculated from the sample values.

NOTE: Parameters are calculated from populations; both begin with *p*.
Statistics are calculated from samples; both begin with *s*.

1.29 a. All individuals who have hypertension and use prescription drugs to control it (a very large group)
b. The 5,000 people in the study
c. The proportion of the population for which the drug is effective
d. The proportion of the sample for which the drug is effective, 80%
e. No, but it is estimated to be approximately 80%

1.31 a. All assembled parts from the assembly line
b. infinite c. the parts checked
d. attribute, attribute (it identifies the assembler), numerical

1.33 a. The population being surveyed is composed of all people suffering from migraine headaches.
b. The sample is the 2,633 people given the drug.
c. The variables are the amount of drug dosage and the side effects encountered.
d. The data is both quantitative and qualitative. The dosage amounts are quantitative. The observed degree of relief from migraines and the type of side effects experienced are qualitative.

1.35 a. numerical b. attribute c. numerical
d. attribute e. numerical f. numerical

1.37 a. The population contains all objects of interest, while the sample contains only those actually studied.
b. convenience, availability, practicality

SECTION 1.3 EXERCISES

1.39 Group 2, the football players, because their weights cover a wider range of values, probably 175 to 300+, while the cheerleaders probably all weigh between 110 and 150.

1.41 By using a standard weight or measure in conjunction with money, prices between competing product brands can be more easily compared, irrespective of purchase quantity. There is a great deal of variability in container sizes between brands and even within brands of the same product. Problems associated with this variability are simplified by showing the standard unit price in addition to the cash register amount at the point of sale.

1.43 a. Answers will vary.
b. Answers will vary.
c. The smaller sample size of 4 shows more variability.
d. The sample size of 10 demonstrated the population average more accurately. There was less variability in the averages.

SECTION 1.4 ANSWER NOW EXERCISES

1.44 Answers will vary but should include reference to an economic bias.

1.45 Volunteer; yes, only those with strong opinions will respond.

1.46 Observational; the experimenter can not control the weather.

1.47 a. Biased, the results were not representative.
b. Experiment

1.48 Landers' survey was a volunteer survey, therefore there is a bias - mostly, only those with strong opinions will respond.

1.49 U.S. Senate

1.50 U.S. House of Representatives

1.51 Election day polls are taken from voting precincts selected because they are believed to be representative of all voters. Each precinct is a cluster and not all precincts are sampled.

1.52
a. American children
b. 16,202 children of participants in Nurses' Health Study II
c. Age, frequency of family dinner, home food, fast food, physical activity, team sports, television
d. Percent of family dinners for each age.
e. No, the sampling was not done randomly. Certain quantities were desired for each age group.
f. Convenience sample.

1.53
a. Answers will vary.
b. Answers will vary.
c. Answers will vary.
d. Answers will vary.

SECTION 1.4 EXERCISES

A convenience sample or volunteer sample, as indicated by their very names, can often result in biased samples.

Data collection can be accomplished with experiments (the environment is controlled) or observational studies (environment is not controlled). Surveys fall under observational studies.

Sample designs can be categorized as judgement samples (believed to be typical) or probability samples (certain chance of being selected is given to each data value in the population).

The random sample (each data value has the same chance) is the most common probability sample.

Methods (simply defined) to obtain a random sample include:

1. Random Number Table - (see Introductory Concepts in the Statistical Tutor)
2. Systematic - every kth element is chosen
3. Stratified - fixed number of elements from each strata (group)
4. Proportional (Quota) - number of elements from each strata is determined by its size
5. Cluster - fixed number or all elements from certain strata.

1.55 a. Observational or volunteer

b. Yes, includes only those people who have internet access, time to answer and, probably strong opinions on the subject.

1.57 a. The set (list) from which a sample is actually drawn

b. A computer list of the full-time students

c. Random Number Table; student numbered 1288 was selected for the sample.

1.59 Judgment sampling - the distributor selected stores in areas he believed would be receptive.

1.61 probability samples

1.63 convenience sampling

1.65 A proportional sample would work best since the area is already divided into 35 (different size) listening areas. The size of the listening area determines the size of the subsample. The total for all subsamples would be 2500.

1.67 Not all adults are registered voters.

SECTION 1.5 ANSWER NOW EXERCISES

1.68 a. probability b. statistics

SECTION 1.5 EXERCISES

Statistics - allows you to make inferences or generalizations about the population based on a sample.

Probability - the chance that something will happen when you know **all** the possibilities (you know all the possibilities when you have the population).

1.69 a. statistics b. probability
c. statistics d. probability

SECTION 1.6 EXERCISES

1.71 Several large comprehensive computer programs (called statistical packages) have been developed that perform many of the computations and tests you will study in this text. In order to have the statistical package perform the computations, you simply enter the data into the computer and the computer does the rest on command. Quickly and easily

1.73 The computer is a remarkable calculating machine, but it is incapable of judging whether or not the data it is working with can be used to assess the truth. Computers can not determine whether or not a study has been conducted properly, or whether the appropriate methodology is being used. The power and speed of the computer in performing calculations needed to analyze data often temps researchers to perform calculations that would never have been performed without careful planning and consideration.

RETURN TO CHAPTER CASE STUDY

1.74 a. Americans, yes
b. Movie- Opinion on the current state of movies.
c. 53% of those people sampled said they used "pens, pencils and paper" during meetings.
d. Answers will vary.

CHAPTER EXERCISES

1.77 Each student's answers will be different. A few possibilities are:
a. color of hair, major, gender, marital status
b. number of courses taken, number of credit hours, number of jobs, height, weight, distance from hometown to college, cost of textbooks

1.79 a. T = 3 is a piece of data - a value from one person.
b. What is the average number of times per week the people in the sample went shopping?
c. What is the average number of times per week that people (all people) go shopping?

1.81 a. All people in the US who died in 1997
b. death from heart disease, state of residence, age, obesity, inactivity
c. mortality rate per 100,000 people in the US
d. death from heart disease - attribute
state of residence - attribute
age at death - numerical
obesity - attribute
inactivity - attribute

1.83 a. Young men who do not go to college and take restaurant jobs.
b. The 2,000 individuals in the study.
c. Most likely, the sample was a judgment sample since "young" was never defined.

1.85 Each will have different examples.

1.87 Each will have different examples.

CHAPTER 2 ∇ DESCRIPTIVE ANALYSIS AND PRESENTATION OF SINGLE-VARIABLE DATA

Chapter Preview

Chapter 2 deals with the presentation of data that were obtained through the various sampling techniques discussed in Chapter 1. The four major areas for presentation and summary of the data are:

1. graphical displays,
2. measures of central tendency,
3. measures of variation, and
4. measures of position.

A study supported by the Stanford Institute for the Quantitative Study of Society on how people utilize the internet is presented in this chapter's case study.

CHAPTER 2 CASE STUDY

2.1 a. Answers will vary. Possibilities might include: sort data, frequency of each value, average.

b. Answers will vary. Possibilities might include: location of my data value with respect to the other values, proximity to average.

SECTION 2.1 ANSWER NOW EXERCISES

2.2

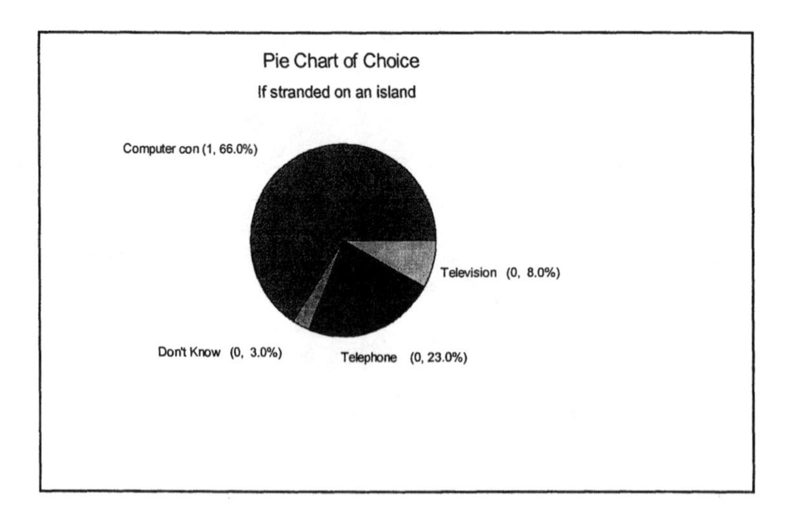

2.3

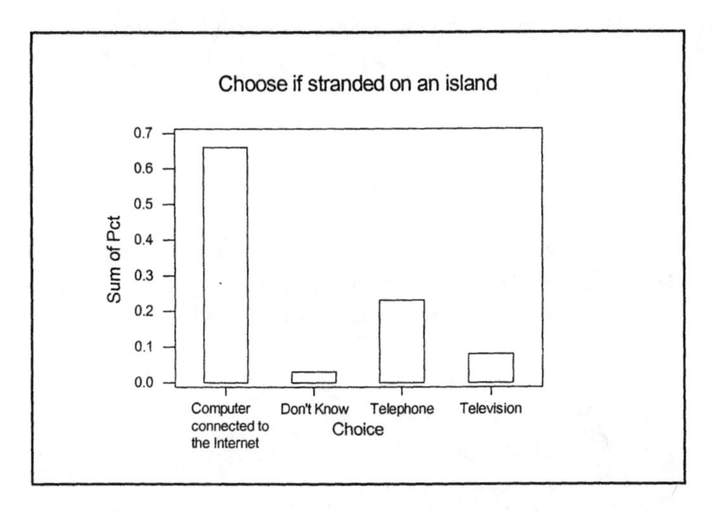

2.4 The circle graph does a better job of representing the relative proportions of the answers to the group as a whole, while the bar graph is more dramatic in representing the relative proportions between the individual answers.

2.5

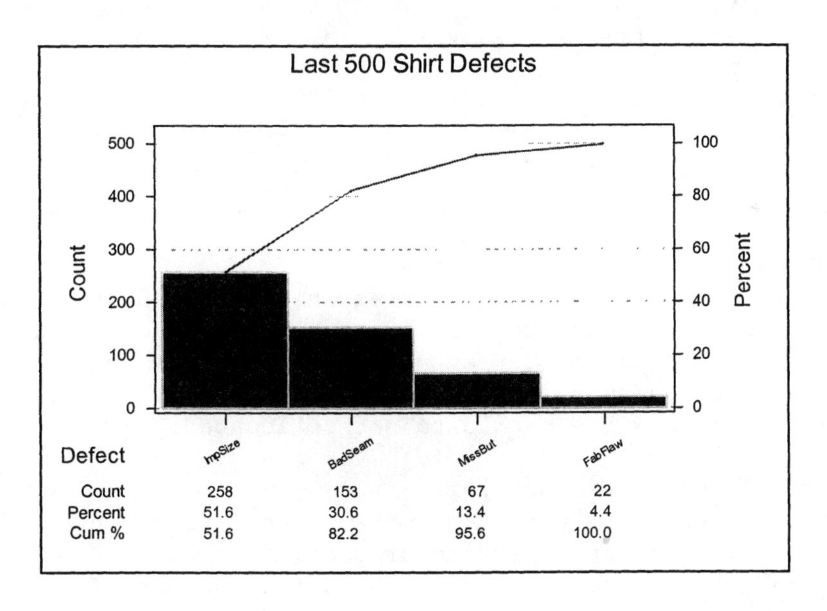

2.6 Points Scored per Game by Basketball Team

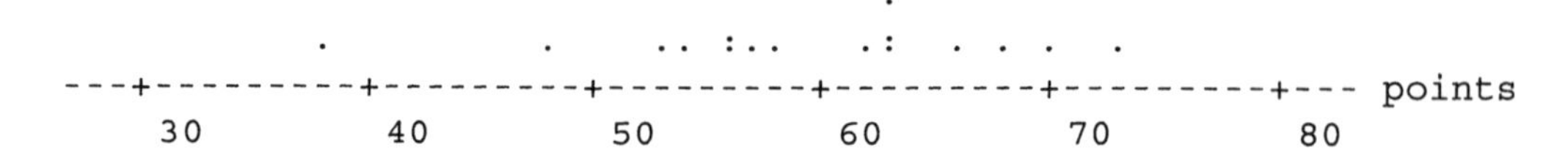

2.7 96: leaf of 6 is placed on the 9 stem
66: leaf of 6 is placed on the 6 stem

2.8 Points scored per game

3	6
4	6
5	6 4 5 4 2 1
6	1 1 8 0 6 1 4
7	1

2.9 Each leaf value is in the ones position of the complete data value. Leaf is 1.0 units wide. For example, with 9|8, the 8 represents the value 8, as in 98.

SECTION 2.1 EXERCISES

MINITAB - Statistical software
Data is entered by use of a spreadsheet divided into columns and rows. Data for each particular problem is entered into its own column. Each column represents a different set of data. Be sure to name the columns in the space provided above the first row, so that you know where each data set is located. (C1 = Column 1)

EXCEL - Spreadsheet software
Data is entered by use of a spreadsheet divided into columns and rows. Data for each particular problem is entered into its own column. Each column represents a different set of data. If needed, use the first row of a column for a title. (A1 = 1st cell of column A)

. . .

TI-83 - Graphing calculator
Data is entered into columns called lists. Data for each particular problem is entered into its own list. Each list represents a different set of data. If needed, use the space provided above the first row for a title. Lists are found under STAT > 1:Edit.
(L1 = List 1)

Partitioning the circle:

1. Divide all quantities by the total sample size and turn them into percents.
2. 1 circle = 100%
   1/2 circle = 50%
   1/4 circle = 25%
   1/8 circle = 12.5%
3. Adjust other values accordingly.
4. Be sure that percents add up to 100 (or close to 100, depending on rounding).

Computer and calculator commands to construct a Pie Chart can be found in ES9-p40.
The TI-83 program 'CIRCLE' and others can be downloaded from the website www.duxbury.com . Select 'Online Book Companions' followed by 'Student Resources' for 'Elementary Statistics, 9th edition'. Select 'IT-83 Programs' under 'Course Resources'. Further details on page 40 in ES9.
*(ES9 denotes the textbook 'Elementary Statistics, 9th edition')

2.11 a.

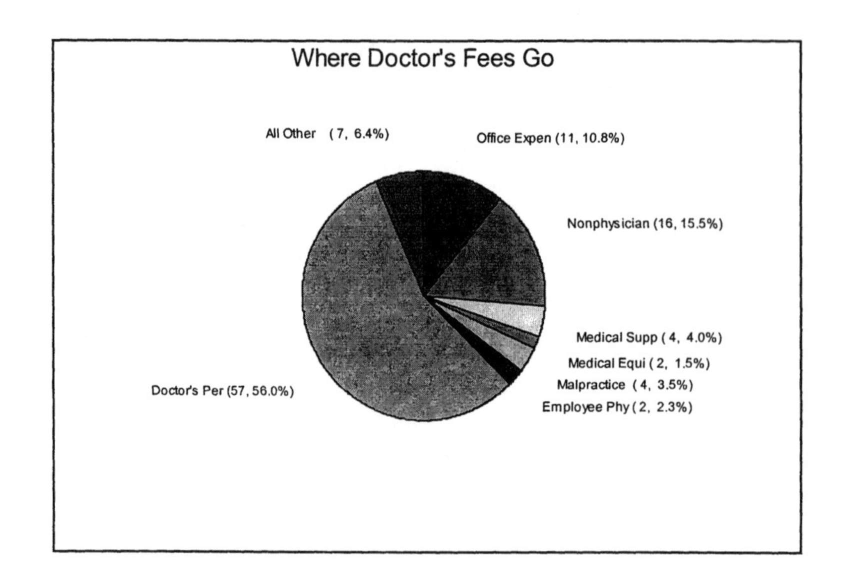

MINITAB commands to construct a bar graph can be found in the GRAPH pull-down menu under CHART. Input the frequencies into C1 and their corresponding categories into C2. EXCEL commands to construct a bar graph can be found under Chart Wizard. Input the categories into column A and the corresponding frequencies into column B. TI-83 commands to construct a bar graph can be found using the STATPLOT command. Input the categories as numbers into list 1 and the corresponding frequencies into list 2. Adjust the x-scale in the WINDOW to 0.5 to allow for gaps between bars.

NOTE: Bar graphs may be vertical (as shown below) or horizontal (as shown in the Chapter Case Study). Information typically on the axes may also be printed inside the bars themselves. There is much flexi-bility in constructing a bar graph. Remember to leave a space between the bars. Be sure to label both axes and give a title to the graph.

b.

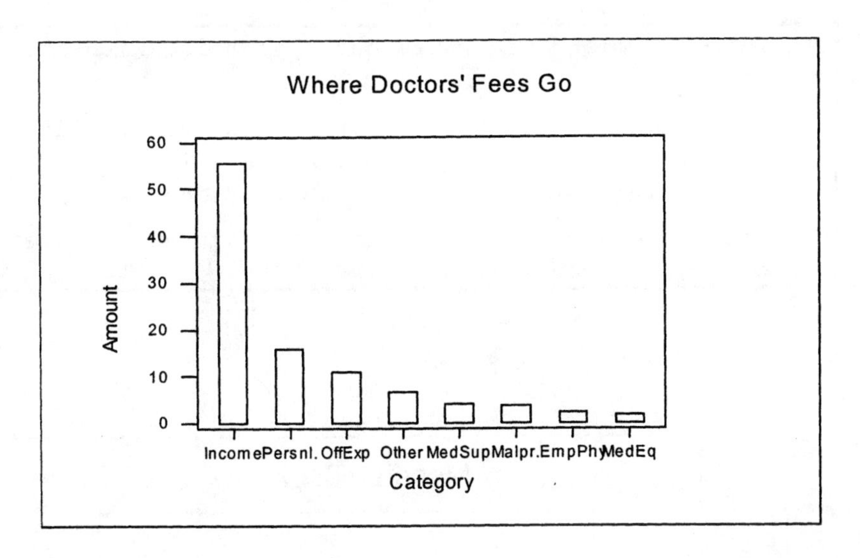

c. The circle graph makes it easy to visually compare the relative sizes of the parts to each other and the size of each part to the whole. The bar graph makes it easy to visually compare the sizes of the parts to each other, but the size of each part relative to the whole is not as obvious. Therefore, most people are able to "read" more from the circle graph.

2.13 a.

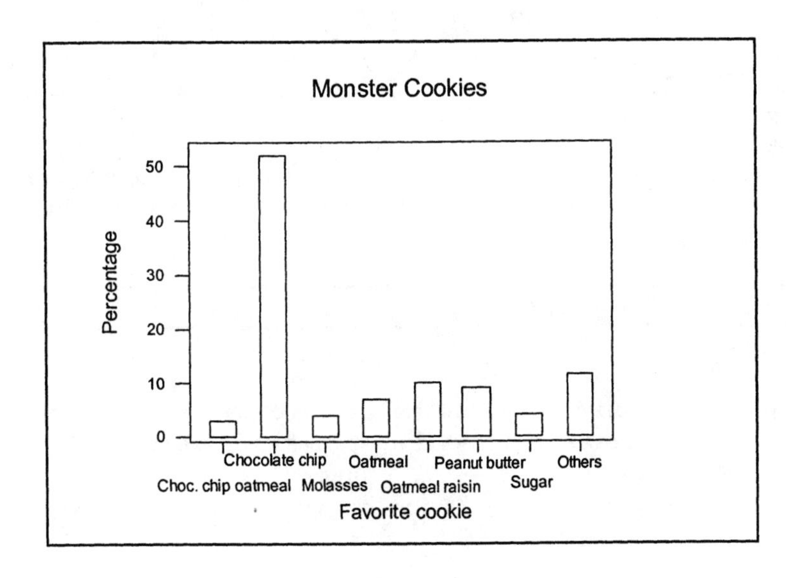

The Pareto command generates bars, starting with the largest category.

NOTE: Pareto diagrams are primarily used for quality control applications and therefore MINITAB's PARETO command identifies the categories as "Defects", even when they may not be defects.

Computer and calculator commands to construct a Pareto diagram can be found in ES9-pp41&42.

b.

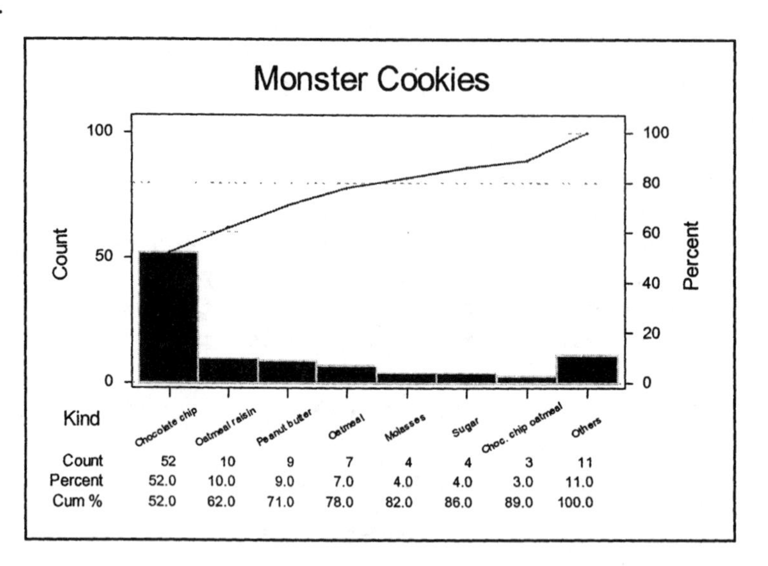

c. They should stock the 4 most popular: Chocolate chip, Oatmeal raisin, Peanut butter, Oatmeal. The Pareto list them first.

Frequency = (total #)(percentage)

d. Approximately 300 multiplied by each percentage: Chocolate chip - 156, Oatmeal raisin - 30, Peanut butter - 27, Oatmeal - 21, Sugar - 12, Molasses - 12, Chocolate chip oatmeal - 9, Others - 33.

2.15 a.

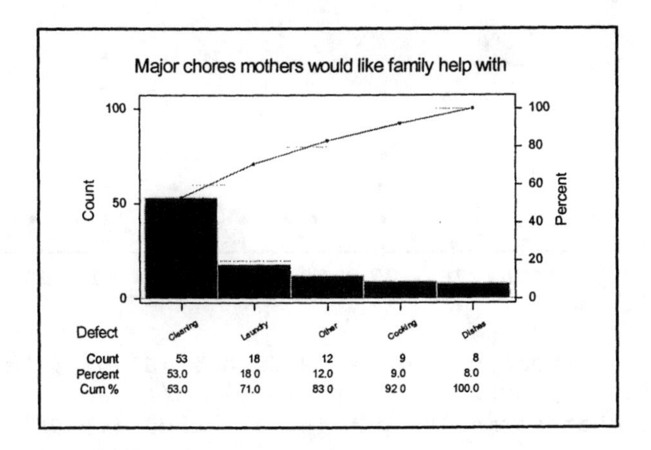

b. The "Other" category is too large, it is a collection of several answers and as such is larger than two of the categories. If it were broken down, then the Pareto diagram would have the categories in order of mother's wishes.

Picking increments (spacing between tick marks) for a dot plot

1. Calculate the spread (highest value minus the lowest value).
2. Divide this value by the number of increments you wish to show (no more than 7 usually).
3. Use this increment size or adjust to the nearest number that is easy to work with (5, 10, etc.).

Computer and calculator commands to construct a dotplot can be found in ES9-pp43&44. Commands for multiple dotplots are also in ES9-pp47-49.

2.17 a.

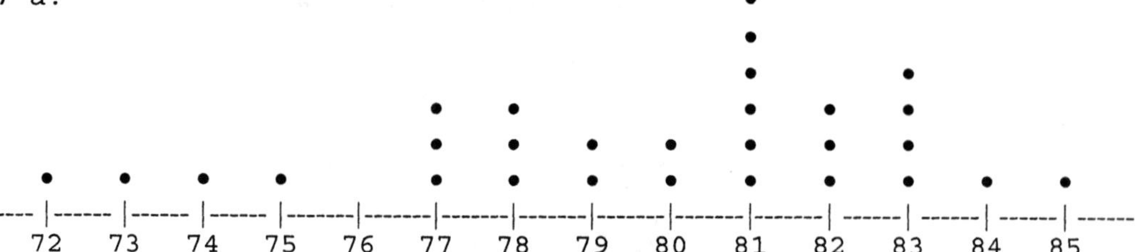

b. The shortest player is 72 inches (6′-0″), and the tallest player is 85 inches (7′-1″).

c. The most common height is 81 inches (6′-9″), and six players were listed at that height.

d. Repeated values show up on the dotplot as points stacked-up. Therefore, the most common height would be represented by the "tallest" stack of points.

2.19 a. 15

b. 11.2, 11.2, 11.3, 11.4, 11.7

c. 15.6

d. 13.7; 3

STEM-AND-LEAF DISPLAYS

1. Find the lowest and highest data values.
2. Decide in what "place value" position, the data values will be split.
3. Stem = leading digit(s)
4. Leaf = trailing digit(s) (if necessary, data is first "rounded" to the desired position)
5. Sort stems and list.
6. Split data values accordingly, listing leaves on the appropriate stem.

Computer and calculator commands to construct a stem-and-leaf display can be found in ES9-pp45&46.

2.21 a.

Stem-and-Leaf Display:
Profitability last 12 mos (pct)
N = 23
Leaf Unit = 0.10

```
  1    10 4
  3    11 37
  8    12 25589
 (4)   13 4689
 11    14 3688
  7    15 17
  5    16
  5    17 9
  4    18 247
  1    19
  1    20 2
```

The column on the left of the stem-and-leaf display is the cumulative count of the data from the top (low-value) down and the bottom (high-value) up until the class containing the median is reached. The number of data values for the median class is in parentheses.

b. The distribution is skewed to the right and appear to be bimodal with a larger cluster centered around 12, 13 and 14 percent, with a smaller cluster centered at 18 percent.

2.23 a. The place value of the leaves is in the hundredths place; i.e., 59|7 is 5.97.

b. 16

c. 5.97, 6.01, 6.04, 6.08

d. Cumulative frequencies starting at the top and the bottom until it reaches the class that contains the median. The number in parentheses is the frequency for just the median class.

SECTION 2.2 ANSWER NOW EXERCISES

2.25

x	f
0	2
1	5
2	3
3	0
4	2
	12

2.26 a. f is frequency, therefore values of 70 or more but less than 80 occurred 8 times.

b. $1 + 3 + 8 + 5 + 2 = 19$

c. The sum represents the sum of all the frequencies, which is the number of data, or the sample size

2.27 a. Time spent cleaning is represented by sectors (pie-shaped piece) of the circle

b. The relative frequency is represented by the size of the angle forming the sector.

c. $2 \leq x < 4$

d.

$0 \leq x < 1$	0.05
$1 \leq x < 2$	0.20
$2 \leq x < 4$	0.33
$4 \leq x$	0.39
don't know	0.03

2.28 a. Class #4; $65 \leq x < 75$

b. This class contains all values greater than or equal to 65 and also less than 75(does not include 75).

c. Difference between upper and lower class boundaries.

i. Subtracting the lower class boundary from the upper class boundary for any one class

ii. Subtracting a lower class boundary from the next consecutive lower class boundary

iii. Subtracting an upper class boundary from the next consecutive upper class boundary

iv. Subtracting a class mark from the next consecutive class mark.

2.29

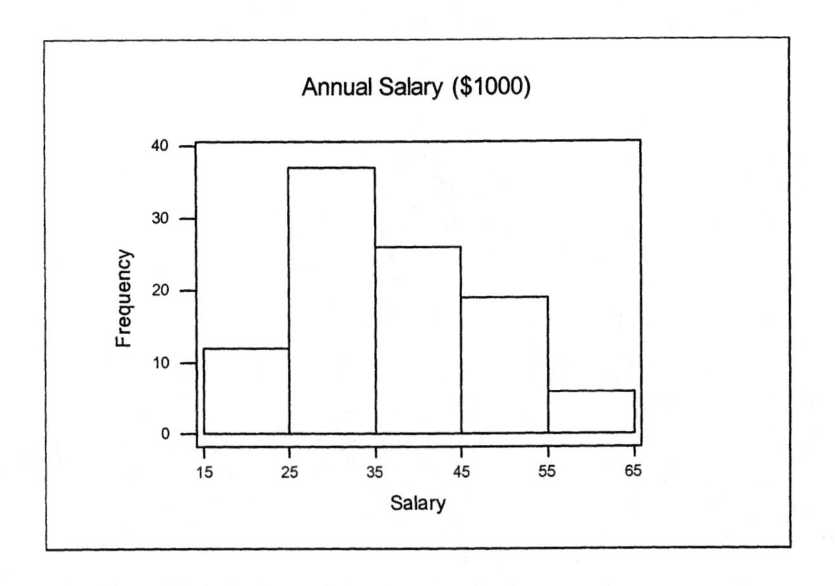

2.30 The shapes are the same but Figure 2.10 uses class midpoints on the horizontal scale, whereas Figure 2.11 uses the class boundaries. Figure 2.10 uses frequency(count) on the vertical axis, whereas Figure 2.11 has turned those counts into percentages.

2.31 Similarities: same classes of data, same shape, Differences: vertical vs. horizontal, individual data points vs. grouped

2.32 Symmetric: weight of dry cereal per box, breaking strength of certain type of string Uniform: rolling a die several hundred times Skewed Right: salaries, high school class sizes Skewed left: hour exam scores Bimodal: heights, weights for groups containing both male and female J-shaped: amount of television watched per day

2.33 a. uniform
b. J-shaped
c. skewed right

2.34

Class Boundaries	Cumulative Frequency
$15 \le x < 25$	12
$25 \le x < 35$	49
$35 \le x < 45$	75
$45 \le x < 55$	94
$55 \le x \le 65$	100

2.35

Class Boundaries	Cum. Rel. Frequency
$15 \le x < 25$	0.12
$25 \le x < 35$	0.49
$35 \le x < 45$	0.75
$45 \le x < 55$	0.94
$55 \le x \le 65$	1.00

2.36

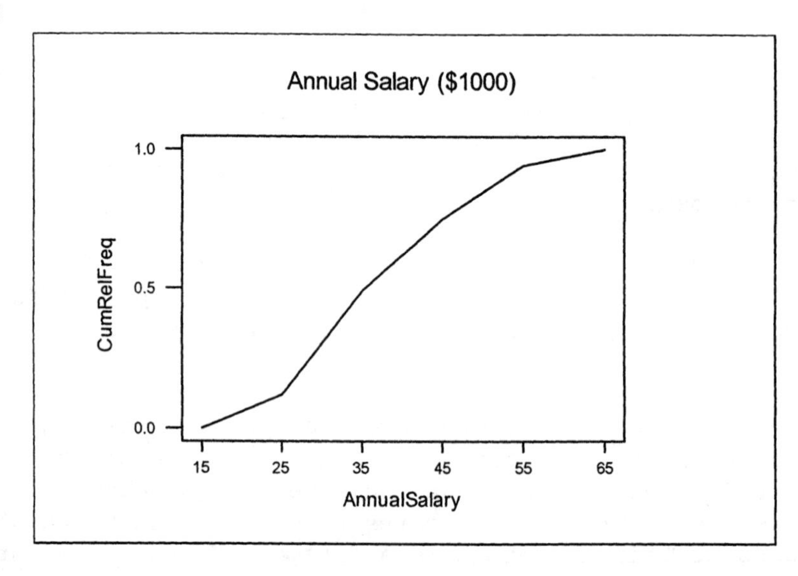

2.37 a. Lower left, upper right   b. Upper left, lower right

SECTION 2.2 EXERCISES

Frequency distributions can be either grouped or ungrouped. Ungrouped frequency distributions have single data values as x values. Grouped frequency distributions have intervals of x values, therefore, use the class midpoints (class marks) as the x values.

Histograms can be used to show either type of distribution graphically. Frequency or relative frequency is on the vertical axis. Be sure the bars touch each other (unlike bar graphs). Increments and widths of bars should all be equal. A title should also be given to the histogram.

Computer or calculator commands to construct a histogram can be found in ES9pp58-60. Note the two methods, depending on the form of your data.
The Student Suite CD has a video clip: "Frequency Distributions & Histograms".

2.39 a.

AP Score	Frequency
1	3
2	18
3	9
4	3
5	4

b.

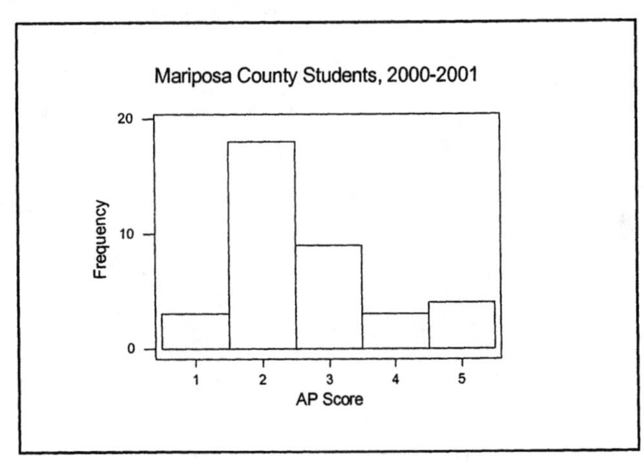

c.

AP Score	Rel. Freq.
1	0.081
2	0.487
3	0.243
4	0.081
5	0.108

d. 0.243 + 0.081 + 0.108 = 0.432 = 43.2%

2.41 a.

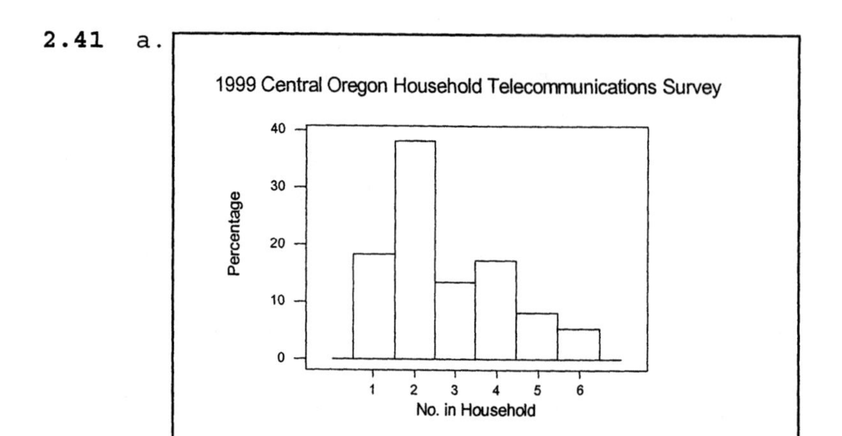

b. Mounded, skewed right

c. Two is the most common number of people per household, with the vast majority of the households being composed of between 1 and 4 people.

Relative frequency = $\frac{\text{frequency}}{\text{sample size}}$

2.43 a.

age	frequency
17	1
18	3
19	16
20	10
21	12
22	5
23	1
24	2
	50

b.

age	rel.freq.
17	0.02
18	0.06
19	0.32
20	0.20
21	0.24
22	0.10
23	0.02
24	0.04
	1.00

d.

age	cum.rel.freq.
17	0.02
18	0.08
19	0.40
20	0.60
21	0.84
22	0.94
23	0.96
24	1.00

CHECKS: sum of frequencies = sample size
sum of relative frequencies = 1.00

c.

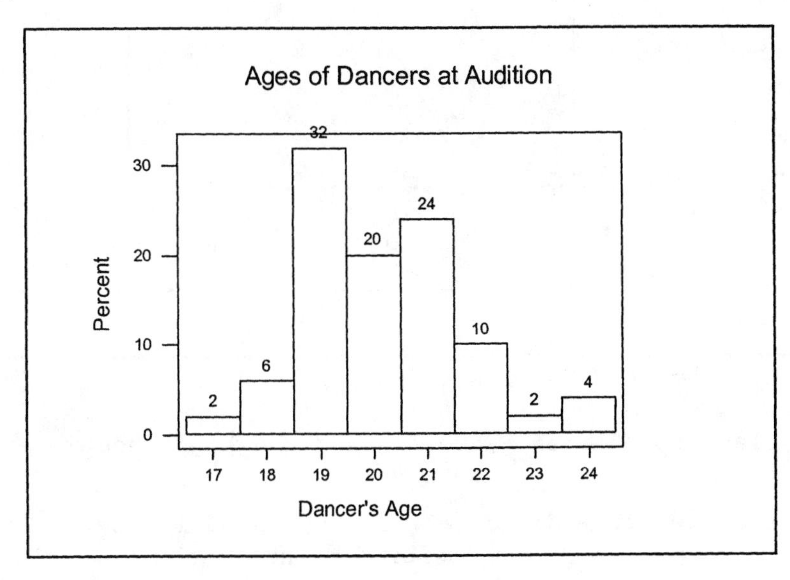

An ogive is a line graph of a cumulative frequency or cumulative relative frequency distribution. Start the line at zero for a class below the smallest class. Plot the upper class boundary points from the remaining values of the cumulative (relative) frequency distribution. Connect all of the points with straight line segments. The last point (class) is at the value of one (vertically).

Computer and calculator commands to construct an ogive can be found in ES9pp64&65.

e.

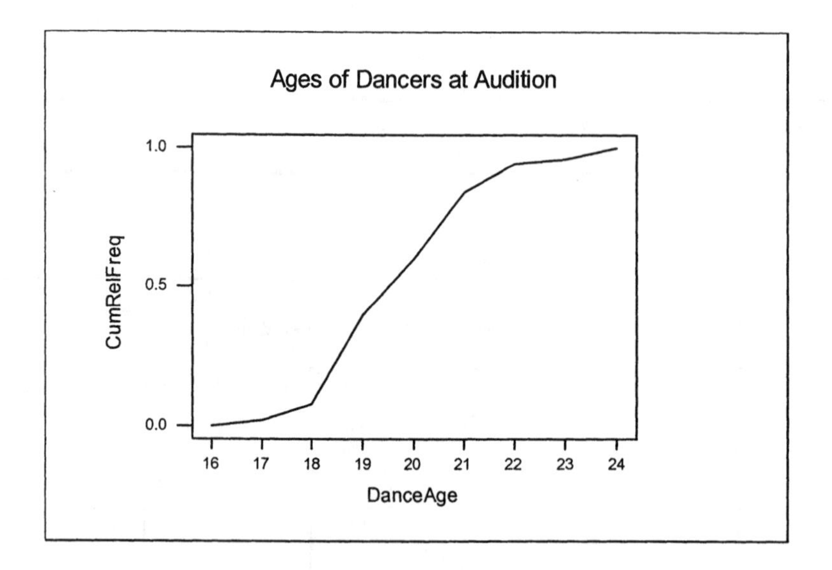

Parts of a grouped frequency distribution -

class boundaries = the low and the high endpoints of the interval

class width = distance from any point in one class to the same position point in the next class or the difference between the upper and lower class boundaries

class midpoint (mark) = (lower boundary + upper boundary)/2, midpoint of the interval

. . .

Example: with respect to the second class interval

30 - 40	form: $(40 \le x < 50)$
40 - 50	lower class boundary = 40
50 - 60	upper class boundary = 50
	class width = 50 - 40 = 10
	class midpoint = (40 + 50)/2 = 45

2.45 a. 12 and 16

b. 2, 6, 10, 14, 18, 22, 26

c. 4.0

d. 0.08, 0.16, 0.16, 0.40, 0.12, 0.06, 0.02

e.

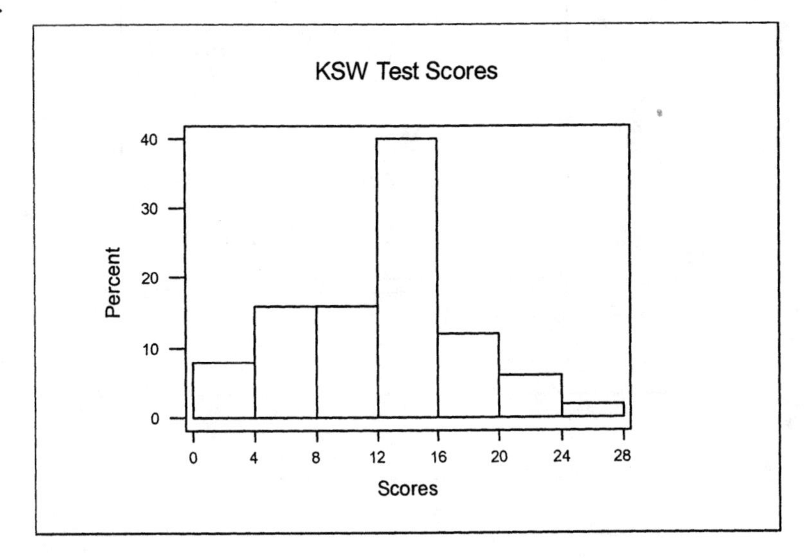

Refer to frequency distribution information before exercise 2.45 if necessary. Either class boundaries or class marks, may be used to determine increments along the horizontal axis for histograms of grouped frequency distributions.

2.47 a.

Class limits	frequency
12 - 18	1
18 - 24	14
24 - 30	22
30 - 36	8
36 - 42	5
42 - 48	3
48 - 54	2

b. class width = 6

c. class midpoint = (24+30)/2
= 27

lower class
boundary = 24

upper class
boundary = 30

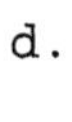

d.

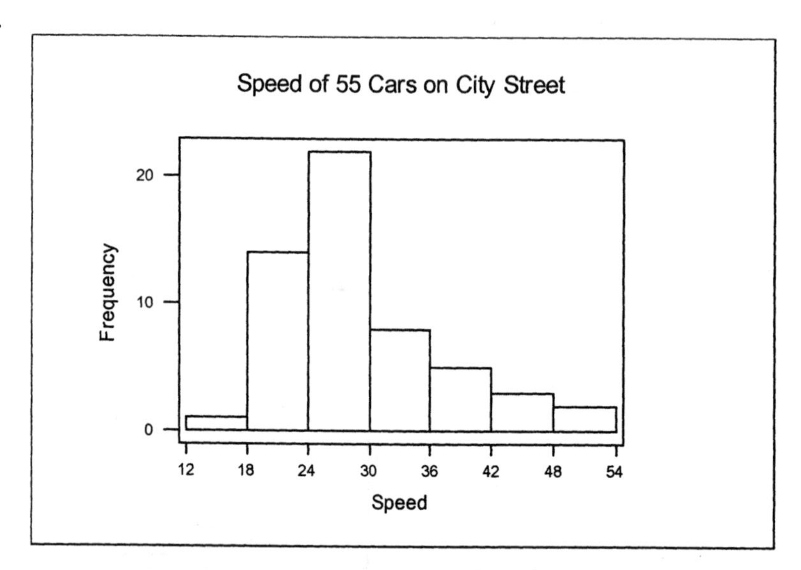

2.49 a.

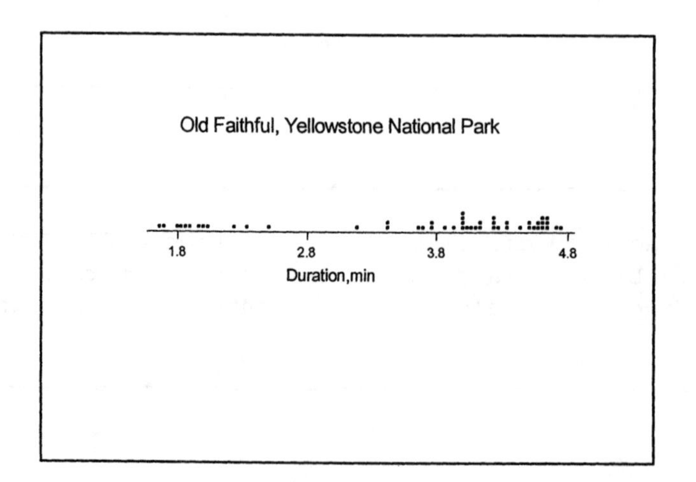

b.

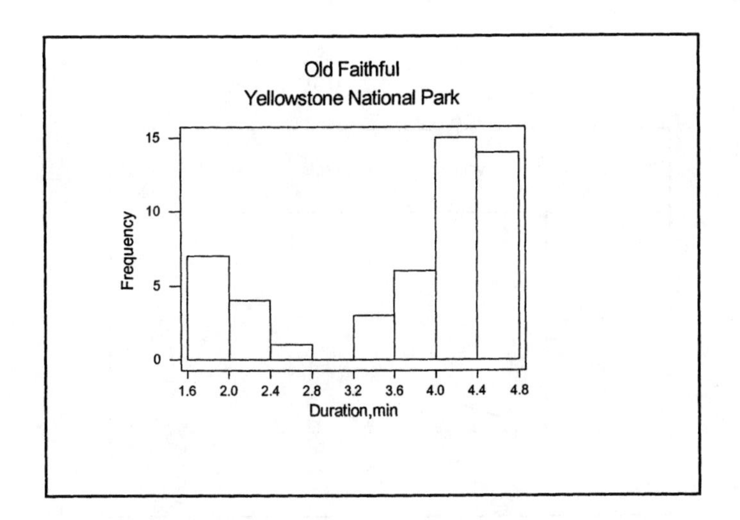

A guideline that can be used for selecting the number of classes is : n(classes) $\approx \sqrt{n}$

c. and d. The histograms will vary as class boundaries and class widths are changed.

e.

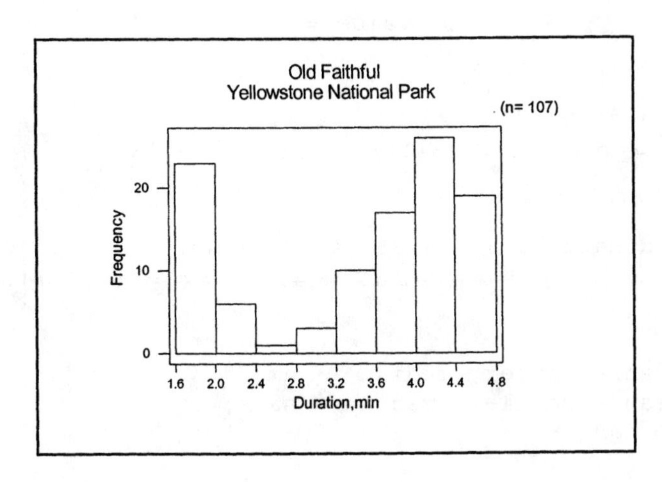

f. Answers will vary. Histogram is easier to read.
g. Answers will vary.

2.51 a.

AP Score	Cum. Rel. Freq.
1	0.081
2	0.568
3	0.811
4	0.892
5	1.000

b.

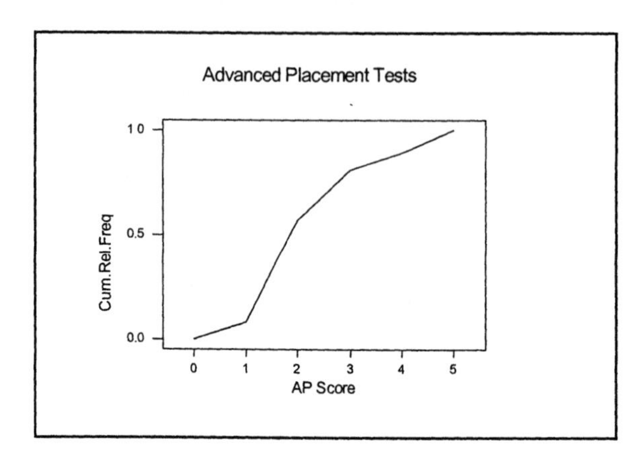

SECTION 2.3 ANSWER NOW EXERCISES

2.53 $\bar{x} = \sum x/n = (1+2+1+3+2+1+5+3)/8 = 18/8 = \underline{2.25}$

2.54 a. 9 b. value = 0

2.55 Ranked data: 70, 72, 73, 74, 76
$d(\tilde{x}) = (n+1)/2 = (5+1)/2 = 3\text{rd}; \ \tilde{x} = \underline{73}$

2.56 Ranked data: 4.15, 4.25, 4.25, 4.50, 4.60, 4.60, 4.75, 4.90
$d(\tilde{x}) = (n+1)/2 = (8+1)/2 = 4.5\text{th}; \ \tilde{x} = (4.50+4.60)/2 = \underline{4.55}$

2.57 a. mean - larger, median - same
b. mean - smaller, median - same;
c. median

2.58 mode = $\underline{2}$

2.59 midrange = (L+H)/2 = (0.20+10.76)/2 = 10.96/2 = <u>$5.48</u>

2.60 $\bar{x} = \sum x/n = (9+6+7+9+10+8)/6 = 49/6 = 8.166 = \underline{8.2}$
Ranked data: 6, 7, 8, 9, 9, 10
$d(\tilde{x}) = (n+1)/2 = (6+1)/2 = 3.5$th; $\tilde{x} = \underline{8.5}$
mode = $\underline{9}$
midrange = (L+H)/2 = (6+10)/2 = 16/2 = $\underline{8.0}$

2.61 a. Ranked data:

25,500	31,500	31,500	31,500	31,500
35,250	36,750	37,500	39,000	54,000

$\bar{x} = \sum x/n = 354{,}000/10 = \underline{35{,}400.00}$
$d(\tilde{x}) = (n+1)/2 = (10+1)/2 = 5.5$th; $\tilde{x} = \underline{33{,}375}$
mode = $\underline{31{,}500}$
midrange = (H+L)/2 = (54,000+25,500)/2 = $\underline{39{,}750}$
The values of these statistics all agree with those in the article.

b. The large value of 54,000 is pulling the mean and midrange towards the larger data value.

c. The higher figures exceed the mean by a total of $25,650.

$54,000 - $35,400 = 18,600
$39,000 - $35,400 = 3,600
$37,500 - $35,400 = 2,100
$36,750 - $35,400 = 1,350
25,650

The lower figures fall short of the mean by a total of $25,650.

$35,400 - $35,250 = 150
$35,400 - $31,500 = 3900
$35,400 - $31,500 = 3900
$35,400 - $31,500 = 3900
$35,400 - $31,500 = 3900
$35,400 - $25,500 = 9900
$25,650

2.62 a. Time of strike
b. One-hour interval
c. Most strikes occur in early afternoon
d. Height of graph for those intervals

SECTION 2.3 EXERCISES

NOTE: A <u>measure of central tendency</u> is a value of the variable. It is that value which locates the "average" value for a set of data. The "average" value may indicate the "middle" or the "center" or the most popular data value.

NOTATION AND FORMULAS FOR MEASURES OF CENTRAL TENDENCY

$\sum x$ = sum of data values
n = # of data values in the sample
$\bar{x}$ = sample mean = $\sum x/n$
$\tilde{x}$ = sample median = middle data value
$d(\tilde{x})$ = depth or position of median = $(n + 1)/2$
mode = the data value that occurs most often
midrange = (highest value + lowest value)/2

NOTE: REMEMBER TO RANK THE DATA BEFORE FINDING THE MEDIAN.
$d(\tilde{x})$ only gives the depth or position, not the value of the median. If n is even, $\tilde{x}$ is the average of the two middle values.

See Introductory Concepts (ST*-p303) for additional information about the Σ (summation) notation.

Computer and calculator commands to find the mean and median can be found in ES9-pp69&72 respectively.
The Student Suite CD has a video clip: "Descriptive Statistics".
*(ST denotes this manual, 'Statistical Tutor')

2.63 The data resulting from a quantitative variable are numbers with which arithmetic (addition, subtraction, etc.) can be performed. The data resulting from a qualitative variable is 'category' type value such as color. It is not possible to add three colors together, and divide by 3, to obtain a value for the mean color.

2.65 a. $\bar{x} = \sum x/n = (3+5+6+7+7+8)/6 = 36/6 = \underline{6.0}$

b. $d(\tilde{x}) = (n+1)/2 = (6+1)/2 = 3.5\text{th}; \quad \tilde{x} = (6+7)/2 = \underline{6.5}$

c. mode = $\underline{7}$

d. midrange = $(H+L)/2 = (8+3)/2 = 11/2 = \underline{5.5}$

2.67 a. $\sum x = 423.00, \quad \bar{x} = 32.538 = 32.5$ hours

b. Ranked data:

24.6	26.9	28.4	29.3	30.5	32.7	33.3	33.4
33.5	34.7	34.9	36.4	44.4			

$d(\tilde{x}) = (13+1)/2 = 7^{th}; \quad \tilde{x} = 33.3$ hours

c. $(24.6 + 44.4)/2 = 34.5$

d. no mode

2.69 a. The only "average" that can be found with this information is the midrange:

midrange = ($3,092+$1,292)/2 = $2,192

b. The only information given are the two extreme values, exactly the information needed to find the midrange.

c. The distribution is most likely skewed to the right, or towards the larger values.

2.71 a. Third Graders at Roth Elementary School

```
                            :
                      . . : .     :     .   . : . : : : . . .
                  . : : : : :   . : : . : . : : : : : : : : : . :
                +---------+---------+---------+---------+---------+--
PhyStren
               0.0       5.0      10.0      15.0      20.0      25.0
```

b. mode = $\underline{9}$

c.

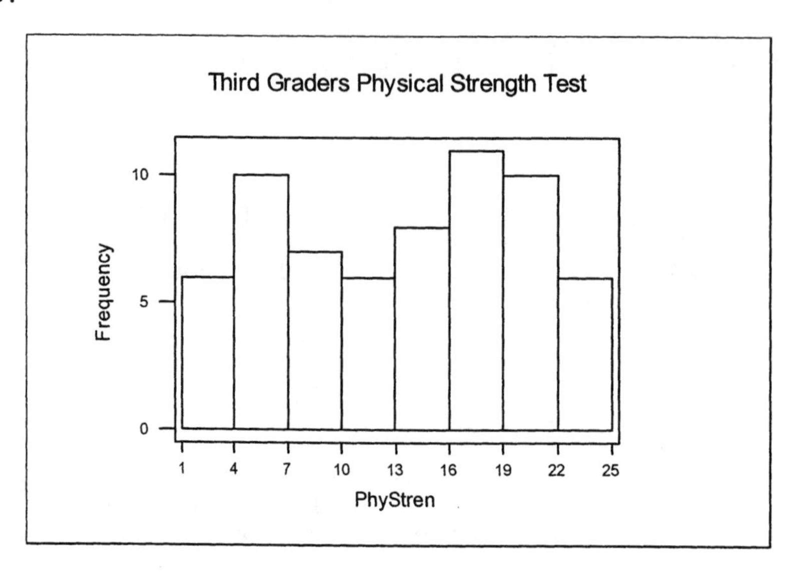

d. The distribution appears to be bimodal. Modal classes are 4-7, and 16-19.

e. Dotplot shows mode to be 9, which is in the 7-10 class; while the histogram shows the two modal classes to be 4-7 and 16-19. The mode is not in either modal class.

f. No. In an ungrouped distribution there is only one numerical value per class.

g. The mode is simply the single data value that occurs most often, while a modal class results from data tending to bunch-up forming a cluster of data values, not necessarily all of one value.

2.73 a. & b.

	Runs at Home	**Runs Away**	**Difference**
Mean	9.77	9.80	−0.03
Median	9.65	9.78	−0.06
Maximum	13.65	11.06	4.89
Minimum	7.64	8.67	−1.74
Midrange	10.65	9.87	1.58

c. Teams playing the Rockies at Coors Field generated the maximum number of runs scored (13.65). On the other hand, while playing their opponents away, the Rockies and their opponents, were able to generate only 8.76 runs, which ranked second from the bottom. Collectively, these two performances produced the greatest spread (4.89) by a considerable margin between runs scored at home and runs scored away by any stadium/team combination in the major leagues. This unusually large value inflates the midrange difference. It appears that playing conditions at Coors Field, therefore, are more responsible for producing the higher combined scores than the strength of either the Rockies' or their opponents' bats or any weakness of the pitching staffs.

2.75 Many different answers are possible.

a. $\sum x$ needs to be 500; therefore, need any three numbers that total 330.
100, 100, 130 [70, 100]
$\bar{x} = \sum x/n = 500/5 = 100$, ck

b. Need two numbers smaller than 70 and one larger.
__, __, 70, ___, 100: 50, 60, 80 [70, 100]
$d(\tilde{x}) = (n+1)/2 = (5+1)/2 = 3rd$; $\tilde{x} = 70$, ck

c. Need multiple 87's. 87, 87, 87 [70, 100]
mode = 87, ck

d. Need any two numbers that total 140 for the extreme values where one is 100 or larger. _, _, _, 70, 100
40, 50, 60 [70, 100]
midrange = $(L+H)/2 = (40+100)/2 = 70$, ck

e. Need two numbers smaller than 70 and one larger than 70 so that their total is 330. _, _, 70, _, 100;
60, 60, 210 [70, 100]
$\bar{x} = \sum x/n = 500/5 = 100$, ck
$d(\tilde{x}) = (n+1)/2 = (5+1)/2 = 3rd$; $\tilde{x} = 70$, ck

f. Need two numbers of 87 and a third number large enough so that the total of all five is 500.
87, 87, 156 [70, 100]
$\bar{x} = \sum x/n = 500/5 = 100$, ck; mode = 87, ck

g. Mean equal to 100 requires the five data to total 500 and the midrange of 70 requires the total of L and H to be 140; 40, _, 70, _, 100; that is a sum of 210, meaning the other two data must total 290. One of the last two numbers must be larger than 145, which would then become H and change the midrange. Impossible.

h. There must be two 87's in order to have a mode, and there can only be two data larger than 70 in order for 70 to be the median. _, 70, 87, 87, 100; Impossible

SECTION 2.4 ANSWER NOW EXERCISES

2.77 range = H - L = 3092 - 1292 = 1800 = $1800

2.78 a. The data value x = 45 is 12 units above the mean; therefore the mean must be 33.
b. The data value x = 84 is 20 units below the mean; therefore the mean must be 104.

2.79 The mean is the 'balance point' or 'center of gravity' to all the data values. Since the weights of the data values on each side of $\overline{x}$ are equal, $\Sigma(x - \overline{x})$ will give a positive amount and an equal negative amount, thereby canceling each other out.

Algebraically: $\Sigma(x - \overline{x}) = \Sigma x - n\overline{x} = \Sigma x - n\cdot(\Sigma x/n) = \Sigma x - \Sigma x = 0$

2.80 1st: find mean, $\overline{x} = \Sigma x/n = 25/5 = 5$

	x	$x - \overline{x}$	$(x - \overline{x})^2$
	1	-4	16
	3	-2	4
	5	0	0
	6	1	1
	10	5	25
Σ	25	0	46

$s^2 = \Sigma(x-\overline{x})^2/(n-1)$

$= 46/4 = 11.5$

2.81

x	x^2
1	1
3	9
5	25
6	36
10	100
25	171

$SS(x) = \sum x^2 - ((\sum x)^2/n)$

$= 171 - ((25)^2/5)$

$= 171 - 125 = 46$

$s^2 = SS(x)/(n-1) = 46/4 = \underline{11.5}$

SECTION 2.4 EXERCISES

NOTE: A measure of dispersion is a value of the variable. It is that value which describes the amount of variation or spread in a data set. A small measure of dispersion indicates data that are closely grouped, whereas, a large value indicates data that are more widely spread.

MEASURES OF DISPERSION - THE SPREAD OF THE DATA

Range = highest value - lowest value

Standard Deviation - s - the average distance a data value is from the mean

$$s = \sqrt{\sum(x - \bar{x})^2 / (n - 1)}$$

Variance - s^2 - the square of the standard deviation (i.e., before taking the square root)

For exercises 2.83-2.86, be sure that the $\sum(x - \bar{x}) = 0$.

NOTE: Standard deviation and/or variance cannot be negative. This would indicate an error in sums or calculations.

See Introductory Concepts (ST-p502) for additional information about Rounding Off.

Computer and calculator commands to find the range and standard deviation can be found in ES9-pp82&83.
If using a non-graphing statistical calculator (one that lets you input the data points) to find the standard deviation of a sample, use the $\sigma(n-1)$ or s_x key. $\sigma(n)$ or σ_x would give the population standard deviation; that is, divide by "n" instead of "n-1".

2.83 a. range = H - L = 9 - 2 = 7

b. 1st: find mean, $\bar{x} = \sum x/n = 30/5 = 6$

	x	$x - \bar{x}$	$(x - \bar{x})^2$
	2	-4	16
	4	-2	4
	7	1	1
	8	2	4
	9	3	9
Σ	30	0	34

$s^2 = \sum(x-\bar{x})^2/(n-1)$

$= 34/4 = 8.5$

c. $s = \sqrt{s^2} = \sqrt{8.5} = 2.915 = 2.9$

2.85 a. 1st: find mean, $\bar{x} = \sum x/n = 72/10 = 7.2$

	x	$x - \bar{x}$	$(x - \bar{x})^2$
	3	-4.2	17.64
	5	-2.2	4.84
	5	-2.2	4.84
	6	-1.2	1.44
	7	-0.2	0.04
	7	-0.2	0.04
	7	-0.2	0.04
	9	1.8	3.24
	10	2.8	7.84
	13	5.8	33.64
Σ	72	0	73.60

$s^2 = \sum(x-\bar{x})^2/(n-1)$

$= 73.60/9$

$= 8.1778 = 8.2$

An easier formula for s - sample standard deviation

1. Calculate "the sum of squares for x", SS(x):

$$SS(x) = \sum x^2 - ((\sum x)^2/n)$$

2. $s = \sqrt{SS(x)/(n-1)}$

This formula eliminates the problem of accumulating round-off errors.

NOTE: SS(x) is formed from the "sum of squared deviations from the mean", $\sum(x - \bar{x})^2$. $\sum x^2$ is the "sum of the squared x's". $SS(x) \neq \sum x^2$.

b.

x	x^2
3	9
5	25
5	25
6	36
7	49
7	49
7	49
9	81
10	100
13	169
$\sum$ 72	592

$SS(x) = \sum x^2 - ((\sum x)^2/n)$

$= 592 - ((72)^2/10)$

$= 592 - 518.4 = 73.6$

$s^2 = SS(x)/(n-1)$

$= 73.6/9 = 8.1778 = \underline{8.2}$

c. $s = \sqrt{s^2} = \sqrt{8.1778} = 2.8597 = \underline{2.9}$

2.87 a. Range = 44.40 - 24.60 = 19.800

b. $n = 13$, $\sum x = 423.00$, $\sum x^2 = 14058$,
$s^2 = 24.51923$

c. $s = 4.96$

2.89 a. Police Recruits

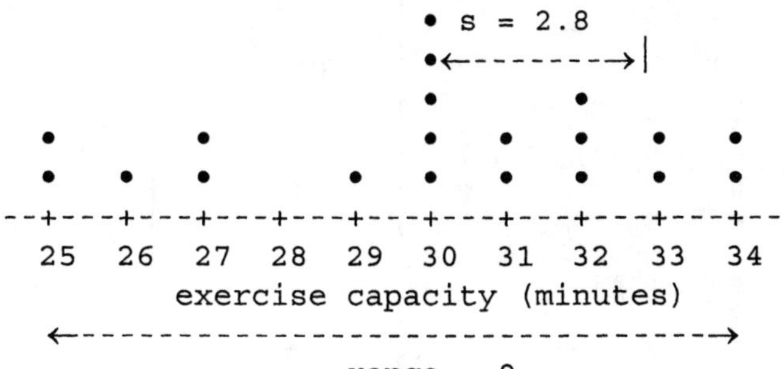

b. $\overline{x} = 601/20 = 30.05$

c. range = H - L = 34 - 25 = $\underline{9}$

d. $n = 20$, $\sum x = 601$, $\sum x^2 = 18{,}209$
$SS(x) = \sum x^2 - ((\sum x)^2/n) = 18{,}209 - (601^2/20) = 148.95$
$s^2 = SS(x)/(n-1) = 148.95/19 = 7.83947 = \underline{7.8}$

e. $s = \sqrt{s^2} = \sqrt{7.83947} = 2.7999 = \underline{2.8}$

f. See graph in (a).

g. Except for the value $x = 30$, the distribution looks rectangular. Range is a little more than 3 standard deviations.

2.91 a. Range = 64.60 - 89.70 = 25.100,

$n = 25$, $\sum x = 2002.0$, $\sum x^2 = 160955$

$$SS(x) = \sum x^2 - ((\sum x)^2/n) = 160955 - (2002.0^2/25)$$
$$= 634.84$$
$$s^2 = SS(x)/(n-1) = 634.84/24 = 26.4516667$$

$$s = \sqrt{s^2} = \sqrt{338.703704} = 5.143118 = \underline{5.14}$$

b.

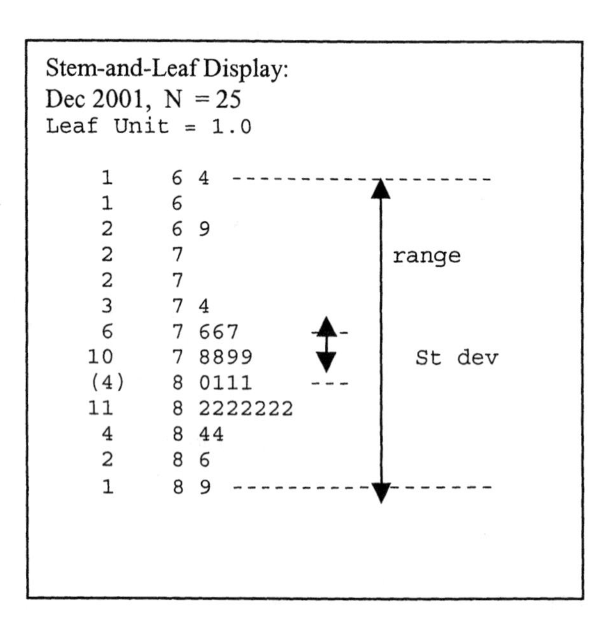

c. Most of the data falls within an interval from one standard deviation below to one standard deviation above the mean. The standard deviation is approximately one-fifth the range.

2.93 Set 1: $\bar{x} = 250/5 = 50$ | Set 2: $\bar{x} = 250/5 = 50$

x	$x - \bar{x}$	$\lvert x - \bar{x} \rvert$	$(x - \bar{x})^2$	x	$x - \bar{x}$	$\lvert x - \bar{x} \rvert$	$(x - \bar{x})^2$
46	-4	4	16	30	-20	20	400
55	+5	5	25	55	+5	5	25
50	0	0	0	65	15	15	225
47	-3	3	9	47	-3	3	9
52	+2	2	4	53	+3	3	9
250	0	14	54	250	0	46	668

	Σx	$\Sigma(x - \bar{x})$	$\Sigma\lvert x - \bar{x} \rvert$	$\Sigma(x - \bar{x})^2$	Range
Set 1:	250	0	14	54	9
Set 2:	250	0	46	668	35

The values of $\Sigma|x - \bar{x}|$, SS(x) [recall SS(x) = $\Sigma(x-\bar{x})^2$], and range all reflect the fact that there is more variability in the data forming set 2 than in the data of set 1.

2.95 a. Since s = 0, all data must be the same value.
100, 100, 100, 100, [100]

Many different answers are possible for (b), (c) and (d). One possible answer is given for each.

b. 99, 99.5, 100.5, 100 [100]; s = 0.57

c. 107, 95, 94, 108, [100]; s = 6.53

d. 75, 78, 123, 124, [100]; s = 23.53

2.97 Answers will vary. One method could utilize the spread of standard deviation first to narrow down the possibilities, then determine the center based on the means given.

SECTION 2.5 ANSWER NOW EXERCISES

2.98 a.

x	f	xf	x^2f
0	1	0	0
1	3	3	3
2	8	16	32
3	5	15	45
4	3	12	48
Σ	20	46	128

b. $\Sigma f = 20$; $\Sigma xf = 46$;

$\Sigma x^2 f = 128$

c. x = 4; 4 is one of the possible data values
f = 8; 8 is the number of times an 'x' value occurred
Σf : sum of the frequencies = sample size
Σxf : sum of the products formed by multiplying a data value by its frequency; the sum of the data

2.99 a. Sum of the x-column has no meaning unless each value occurred only once.

b. Each data value is multiplied by how many times it occurred. Summing these products will give the same sum if all data values were listed individually.

Note:

x	f	xf
3	5	15

xf = 15 or 3+3+3+3+3=15

2.100 $\bar{x} = \Sigma xf/\Sigma f = 46/20 = \underline{2.3}$

2.101 $SS(x) = \Sigma x^2 f - ((\Sigma xf)^2/\Sigma f)$
$= 128 - (46^2/20) = 128 - 105.8 = 22.2$

$s^2 = SS(x)/(\Sigma f-1)$
$= 22.2/19 = 1.16842 = \underline{1.2}$

2.102 $s = \sqrt{s^2} = \sqrt{1.16842} = 1.080935 = \underline{1.1}$

2.103

Class limits	x	f	xf	x^2f
2 - 6	4	2	8	32
6 - 10	8	10	80	640
10 - 14	12	12	144	1728
14 - 18	16	9	144	2304
18 - 22	20	7	140	2800
	Σ	40	516	7504

$\overline{x} = \Sigma xf/\Sigma f = 516/40 = \underline{12.9}$

$SS(x) = \Sigma x^2f - ((\Sigma xf)^2/\Sigma f) = 7504 - (516^2/40) = 847.6$

$s^2 = SS(x)/(\Sigma f-1) = 847.6/39 = 21.7333 = \underline{21.7}$

$s = \sqrt{s^2} = \sqrt{21.7333} = 4.6619 = \underline{4.7}$

SECTION 2.5 EXERCISES

MEAN AND STANDARD DEVIATION OF FREQUENCY DISTRIBUTIONS

Mean - $\overline{x} = \Sigma xf/\Sigma f$

Standard Deviation - s

1. $SS(x) = \Sigma x^2f - ((\Sigma xf)^2/\Sigma f)$
2. $s^2 = SS(x)/(\Sigma f-1)$
3. $s = \sqrt{s^2}$

NOTE: in grouped frequency distributions, the calculated statistics are approximations.

There are no* "mean of grouped data" nor "standard deviation of grouped data" computer or calculator commands. Therefore, the computer and calculator commands needed to form the *xf* and x^2f columns and to work through the formulas to calculate the mean and standard deviation are shown in ES9-pp90&91.
* Grouped data techniques are for "hand" calculations. Usually a computer has all of the data and has no problem working with large sets of data.

2.105

x	f	xf	x^2f
0	15	0	0
1	12	12	12
2	26	52	104
3	14	42	126
4	4	16	64
6	2	12	72
Σ	73	134	378

$\overline{x} = \Sigma xf/\Sigma f = 134/73 = 1.836 = \underline{1.8}$

$SS(x) = \Sigma x^2f - ((\Sigma xf)^2/\Sigma f)$
$= 378 - (134^2/73)$
$= 378 - 245.97260 = 132.0274$

$s^2 = SS(x)/(\Sigma f-1)$
$= 132.0274/72 = 1.8337 = \underline{1.8}$

$s = \sqrt{s^2} = \sqrt{1.8337} = 1.354 = \underline{1.4}$

In exercises 2.107 through 2.111, the calculated means, variances, and standard deviations of the grouped frequency distributions will be approximations. This is due to the use of the class midpoints versus the actual data values.

For example, suppose the class limits for a particular class are 0-6, and that 5 data values fall in that class interval. The class midpoint of 3 would be used in the calculations, thereby treating all 5 data values as 3s, when they each could be any numbers from 0 through 6 (even all 0's).

2.107

Class limits	x	f	xf	x^2f
3 - 6	4.5	2	9	40.50
6 - 9	7.5	10	75	562.50
9 - 12	10.5	12	126	1323.00
12 - 15	13.5	9	121.5	1640.25
15 - 18	16.5	7	115.5	1905.75
	Σ	40	447.0	5472.00

$\overline{x} = \Sigma xf/\Sigma f = 447.0/40 = 11.175 = \underline{11.2}$

$SS(x) = \Sigma x^2f-((\Sigma xf)^2/\Sigma f) = 5472 - (447^2/40) = 476.775$

$s^2 = 476.775/39 = 12.225 = \underline{12.2}$

$s = \sqrt{s^2} = \sqrt{12.225} = 3.496 = \underline{3.5}$

2.109

Class limits	x	f	xf	x^2f
1.0 - 3.0	2	2	4	8
3.0 - 5.0	4	6	24	96
5.0 - 7.0	6	12	72	432
7.0 - 9.0	8	50	400	3200
9.0 - 11.0	10	35	350	3500
11.0 - 13.0	12	15	180	2160
13.0 - 15.0	14	5	70	980
	Σ	125	1100	10,376

$\bar{x} = \Sigma xf/\Sigma f = 1100/125 = \underline{8.8}$

$SS(x) = \Sigma x^2 f - ((\Sigma xf)^2/\Sigma f) = 10,376 - (1100^2/125)$
$= 696.0000$

$s^2 = SS(x)/(\Sigma f-1) = 696.00/124 = 5.6129$

$s = \sqrt{s^2} = \sqrt{5.6129} = 2.36916 = \underline{2.37}$

2.111

Class limits	x	f	xf	x^2f
15 - 25	20	13	260	5200
25 - 35	30	20	600	18000
35 - 45	40	28	1120	44800
45 - 55	50	20	1000	50000
55 - 65	60	10	600	36000
65 - 75	70	9	630	44100
	Σ	100	4210	198100

$\bar{x} = \Sigma xf/\Sigma f = 4210/100 = \underline{42.1}$

$SS(x) = \Sigma x^2 f - ((\Sigma xf)^2/\Sigma f) = 198100 - (4210^2/100)$
$= 20859$

$s^2 = SS(x)/(\Sigma f-1) = 20859/99 = 210.6969697$

$s = \sqrt{s^2} = \sqrt{210.6969697} = 14.5154 = \underline{14.5}$

2.113 a. 23 is the number of students tested at Caruthers Unified, and the 441 is the average verbal score for those 23 students for the year 2000-2001.

b. 23 x 441 = 10143, assuming that the average is the mean.

c. Total number of students is 3199

d. The total of all 3199 test scores is 1508912

e. mean = 1508912/3199 = 471.68 = 471.7

OTHER MEASURES OF CENTRAL TENDENCY FOR FREQUENCY DISTRIBUTIONS

NOTE: Data are already ranked.

Median - $\tilde{x}$ - find the depth and count down the frequency column until you include that position number. This is the median class. In an ungrouped frequency distribution, the median equals the x value of that class. In a grouped distribution, the data must be ranked in that particular class, then count to the appropriate position. If the original data are not given, use the class mark.

Mode - class mark of the highest frequency class

Modal Class - interval bounded by the class boundaries of the class with the highest frequency

Midrange = (highest value + lowest value)/2. If the original data is not given, use the lowest class boundary and the highest class boundary from the entire distribution, or the lowest and highest class marks.

2.115 a. $n = 997$, $\Sigma xf = 63,665$, $\sum x^2 f = 4,302,725$

mean = 63.857 = 63.9, median = 65, mode = 65,
midrange = (20 + 90)/2 = 55

b. variance = 238.25, st.dev. = 15.435 = 15.4

2.117 a. and b.

Class	x	f	xf	xsqf
9 - 11	10	2	20	200
11 - 13	12	0	0	0
13 - 15	14	2	28	392
15 - 17	16	1	16	256
17 - 19	18	1	18	324
19 - 21	20	2	40	800
21 - 23	22	0	0	0
23 - 25	24	4	96	2304
25 - 27	26	4	104	2704
27 - 29	28	5	140	3920
29 - 31	30	8	240	7200
31 - 33	32	9	288	9216
33 - 35	34	8	272	9248
35 - 35	36	4	144	5184

Grouped:

$\Sigma f = 50$, $\Sigma xf = 1406.0$, $\sum x^2 f = 41748$

$\overline{x} = \Sigma xf/\Sigma f = 1406/50 = 28.12$

$s^2 = 45.128$, $s = 6.7177$

Ungrouped: using 50 data:

$n = 50$, $\Sigma x = 1393.0$, $\Sigma x^2 = 41057$

$\overline{x} = \Sigma x/n = 1393/50 = 27.86$

$s^2 = 45.878$, $s = 6.773$

c. Error for mean = (28.12-27.86)/27.86 = 0.009 = 0.9%
Error for variance = (45.128-45.877959)/45.877959 = -0.016 = -1.6%
Error for standard deviation = (6.7177-6.773)/6.773 = -0.008 = -0.8%

SECTION 2.6 ANSWER NOW EXERCISES

2.118 91 is in the 44th position from the Low value of 39
91 is in the 7th position from the High value of 98

2.119 $nk/100 = (50)(20)/100 = 10.0$; therefore $d(P_{20}) = 10.5$th from L
$P_{20} = (64+64)/2 = \underline{64}$
$nk/100 = (50)(35)/100 = 17.5$; therefore $d(P_{35}) = 18$th from L
$P_{35} = \underline{70}$

2.120 $nk/100 = (50)(20)/100 = 10.0$; therefore $d(P_{80}) = 10.5$th from H
$P_{80} = (88+89)/2 = \underline{88.5}$
$nk/100 = (50)(5)/100 = 2.5$; therefore $d(P_{95}) = 3$rd from H
$P_{95} = \underline{95}$

2.121 The distribution needs to be symmetric about the mean.

2.122

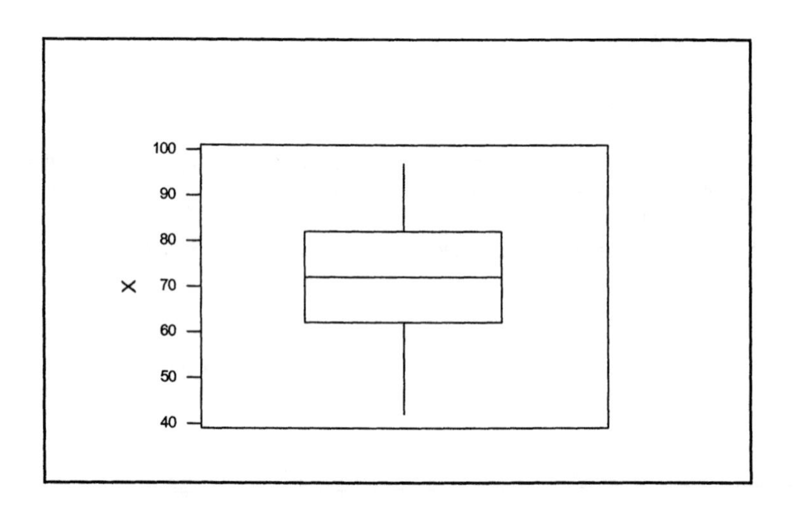

2.123 $z = (x - \text{mean})/\text{st.dev.}$

for $x = 92$, $z = (92 - 72)/12 = \underline{1.67}$
for $x = 63$, $z = (63 - 72)/12 = \underline{-0.75}$

SECTION 2.6 EXERCISES

NOTE: A measure of position is a value of the variable. It is that value which divides the set of data into two groups: those data smaller in value than the measure of position, and those larger in value than the measure of position.

To find any measure of position:

1. Rank the data - DATA MUST BE RANKED LOW TO HIGH
2. Determine the depth or position in two separate steps:
   a. Calculate nk/100, where n = sample size,
   k = desired percentile
   b. Determine $d(P_k)$:
   If nk/100 = integer ⇒ add .5 (value will be halfway between 2 integers)
   If nk/100 = decimal ⇒ round up to the nearest whole number
3. Locate the value of P_k

REMEMBER:
$Q_1 = P_{25}$ = 1st quartile - 25% of the data lies below this value
$Q_2 = P_{50} = \tilde{x}$ = 2nd quartile - 50% of the data lies below this value
$Q_3 = P_{75}$ = 3rd quartile - 75% of the data lies below this value

2.125 Ranked data:

2.6	2.7	3.4	3.6	3.7	3.9	4.0	4.4	4.8	4.8
4.8	5.0	5.1	5.6	5.6	5.6	5.8	6.8	7.0	7.0

a. nk/100 = (20)(25)/100 = 5.0; therefore $d(P_{25})$ = 5.5th
$Q_1 = P_{25} = (3.7 + 3.9)/2 = \underline{3.8}$
nk/100 = (20)(75)/100 = 15.0; therefore $d(P_{75})$ = 15.5th
$Q_3 = P_{75} = (5.6 + 5.6)/2 = \underline{5.6}$

b. midquartile = $(Q_1 + Q_3)/2 = (3.8 + 5.6)/2 = \underline{4.7}$

c. nk/100 = (20)(15)/100 = 3.0; therefore $d(P_{15})$ = 3.5th
$P_{15} = (3.4+3.6)/2 = \underline{3.5}$
nk/100 = (20)(33)/100 = 6.6; therefore $d(P_{33})$ = 7th
$P_{33} = \underline{4.0}$
nk/100 = (20)(90)/100 = 18.0; therefore $d(P_{90})$ = 18.5th
$P_{90} = (6.8+7.0)/2 = \underline{6.9}$

Box-and-whisker displays may be drawn horizontal or vertical. The Student Suite CD contains the Excel macro, Data Analysis Plus, for constructing box-and-whisker displays.

2.127 ranked data:

1.4	2.3	2.4	2.6	2.6	2.7	2.7	2.8	2.8
2.9	2.9	2.9	3.0	3.1	3.1	3.2	3.3	3.4
3.5	3.5	3.6	3.7	3.7	3.9	3.9	4.0	4.0
4.0	4.1	4.1	4.2	4.2	4.2	4.4	4.4	4.5
4.6	4.6	4.6	4.7	4.8	4.8	4.8	4.9	5.2
5.2	5.5	5.6	5.7	6.5	7.0	13.3		

a. nk/100 = (52)(25)/100 = 13; therefore $d(Q_1)$ = 13.5^{th}
   Q1 = (3.0+3.1)/2 = 3.05

b. d(median)=(52+1)/2 = 26.5^{th}, Q_2 = median = (4.0+4.0)/2= 4.0

c. Q_3 = (4.6+4.7)/2 = 4.65

d. midquartile = (3.05+4.65)/2 = 3.85

e. nk/100 = (52)(30)/100 = 15.6; therefore $d(P_{30})$ = 16th
   P_{30} = 3.2

f. 5-number summary: 1.4, 3.05, 4.0, 4.65, 13.3

g.

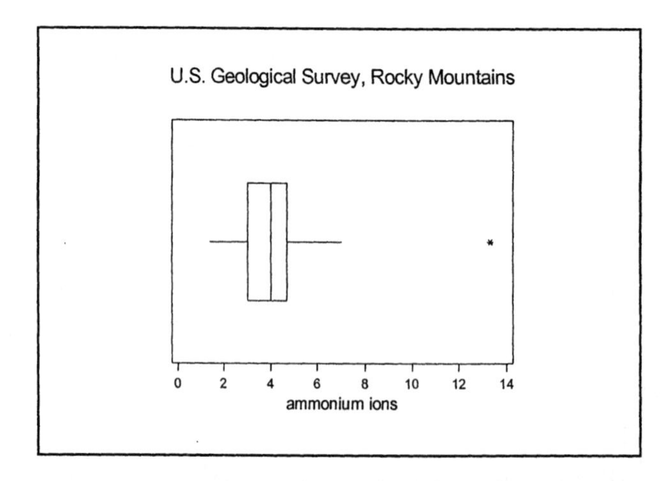

z is a measure of position. It gives the number of standard deviations a piece of data is from the mean. It will be positive if x is to the right of the mean (larger than the mean) and negative, if x is to the left of the mean (smaller than the mean). Keep 2 decimal places. (hundredths)

$z = (x - \text{mean})/\text{st. dev.}$ $z = (x - \bar{x})/s$

2.129 $z = (x - \text{mean})/\text{st.dev.}$

a. for $x = 54$, $z = (54 - 50)/4.0 = \underline{1.0}$
b. for $x = 50$, $z = (50 - 50)/4.0 = \underline{0.0}$
c. for $x = 59$, $z = (59 - 50)/4.0 = \underline{2.25}$
d. for $x = 45$, $z = (45 - 50)/4.0 = \underline{-1.25}$

2.131 If $z = (x - \text{mean})/\text{st.dev}$; then $x = (z)(\text{st.dev}) + \text{mean}$

for $z = 1.8$, $x = (1.8)(100) + 500 = \underline{680}$

2.133 a. 152 is one and one-half standard deviations above the mean
b. the score is 2.1 standard deviations below the mean
c. the number of standard deviations from the mean

2.135 for A: $z = (85 - 72)/8 = 1.625$
for B: $z = (93 - 87)/5 = 1.2$
Therefore, A has the higher relative position.

SECTION 2.7 ANSWER NOW EXERCISES

2.137 From 175 through 225 words, inclusive.

2.138 a. 50% b. ≈68% c. ≈84%

2.139 $1 - (1/k^2) = 1 - (1/4^2) = 1 - (1/16) = 15/16 = 0.9375$;
at least 93.75%

SECTION 2.7 EXERCISES

The empirical rule applies to a normal distribution.
Approximately 68% of the data lies within 1 standard deviation of the mean.
Approximately 95% of the data lies within 2 standard deviations of the mean.
Approximately 99.7% of the data lies within 3 standard deviations of the mean.

Chebyshev's theorem applies to any shape distribution.

At least 75% of the data lies within 2 standard deviations of the mean.

At least 89% of the data lies within 3 standard deviations of the mean.

2.141 Nearly all of the data, 99.7%, lies within 3 standard deviations of the mean.

2.143 a. 97.6 is 2 standard deviations above the mean $\{z = (97.6-84.0)/6.8 = 2.0\}$, therefore 2.5% of the time more than 97.6 hours will be required.

b. 95% of the time the time to complete will fall within 2 standard deviations of the mean, that is $84.0 \pm 2(6.8)$ or from 70.4 to 97.6 hours.

2.145 a. Range should be approximately equal to 6 times the standard deviation for data with a normal distribution.

b. An approximation of the standard deviation can be found by dividing the range by 6.

Chebyshev's theorem

At least $\left(1-\frac{1}{k^2}\right)\%$ of the data lies within k standard deviations of the mean. $(k > 1)$

2.147 a. at most 11% b. at most 6.25%

2.149 Summary: $n = 50$, $\sum x = 5917489$, $\sum x^2 = 810941217297$

a. $\bar{x} = \$118,350$ $s = \$47,511$

b. $\bar{x} - s = 118,350 - 47,511 = \$70,839$ and

$\bar{x} + s = 118,350 + 47,511 = \$165,861$

c. 40, 40/50 = 0.80 = 80%

d. $\overline{x} - 2s = 118,350 - 2(47,511) = \$23,328$ and

$\overline{x} + 2s = 118,350 + 2(47,511) = \$213,372$

e. 46, $46/50 = 0.92 = 92\%$

f. $\overline{x} - 3s = 118,350 - 3(47,511) = -\$24,183$ and
$\overline{x} + 3s = 118,350 + 3(47,511) = \$260,883$

g. $49/50 = 0.98 = 98\%$

h. They agree with Chebyshev's theorem, both percentages exceed the values cited.

i. 80%, 92% and 98% as a set are not close to the 68%, 95% and 99.7% cited by the empirical rule. The results suggest the distribution is not approximately normal.

j.

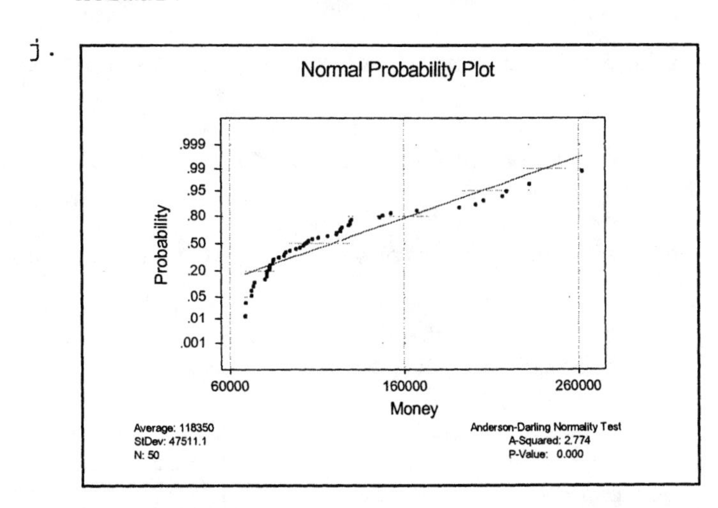

Helpful hint for use when expecting to count data on histogram:

Minitab: While on the Histogram dialogue box,
Select: **Annotations > Data Labels...**
Select: **Show data labels**

This will direct the computer to print the frequency of each class above its corresponding bar.

Excel: Returns a frequency distribution with the histogram.
TI-83: Use the TRACE and arrow keys.

2.151 a. Answers will vary.

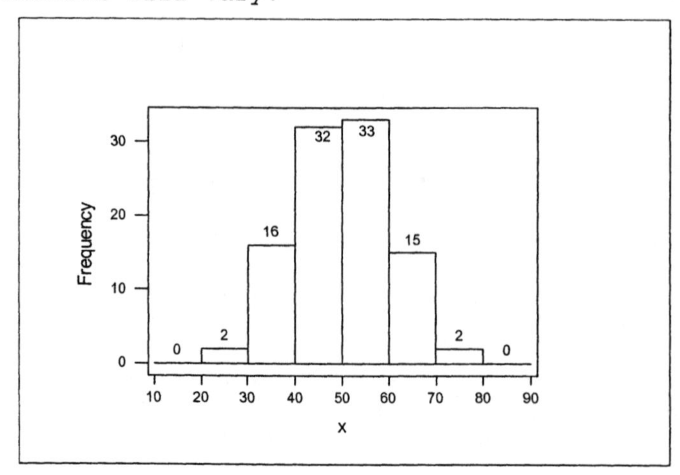

Within one standard deviation, 40 to 60, is 33 + 32 or 65 of the 100 data. 65%

Within two standard deviations, 30 to 70, is 16+32+33+15 or 96 of the 100 data. 96%

Within three standard deviations, 20 to 80, is all 100 of the data, or 100%.

The above results are extremely close to what the empirical rule claims will happen.

b, c, d. Not all sets of 100 data will result in percentages this close. However, expect very similar results to occur most of the time.

SECTION 2.8 ANSWER NOW EXERCISES

2.153 Yes, if all 8 employees earned $300 each, the mean would be $405.56 and if all 8 employees earned $350 each, the mean would be $450. $430 falls within this interval.

Or, if the mean of 9 employees is 430, then the total is 3870. The 8 employees then would need to make 2620 (3870-1250) and their average earnings would be 2620/8 or 327.50, which is within the interval.

2.154 a. Here's a few of the more obvious ones. If you research and find the original information, you'll find more.

1. The ranking information covers an 11-year period while the tuition information covers 35 years. Yet they are shown horizontally as being the same.
2. The units for tuition (share of median income) and ranking (rank number) are totally different, yet the vertical axis treats them as having common units.
3. The ranking graph is placed below the tuition graph creating the impression that cost exceeds quality. Since the vertical scale is meaningless, either line could have been "on top".
4. The sharp "drop" in the ranking graph actually represents an improved ranking. A ranking of "15th best" is not better than a ranking of "6th best", however the vertical scale used makes it look like it is.

b.

1. The caption under the graph suggests that Cornell's rank has been erratic by varying from 6th to 15th on the national ranking over the 12 years reported. With the hundreds of colleges and universities that exist, to consistently hold a rank like this is quite good.
2. The "upside-down" scale with the best ranking at the bottom is totally misleading.

2.155 a. Here are a few: Answers will vary.

1. The response "Take too long" is twice as likely as "Messy"
2. The response "Have to come back" is three times as likely as "Messy"
3. The response "Show up late" is four times as likely as "Messy"
4. The response "Show up late" is twice as likely as "Take too long"
5. The percentage answering with each response decreases the same amount from response to response as you read down the Snapshot.

b.

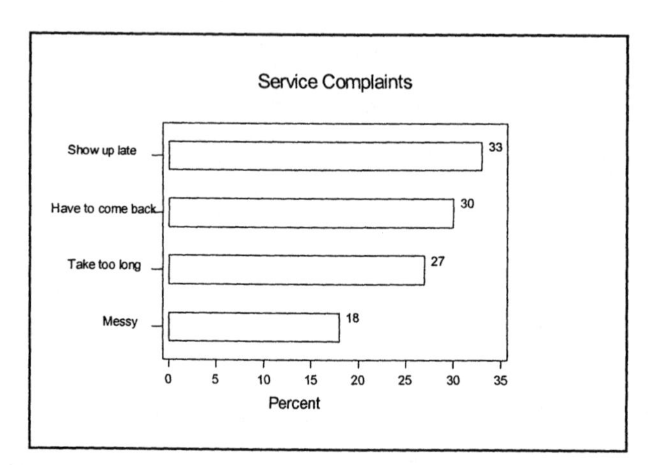

c. Comments - 'Messy' is a substantial percent when displayed against the other complaints in the proper format. 'Show up late' is not even twice that of 'Messy'. The difference between 'Show up late' and 'Take too long' is minimal.

2.156 a.

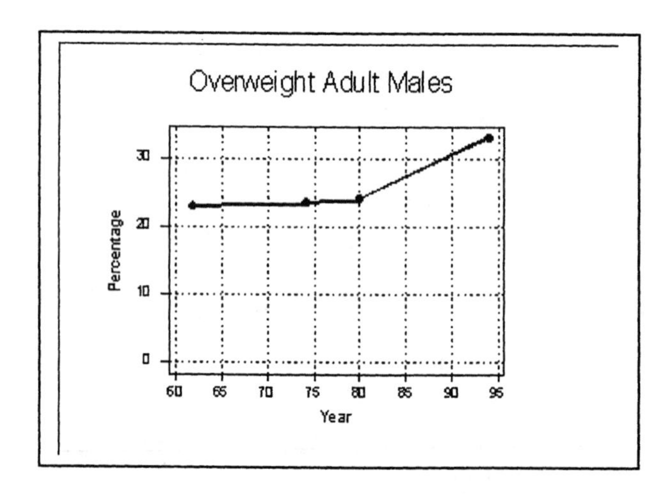

b. The graph in (a) gives the truer picture with respect to time. 1974 to 1980 is a 6 year interval versus the 12 and 14 year intervals at the beginning and end of the graph. The percentage has increased greatly, however the graph in the case study seems to overstate the increase.

SECTION 2.8 EXERCISES

2.157 The class width is not uniform.

2.159 Answers will depend on the article selected.
Answers will vary.

RETURN TO CHAPTER CASE STUDY

2.160 a, b, c. Answers will vary.

CHAPTER EXERCISES

2.163 a.

Corporate Area	relative frequency
1. Sales	0.15
2. Research & Development	0.34
3. Finance	0.18
4. Administration	0.24
5. Other	0.09

b.

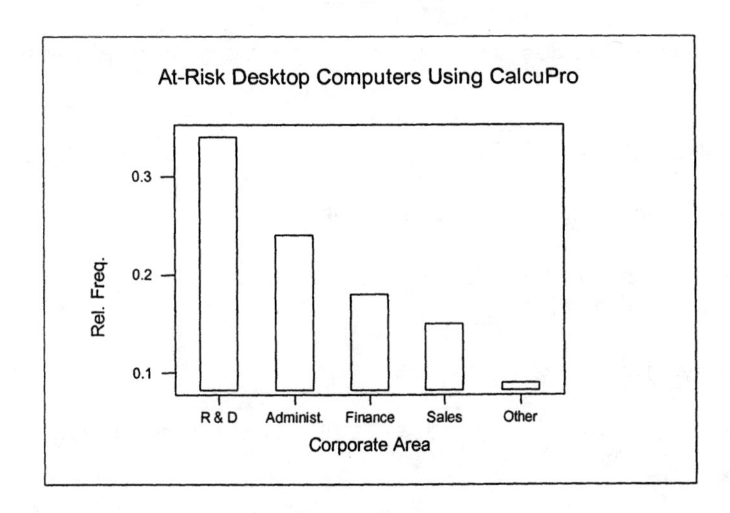

c. Histograms are used when the frequencies correspond to numerical data rather than attribute data. The bars are touching in a histogram emphasizing the sequence of the data.

2.165 a. Mean increased; when one data increases, the sum increases.

b. Median is unchanged; the median is affected only by the middle value(s).

c. Mode is unchanged.

d. Midrange increased; an increase in either extreme value increases the sum H+L.

e. Range increased; difference between high and low values increased.

f. Variance increased; data are now more spread out.

g. Standard deviation increased; data are now more spread out.

2.167 Data summary: $n = 8$, $\sum x = 36.5$, $\sum x^2 = 179.11$

a. $\bar{x} = \sum x/n = 36.5/8 = 4.5625 = \underline{4.56}$

b. $s = \sqrt{[\sum x^2 - ((\sum x)^2/n)]/(n-1)}$
$= \sqrt{[179.11 - (36.5^2/8)]/7}$
$= \sqrt{1.79696} = 1.3405 = \underline{1.34}$

c. These percentages seem to average very closely to 4%.

2.169 Data summary: $n = 118$, $\sum x = 2364$

a. $\bar{x} = \sum x/n = 2364/118 = 20.034 = \underline{20.0}$

b. $d(\tilde{x}) = (n+1)/2 = (118+1)/2 = 59.5$th;
$\tilde{x} = (17+17)/2 = \underline{17}$

c. mode $= \underline{16}$

d. $nk/100 = (118)(25)/100 = 29.5$; therefore $d(P_{25}) = 30$th
$Q_1 = P_{25} = \underline{15}$

$nk/100 = (118)(75)/100 = 88.5$; therefore $d(P_{75}) = 89$th
$Q_3 = P_{75} = \underline{21}$

e. $nk/100 = (118)(10)/100 = 11.8$; therefore $d(P_{10}) = 12$th
$P_{10} = \underline{14}$

$nk/100 = (118)(95)/100 = 112.1$; therefore $d(P_{95}) = 113$
$P_{95} = \underline{43}$

2.171 Data:

63	67	66	63	69	74	72	70	71	71
72	70	75	85	84	85	85	86	94	91
90	90	95	105	104					

Data summary: $n = 25$, $\sum x = 1{,}997$, $\sum x^2 = 163{,}205$

$\bar{x} = \sum x/n = 1997/25 = 79.88 = \underline{79.9}$

$SS(x) = \sum x^2 - ((\sum x)^2/n)$
$= 163{,}205 - (1{,}997^2/25) = 3684.64$

$s^2 = SS(x)/(n-1) = 3684.64/24 = 153.5267$

$s = \sqrt{s^2} = \sqrt{153.5267} = 12.3906 = \underline{12.4}$

2.173 a. The population would be all airline passengers; the variable would be passenger luggage status, lost or not. Other variables needed: airline, date of flight.
b. Statistics; they summarize the data collected for each airline on the given day.
c. Statistic; it summarizes all the data. It is used to estimate the parameter, the value for the whole population.
d. 5.09 is not the midrange. Midrange = (3.50+8.64)/2 = 6.07
No

2.175

CredHrs	Freq	xf	x^2f
3	75	225	675
6	150	900	5400
8	30	240	1920
9	50	450	4050
12	70	840	10080
14	300	4200	58800
15	400	6000	90000
16	1050	16800	268800
17	750	12750	216750
18	515	9270	166860
19	120	2280	43320
20	60	1200	24000
Σ	3570	55155	890655

Summary: $n = 3570$; $\Sigma xf = 55{,}155$, $\Sigma x^2 f = 890{,}655$

a.

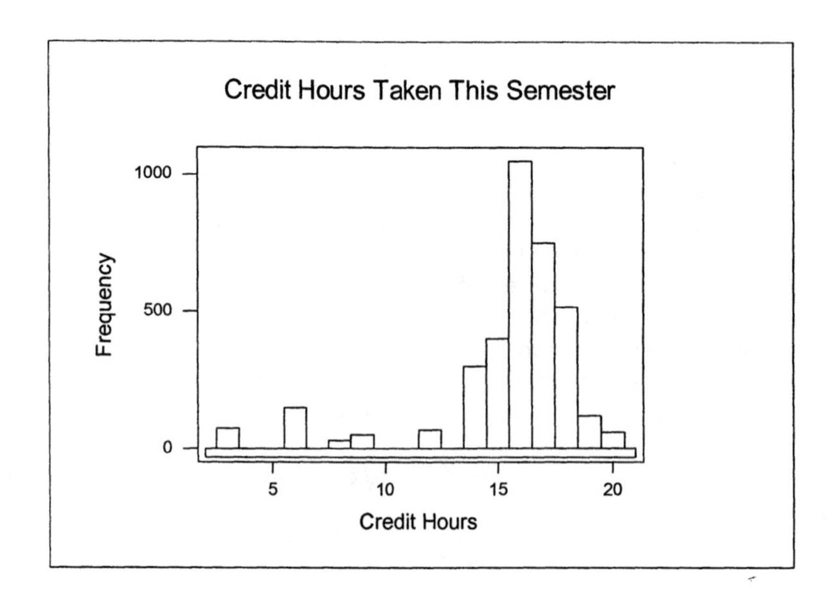

b. mean: $\bar{x} = \Sigma xf/\Sigma f = 55{,}155/3570 = 15.449 = \underline{15.4}$

median: $d(\tilde{x}) = (\Sigma f+1)/2 = (3570+1)/2 = 1785.5^{th}$; $\tilde{x} = \underline{16}$

mode: mode = $\underline{16}$

midrange: midrange = $(H+L)/2 = (3+20)/2 = \underline{11.5}$

midquartile: midquartile = $(Q1+Q3)/2 = (15+17)/2 = \underline{16}$

c. $nk/100 = (3570)(25)/100 = 892.5$; $d(Q_1) = 893^{rd}$;
$Q_1 = \underline{15}$ and $Q_3 = \underline{17}$

d. $nk/100 = (3570)(15)/100 = 535.5$; $d(P_{15}) = 536$th
   $P_{15} = \underline{14}$
   $nk/100 = (3570)(12)/100 = 428.4$; $d(P_{12}) = 429$th
   $P_{12} = \underline{14}$

e. range: range = H - L = 20 - 3 = $\underline{17}$
   variance: $SS(x) = \sum x^2 - ((\sum x)^2/n)$
   $= 890655 - (55155^2/3570) = 38533.42437$
   $s^2 = SS(x)/(n-1) = 38533.42437/3569 = \underline{10.7967}$
   standard deviation:
   $s = \sqrt{s^2} = \sqrt{10.7967} = 3.2858 = \underline{3.3}$

2.177 a.

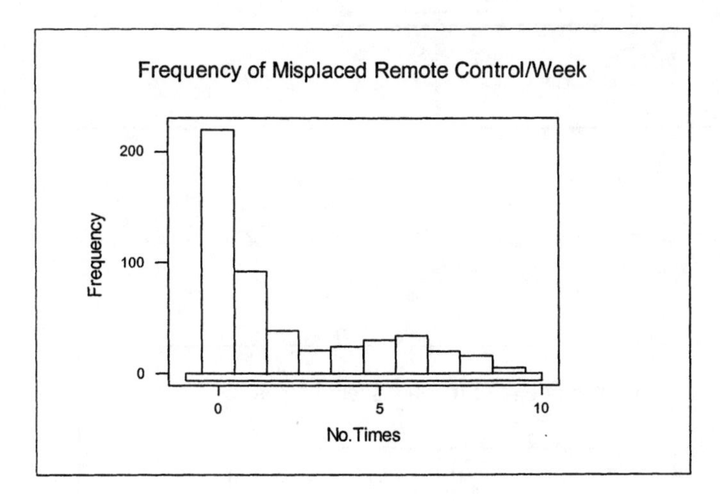

b. $n = \sum f = 500$, $\sum xf = 994$, $\sum x^2 f = 5200$
   mean = 1.988, median = 1, mode = 0, midrange = 4.5

c. $SS(x) = \sum x^2 f - ((\sum xf)^2/n)$
   $= 5200 - (994^2/500) = 3223.928$

   $s^2 = SS(x)/(\sum f-1) = 3223.928/499 = 6.46078 = \underline{6.46}$

   $s = \sqrt{s^2} = \sqrt{6.46078} = 2.5418 = \underline{2.5}$

d. $Q_1 = 0$, $Q_3 = 4$, $P_{90} = 6$

e. midquartile = 2

f. 5-number summary: 0, 0, 1, 4, 9

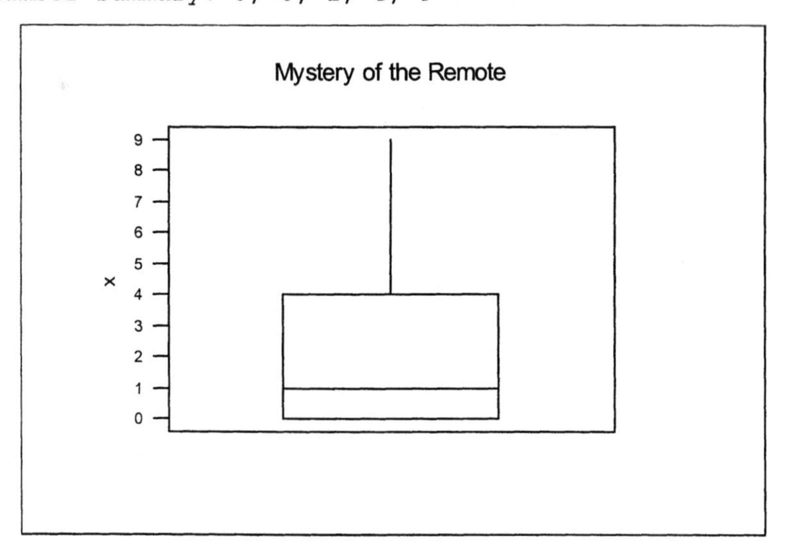

2.179 Summary: $n = 220$; $\Sigma xf = 219,100$; $\Sigma x^2 f = 224,470,000$

a.

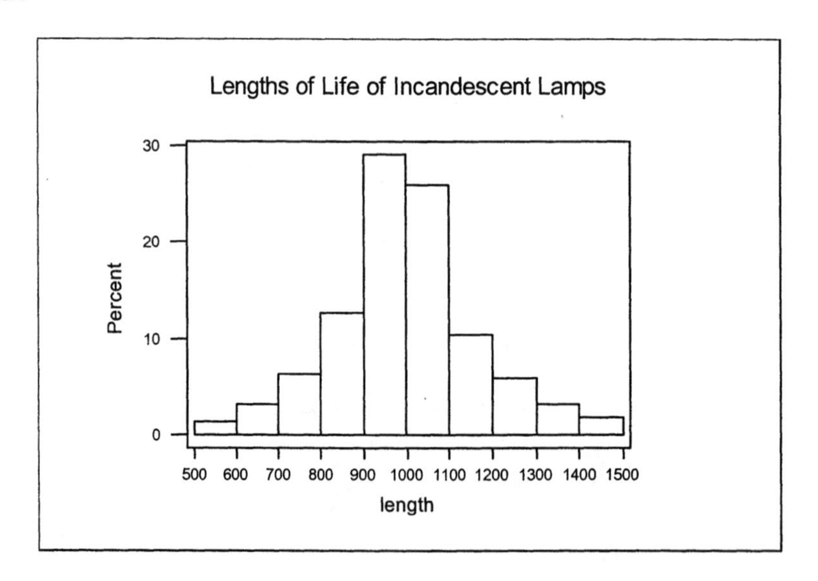

b. $\bar{x} = \Sigma xf/\Sigma f = 219,100/220 = 995.90909 = \underline{995.9}$

c. $SS(x) = \sum x^2 f - ((\sum xf)^2/\sum f)$
$= 224{,}470{,}000 - (219{,}100^2/220) = 6{,}266{,}318.182$

$s^2 = SS(x)/(\sum f - 1) = 6{,}266{,}318.182/219 = 28{,}613.32503$

$s = \sqrt{s^2} = \sqrt{28613.32503} = 169.15473 = \underline{169.2}$

2.181 a.

Class limits	f
-1.00-0.00	1
0.00-1.00	6
1.00-2.00	10
2.00-3.00	7
3.00-4.00	6
4.00-5.00	3
5.00-6.00	3
6.00-7.00	1
7.00-8.00	2
8.00-9.00	0
9.00-10.0	1
$\sum$	40

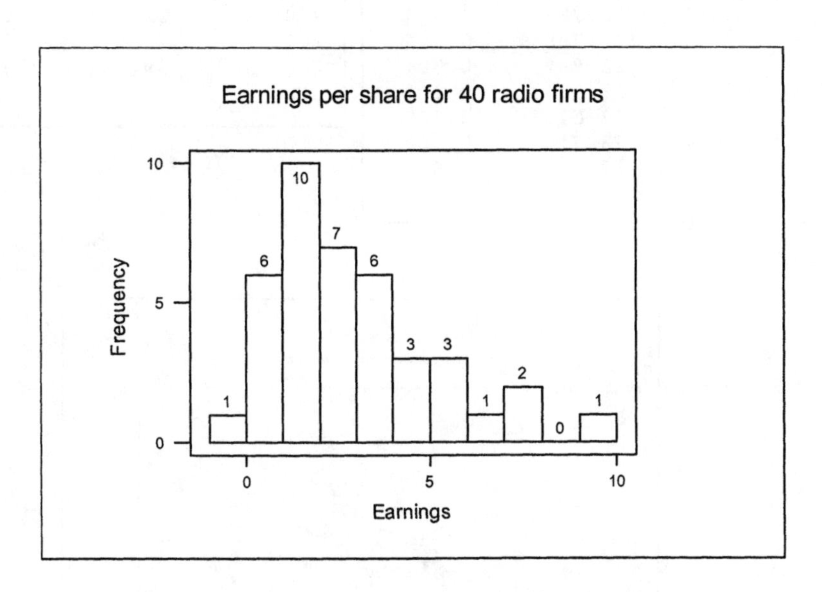

b. $d(\tilde{x}) = (n+1)/2 = 20.5$th; median is in the class \$2.00-\$3.00.

2.183 a. Using the 5 years, 1998 - 2002

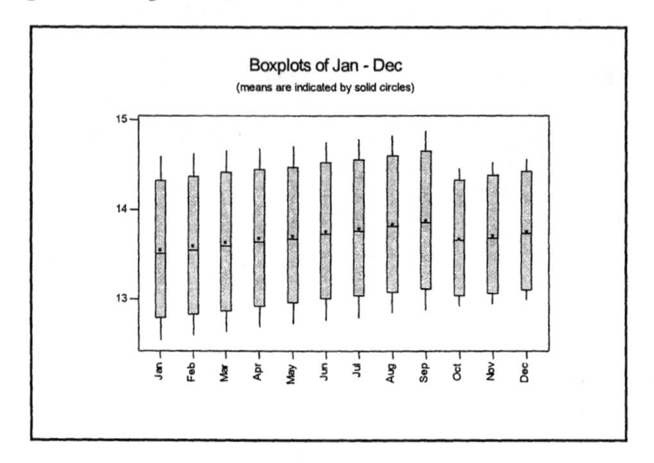

Month	n	Mean	StDev
Jan	5	13.546	0.808
Feb	5	13.590	0.807
Mar	5	13.630	0.805
Apr	5	13.674	0.797
May	5	13.706	0.790
Jun	5	13.756	0.793
Jul	5	13.788	0.796
Aug	5	13.832	0.794
Sep	5	13.878	0.801
Oct	4	13.673	0.668
Nov	4	13.710	0.682
Dec	4	13.755	0.685

Year	n	Mean	StDev
1998	12	12.779	0.144
1999	12	13.244	0.136
2000	12	13.754	0.170
2001	12	14.313	0.162
2002	9	14.718	0.094

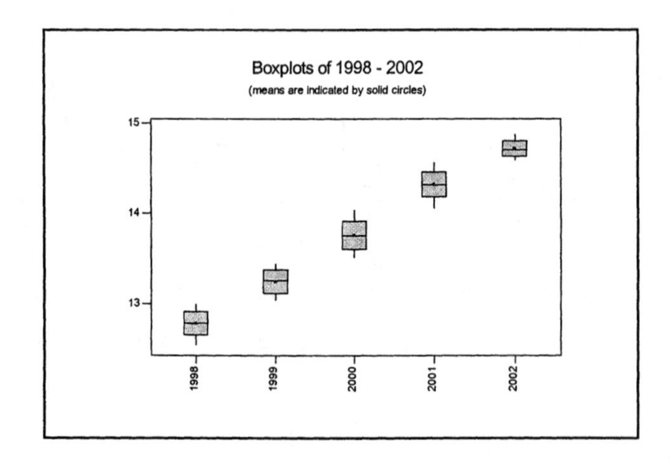

Hourly earnings are definitely increasing annually as seen by the regular increase in the annual mean hourly earnings. The standard deviation seems quite stable for both the monthly and yearly calculations. However on closer inspection of the data, it can be seen that the hourly earnings is increasing at a very steady rate, approximately $0.04, every month over the 5-year period.

b. Using all 11 years, 1992 - 2002

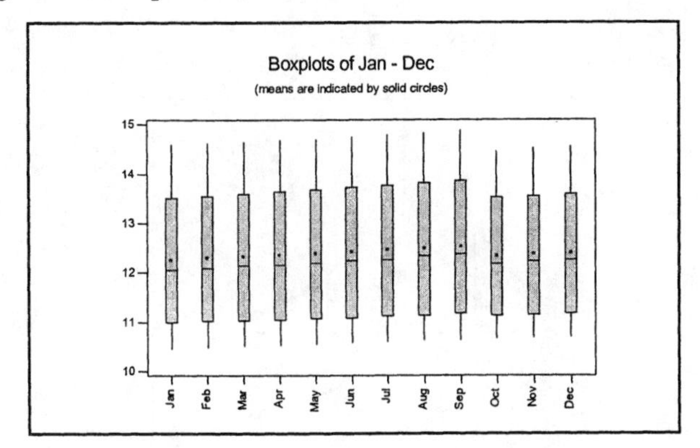

Month	n	Mean	StDev
Jan	11	12.259	1.399
Feb	11	12.292	1.409
Mar	11	12.327	1.413
Apr	11	12.359	1.423
May	11	12.386	1.426
Jun	11	12.429	1.436
Jul	11	12.457	1.440
Aug	11	12.495	1.448
Sep	11	12.531	1.460
Oct	10	12.336	1.310
Nov	10	12.368	1.320
Dec	10	12.404	1.329

Year	n	Mean	StDev
1992	12	10.574	0.079
1993	12	10.835	0.077
1994	12	11.113	0.086
1995	12	11.428	0.106
1996	12	11.812	0.131
1997	12	12.272	0.151
1998	12	12.779	0.144
1999	12	13.244	0.136
2000	12	13.754	0.170
2001	12	14.313	0.162
2002	9	14.718	0.094

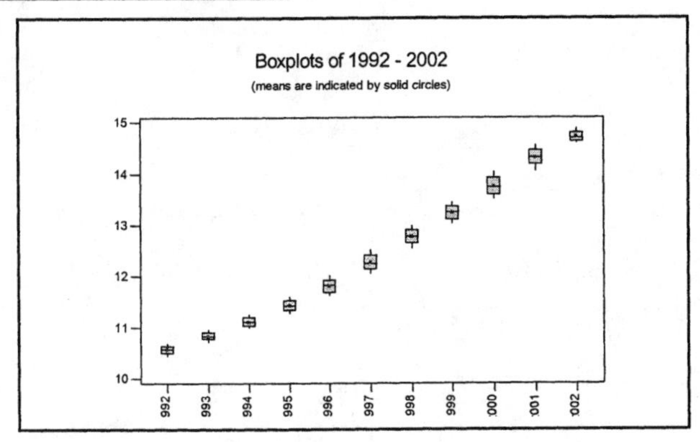

Draw a diagram of a normal curve with its corresponding percentages for standard deviations away from the mean (see Figure 2.32 in ES9-p107). Add the percentages from the left to the right, until the desired z-value is reached. The sum equals the percentile.

2.185 a. 2.5+13.5+34+34+13.5 = 97.5%; therefore, P_{98}

b. 2.5+13.5 = 16%; therefore, P_{16}

c. need curve

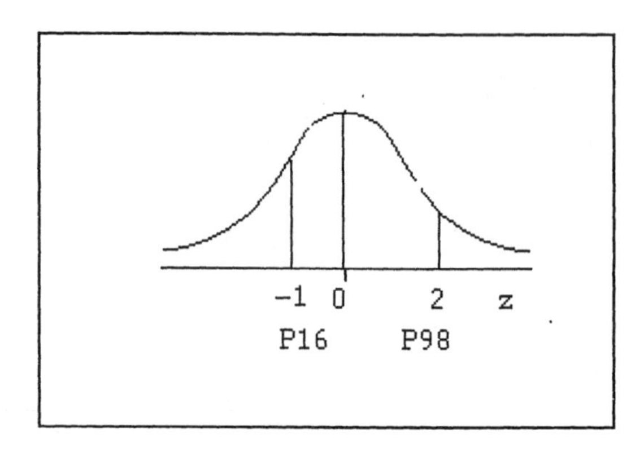

2.187 x = (z)(st.dev) + mean

Sit-ups: x = (-1)(12) + 70 = 58
Pull-ups: x = (-1.3)(6) + 8 = 0.2 = 0
Shuttle Run: x = (0)(0.6) + 9.8 = 9.8
50 yd. dash: x = (1)(.3) + 6.6 = 6.9
Softball: x = (0.5)(16) + 173 = 181

2.189 Data summary: n = 8, $\sum x = 31,825$, $\sum x^2 = 126,894,839$

a. $\bar{x} = \sum x/n = 31,825/8 = \underline{3978.1}$

b. $SS(x) = \sum x^2 - ((\sum x)^2/n)$
$= 126,894,839 - (31,825^2/8) = 291,010.88$

$s^2 = SS(x)/(n-1) = 291,010.88/7 = 41,572.982$

$s = \sqrt{s^2} = \sqrt{41572.98} = \underline{203.9}$

c. $\bar{x} \pm 2s = 3978.1 \pm 2(203.9)$
$= 3978.1 \pm 407.8$ or $\underline{3570.3}$ to $\underline{4385.9}$

2.191 a. Summary of calories data: $n=50, \Sigma x=5805$, $\Sigma x^2 = 781425$

$\overline{x} = \Sigma x/n = 5805/50 = \underline{116.1}$

$SS(x) = \Sigma x^2 - ((\Sigma x)^2/n) = 781425 - (5805^2/50) = 107464.5$

$s^2 = SS(x)/(n-1) = 107464.5/49 = 2193.15306$

$s = \sqrt{s^2} = \sqrt{2193.15306} = \underline{46.83}$

Summary of Sodium data: $n= 50, \Sigma x= 28990, \Sigma x^2 = 19880100$

$\overline{x} = \Sigma x/n = 28990/50 = \underline{579.8}$

$SS(x) = \Sigma x^2 - ((\Sigma x)^2/n) = 19880100 - (28990^2/50) = 3071698$

$s^2 = SS(x)/(\Sigma f-1) = 3071698/49 = 62687.71429$

$s = \sqrt{s^2} = \sqrt{62687.71429} = \underline{250.4}$

b. Calories: $116.10 \pm 2(46.83)$ or between 22.44 and 209.76

Sodium: $579.80 \pm 2(250.38)$ or between 79.04 and 1,080.56 mg

Yes. Only 3 of the brands of soups fall outside this calorie interval, so 94% are included within the interval. Only 1 brand falls outside the sodium content interval, so 98% are included within the interval.

c. Sodium: 579.80 ± 250.38 or between 329.42 and 830.18 mg

The empirical rule predicts that 68% of the brands of soups' sodium content will fall between 329.42 and 830.18, provided the distribution is normally distributed. In fact, these limits include 32 of the brands out of 50 or 64%. The empirical rule predicts that 95% of the player's earnings will fall between 79.04 and 1,080.56, if normally distributed. In fact, these limits include 49 of the brands out of 50, or 98%. Therefore, the results suggest the sodium content of the soups to be an approximately normal distribution.

2.193 $n = 40$, $\Sigma x = 103545$, $\Sigma x^2 = 281671245$

a. $\overline{x} = \Sigma x/n = 103545/40 = \underline{2588.6}$

$SS(x) = \Sigma x^2 - ((\Sigma x)^2/n)$

$= 281671245 - (103545^2/40) = 13632069.38$

$s^2 = SS(x)/(n-1) = 13632069.38/39 = 349540.2404$

$s = \sqrt{s^2} = \sqrt{349540.2404} = \underline{591.22}$

b.

1393	1448	1656	1734	1781	1837	1889
1895	2045	2064	2191	2230	2345	2377
2388	2440	2511	2603	2604	2611	2632
2637	2731	2773	2827	2856	2915	2948
3078	3094	3168	3191	3191	3255	3293
3321	3357	3367	3430	3439		

$d(\tilde{x}) = (n+1)/2 = (40+1)/2 = 20.5^{th}$
$\tilde{x} = (2611+2632)/2 = 2621.5$

$Q1 = (2064+2191)/2 = 2127.5$
$Q3 = (3094+3168)/2 = 3131$

Answer: 1393, 2127.5, 2621.5 3131, 3439

c. $2588.6 \pm 2(591.2)$ or between 1406.2 and 3771. Yes, 39 of the 40, or 97.5%, fall within this interval.

d. 2588.6 ± 591.2 or between 1997.4 and 3179.8. No, 23 of the 40, or 57.5%, fall within this interval.

e. The points distribution for the top 40 is part of a skewed left distribution. There is a "wide" cluster near the top and then the distribution tails out to the left. By using the top 40, much of the left tail has been chopped off. Thus the percentages from the empirical rule do not hold true.

2.195 There are many possible answers for this question; only one of those possibilities is shown.

a. 70, 77.5, 77.5, 77.5, 85 yields $s = 5.30$, which is the smallest standard deviation for a sample of 5 data with 70 and 85.

b. 70, 76, 85, 89, 95 yields $s = 10.02$.

c. 70, 85, 90, 99, 110 yields $s = 15.02$.

d. In order to increase the standard deviation the data had to become more dispersed.

2.197

a.

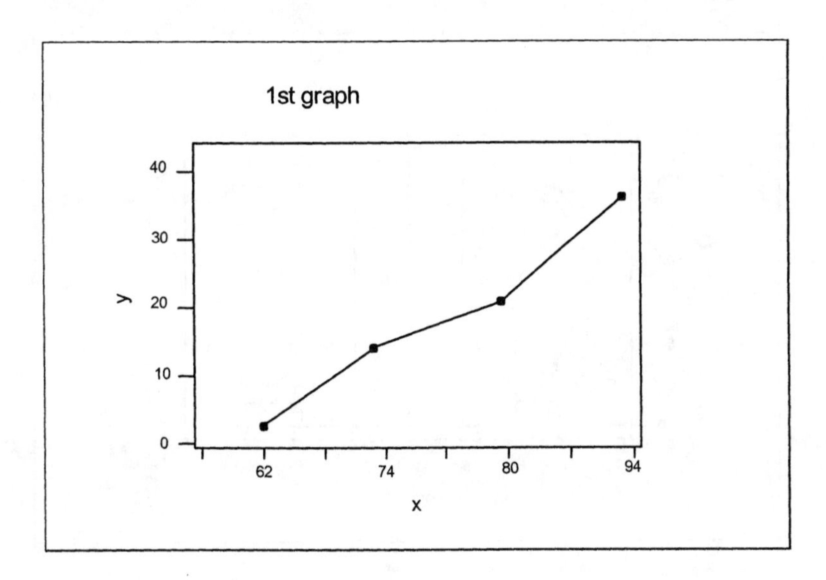

b.

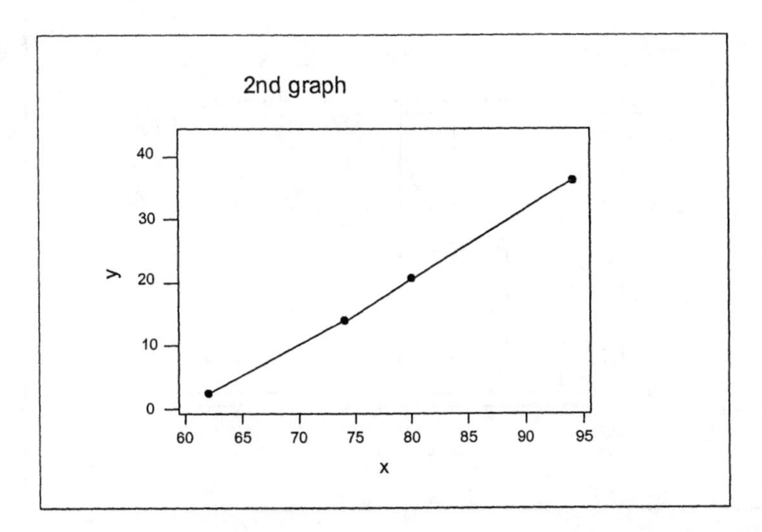

c. The line graph in (a) suggests an accelerated rate of increase from 1980 to 1995, while the line graph in (b) suggests that the rate of increase has been constant from 1962 to 1995.

d. By adjusting the horizontal intervals, it is possible to cause the line to show different slopes in those different intervals.

2.199 a.

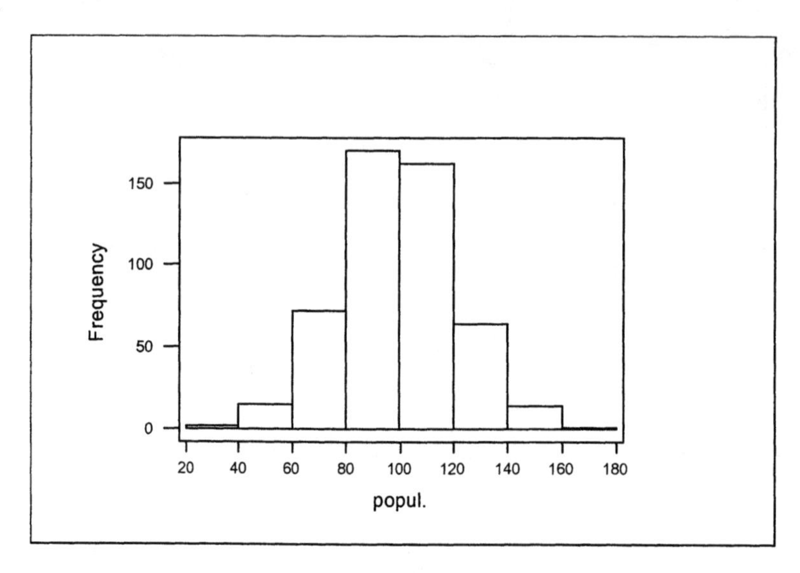

Frequency
150
100
50
0
20
40
60
80
100
120
140
160
180
popul.

b.

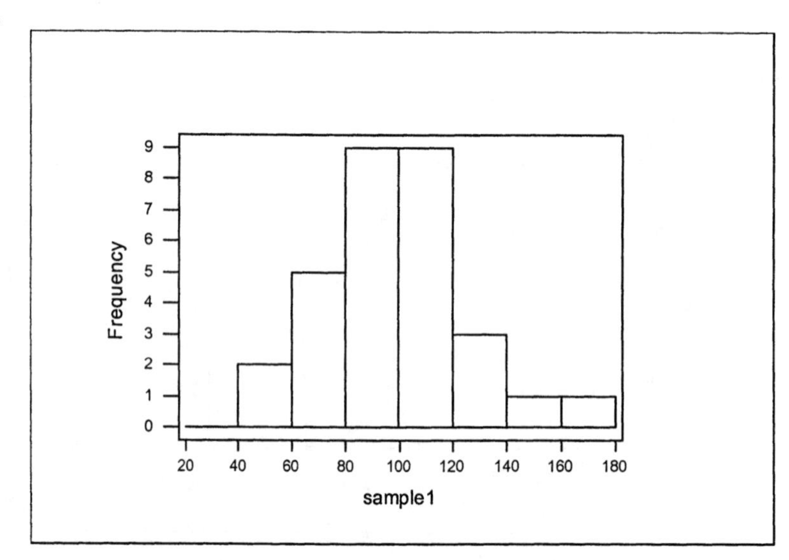

Frequency
9
8
7
6
5
4
3
2
1
0
20
40
60
80
100
120
140
160
180
sample1

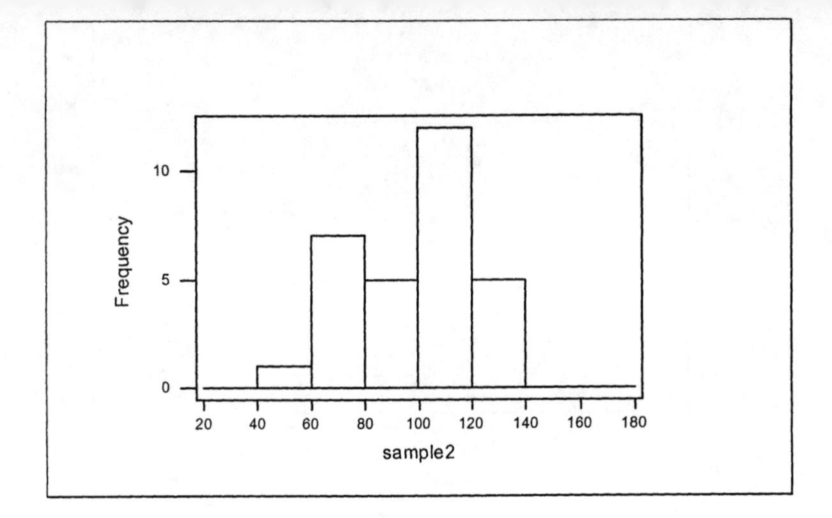

Frequency
10
5
0
20
40
60
80
100
120
140
160
180
sample2

c.

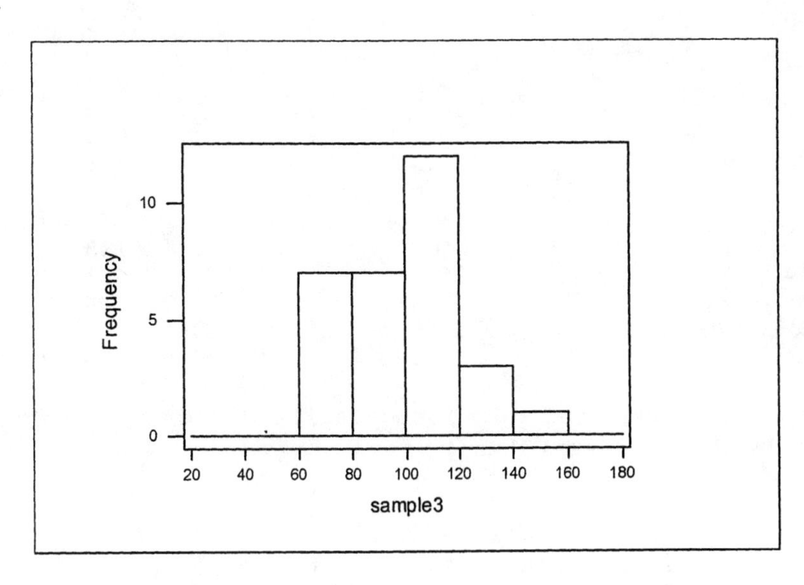

Frequency
10
5
0
20
40
60
80
100
120
140
160
180
sample3

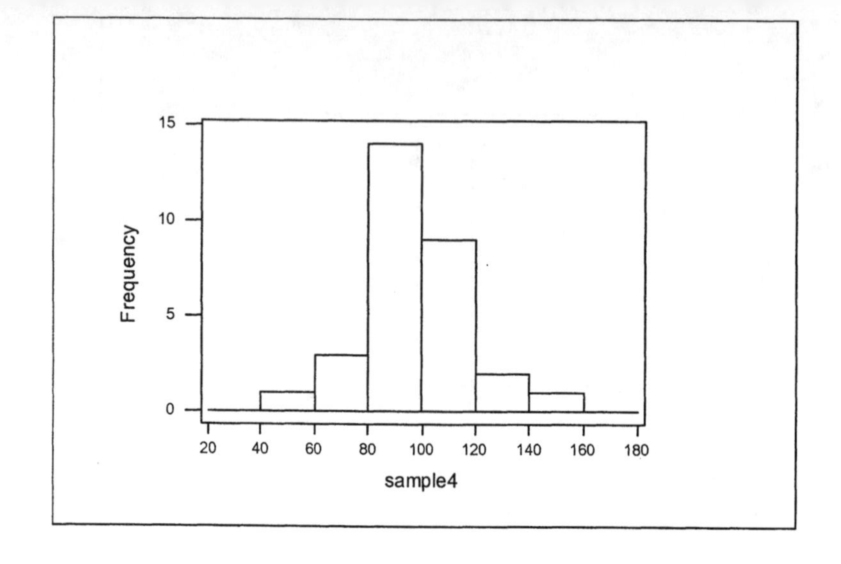

d.

Variable	N	Mean	Median	StDevpopul.
	500	98.932	99.190	20.915
sample1	30	98.35	96.19	25.53
sample2	30	96.84	101.26	21.12
sample3	30	99.25	100.75	20.01
sample4	30	97.45	95.28	18.83

Variable	Minimum	Maximum	Q1	Q3popul.
	31.792	162.786	84.358	113.016
sample1	53.65	162.79	83.17	114.72
sample2	53.18	128.83	77.83	112.62
sample3	65.26	141.59	80.76	115.00
sample4	57.90	151.98	88.52	107.45

e. Yes, the sample statistics calculated closely resemble the population parameters.

2.201 Samples of size 30 usually demonstrated some of the properties of the population. As the sample size was increased, more of the properties of the population were shown. The suggested distributions in this exercise seem to require sample sizes greater than 30 for a closer match to the population.

CHAPTER 3 ∇ DESCRIPTIVE ANALYSIS AND PRESENTATION OF BIVARIATE DATA

Chapter Preview

Chapter 3 deals with the presentation and analysis of bivariate (two variables) data. There are three main categories of bivariate data.

1. Two Qualitative Variables

This type of data is best presented in a contingency table and/or bar graph. Variations of the contingency table are given in Section 1 of Chapter 3.

2. One Qualitative Variable and One Quantitative Variable

This type of data can be presented and/or summarized in table form or graphically. More statistical techniques are available because of the one quantitative variable. Dot plots, box plots, and stem-and-leaf diagrams can represent the data for each different value of the qualitative variable.

3. Two Quantitative Variables

Initially, this type of data is best presented in a scatter diagram. If a relationship seems to exist, based on the scatter plot, then linear correlation and regression techniques will be performed.

An article reported by ESPN on Tim Duncan of the San Antonio Spurs basketball team is presented in this chapter's case study.

CHAPTER 3 CASE STUDY

3.1 a. Yes. A relationship seems to exist, the higher number of personal fouls go with the higher values of points scored per game.

b. Somewhat. Explanations will vary.

SECTION 3.1 ANSWER NOW EXERCISES

3.2

	On Airplane	Hotel Room	All Other	Marginal total
Business	35.5%	9.5%	5.0%	50%
Leisure	25.0%	16.5%	8.5%	50%
Marginal total	60.5%	26.0%	13.5%	100%

3.3

	On Airplane	Hotel Room	All Other	Marginal total
Business	71.0%	19.0%	10.0%	100%
Leisure	50.0%	33.0%	17.0%	100%
Marginal total	60.5%	26.0%	13.5%	100%

The table shows the distribution of ratings for business and leisure separately. For example, 71% of business travelers would like more space on the airplanes while 50% of leisure travelers would like more space on the airplanes.

3.4

	On Airplane	Hotel Room	All Other	Marginal total
Business	58.7%	36.5%	37.0%	50%
Leisure	41.3%	63.5%	63.0%	50%
Marginal total	100%	100%	100%	100%

The table shows the distribution of business travelers and leisure travelers for each of the categories. For example, for more space in the hotel room, 36.5% of the responses were from business travelers and 63.5% from leisure travelers.

3.5

```
   .  .  .         .                          .
-----+---------+---------+---------+---------+---------+-
Eastrate


      .      .            .            .               .
-----+---------+---------+---------+---------+---------+-
Westrate
       5.40      5.70      6.00      6.30      6.60      6.90
```

East coast cities: $\overline{x} = \Sigma x/n = 28.6/5 = 5.72$

$d(\tilde{x}) = (n+1)/2 = (5+1)/2 = 3\text{rd};\ \tilde{x} = 5.5$

West coast cities: $\overline{x} = \Sigma x/n = 30.3/5 = 6.06$

$d(\tilde{x}) = (n+1)/2 = (5+1)/2 = 3\text{rd};\ \tilde{x} = 6.0$

3.6 The input variable most likely would be height. Based on height, weight is often predicted or given in a range of acceptable values depending on the size of a person's frame.

3.7

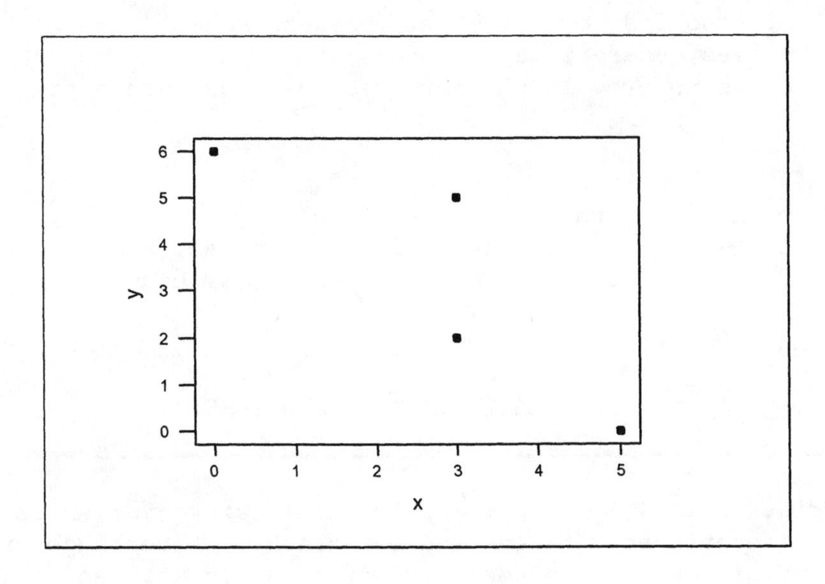

3.8 a.

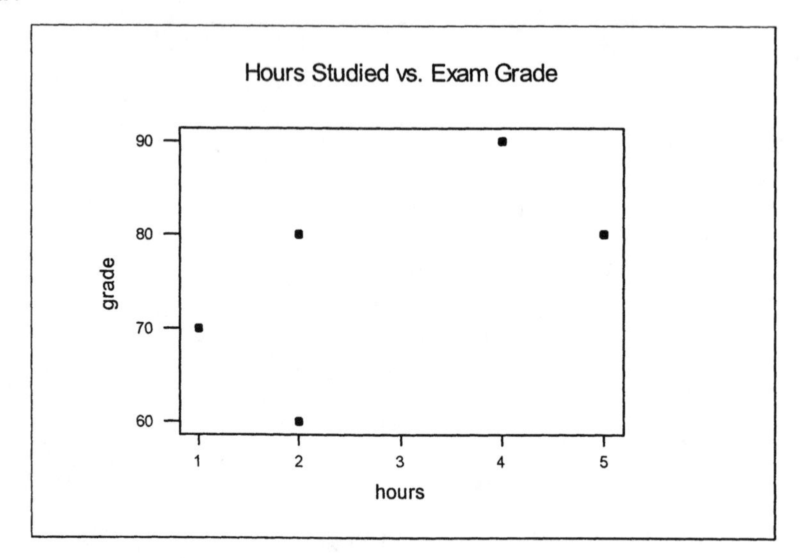

b. As hours studied increased, there seems to be a trend for the exam grades to also increase.

3.9 a. school's poverty level, passage rate;
b. Yes, there is a pattern.
c. As poverty level increased, passage rate decreased.

3.10 a. age, height,
b. Age = 3 yrs., height = 87 cm.
c. Sarah's growth was well below normal,
d. Sarah's information is below normal band

SECTION 3.1 EXERCISES

Exercises 3.11 & 3.13 present two qualitative variables in the form of contingency tables and bar graphs. A contingency table is made up of rows and columns. Rows are horizontal and columns are vertical. Adding across the rows gives marginal row totals. Adding down the columns gives marginal column totals. The sum of the marginal row totals should be equal to the sum of the marginal column totals, which in turn, should be equal to the sample size.

Computer and/or calculator commands to construct a cross-tabulation table can be found in ES9-p134.

3.11 a. Population: Employed college graduates age 30 - 55 who had been out of college 10 or more years
Variables: Type of worker: Tech worker or Other worker
Reason for going to school: Professional, Personal, Both

b.

	Reason for going to school			
Worker	Professional	Personal	Both	Marginal Total
Tech	28%	31%	41%	100%
Other	47%	20%	33%	100%

3.13 a. Cross tabulation of vehicle type and maximum interstate highway speed limit (frequencies).

	Interstate Highway Speed Limits (mph)					
Vehic	***75***	***70***	***65***	***60***	***55***	***Row Totals***
Cars	*10*	*16*	*22*	*0*	*2*	*50*
Truck	*9*	*11*	*20*	*3*	*7*	*50*
Col Total	*19*	*27*	*42*	*3*	*9*	*100*

For percentages based on the grand total, divide each count by 100, the grand total.

b. Cross tabulation of vehicle type and maximum interstate highway speed limit (relative frequencies, % of grand total).

	Interstate Highway Speed Limits (mph)					
Vehic	**75**	**70**	**65**	**60**	**55**	**Row Totals**
Cars	10%	16%	22%	0%	2%	50%
Truck	9%	11%	20%	3%	7%	50%
Colum Total	19%	27%	42%	3%	9%	100%

Be sure to leave spaces between the bars of different categories in the bar graph.

c. Bar graph of (b):

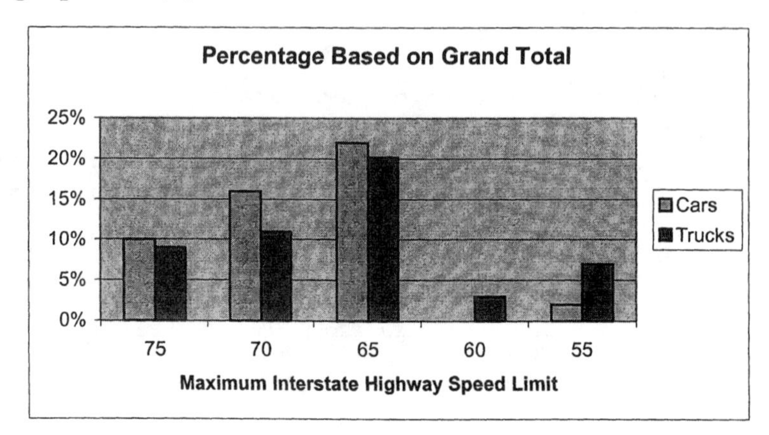

Before answering any questions concerning data in contingency tables, add all of the rows and columns. Be sure the sum of the row totals = the sum of the column totals = the grand total. Now you are ready to answer all questions easily.

3.15 a. 3000

b. Two variables, political affiliation and news information preferred, are paired together. Both variables are qualitative.

c. 950

d. 50% [1500/3000]

e. 25% [200/800]

Exercise 3.17 demonstrates the statistical methods that can be used on "one qualitative, one quantitative" type data. Be sure to split the data based on the qualitative variable. The effect is a side-by-side comparison of the quantitative variable for each different value of the qualitative variable.

Computer and/or calculator commands to construct multiple dotplots or boxplots can be found in ES9-p136.

3.17 a.

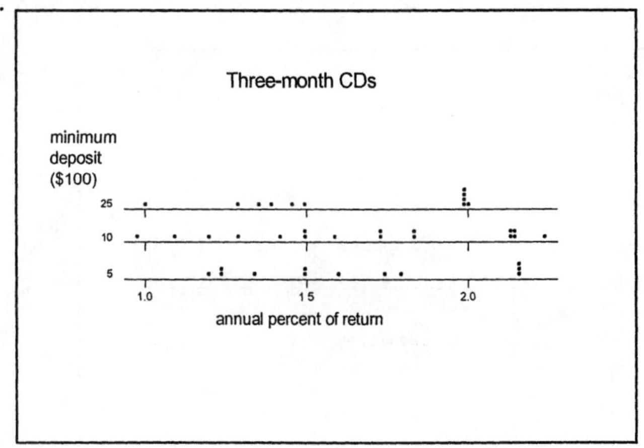

b.

	Low	Q_1	Median	Q_3	High
5:	1.19	1.265	1.545	2.068	2.16
10:	0.98	1.35	1.73	2.13	2.23
25:	1.00	1.35	1.490	1.990	2.00

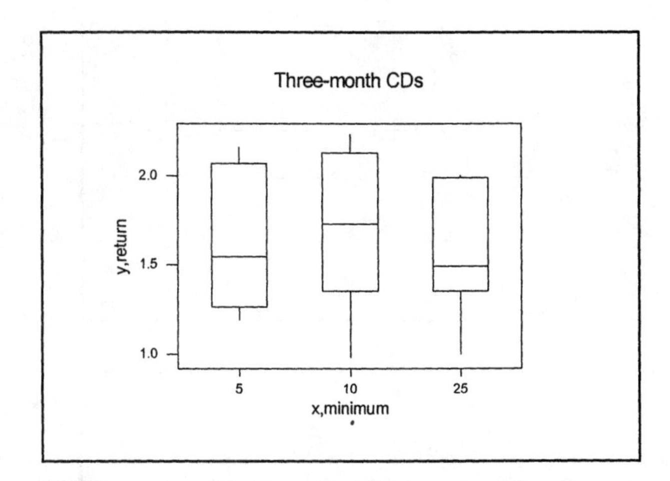

c. The $500 minimum does not have any of the very low return rates like the other two have.
   1. All three have similar interquartile range values.
   2. The $2500 minimum does not have any of the higher return rates.

3.19 a.

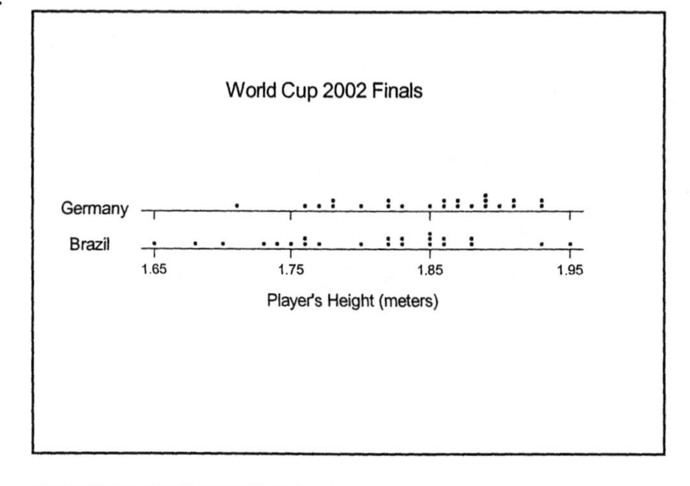
World Cup 2002 Finals
Germany
Brazil
1.65
1.75
1.85
1.95
Player's Height (meters)

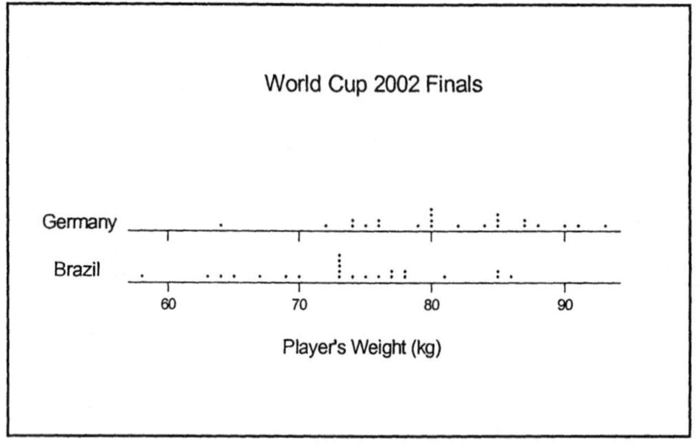
World Cup 2002 Finals
Germany
Brazil
60
70
80
90
Player's Weight (kg)

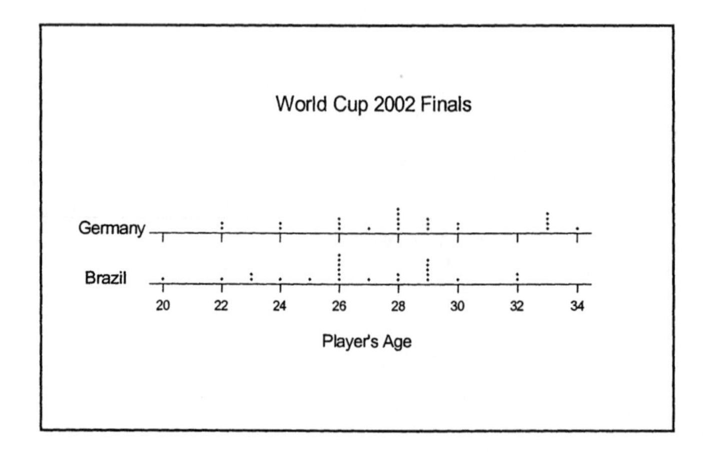
World Cup 2002 Finals
Germany
Brazil
20
22
24
26
28
30
32
34
Player's Age

b. It appears that the German team is made up of players that are slightly taller and slightly heavier, and perhaps a bit older than the Brazilian team.

c. Each variable was used separately and the players on each team are different - no pairing took place.

Exercises 3.21 - 3.25 demonstrate the numerical approach that can be taken now that we have two quantitative variables. A scatter diagram is the first tool we use in determining whether a linear relationship exists between the two variables. Decide which variable is to be predicted. This variable will be the dependent variable.* Let x be equal to the independent variable (input variable) and y be equal to the dependent variable (output variable).

How to construct a scatter diagram:

1. Find the range of the x values and the range of the y values.
2. Based on these, choose your increments for the x-axis (horizontal axis) and then for the y-axis (vertical axis). They will not always be the same.
3. Each point on the scatter diagram is made up of an ordered pair(x,y). (x,y) is plotted at the point that is x units on the x-axis and y units on the y-axis.
4. Label both axes and give a title to the diagram.

*(ex. the age of a car and the price of a car; price would be the dependent variable if we wish to predict the price of a car based on its age)

Computer and/or calculator commands to construct a scatter diagram can be found in ES-pp139&140.
The Student Suite CD has a video clip: "Bivariate Data".

3.21

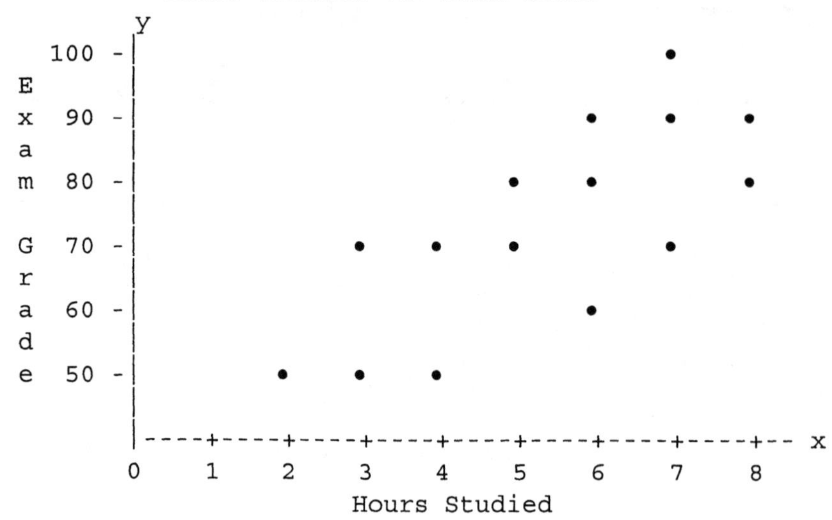
Hours Studied vs. Exam Grade
y
100
90
80
70
60
50
Exam Grade
x
0
1
2
3
4
5
6
7
8
Hours Studied

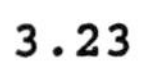
3.23

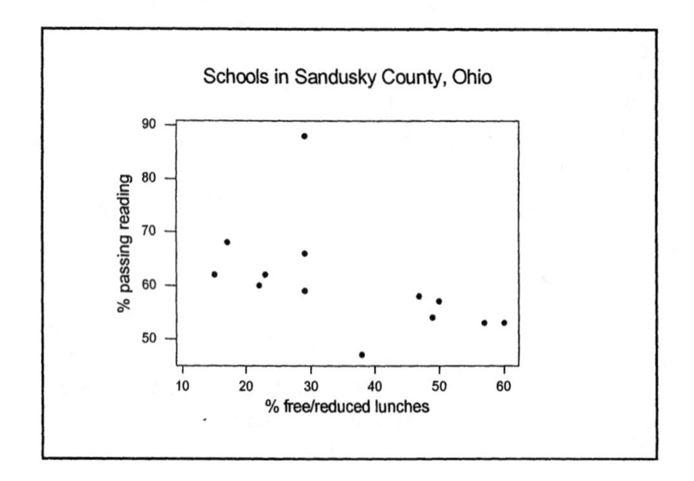
Schools in Sandusky County, Ohio
90
80
70
60
50
% passing reading
10
20
30
40
50
60
% free/reduced lunches

3.25 a.

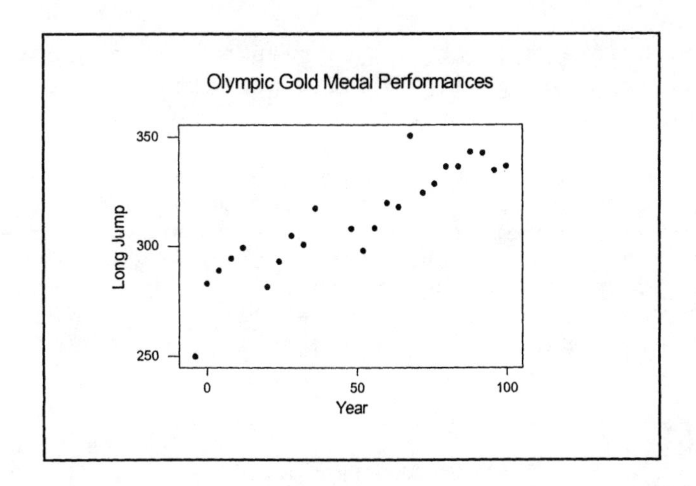

Olympic Gold Medal Performances
Long Jump
350
300
250
0
50
100
Year

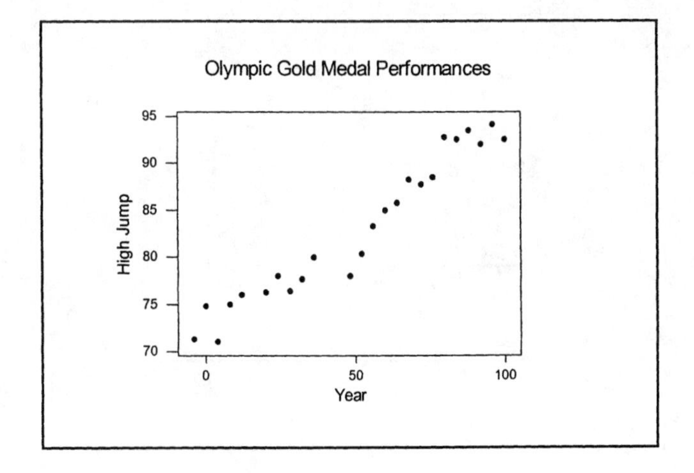

Olympic Gold Medal Performances
High Jump
95
90
85
80
75
70
0
50
100
Year

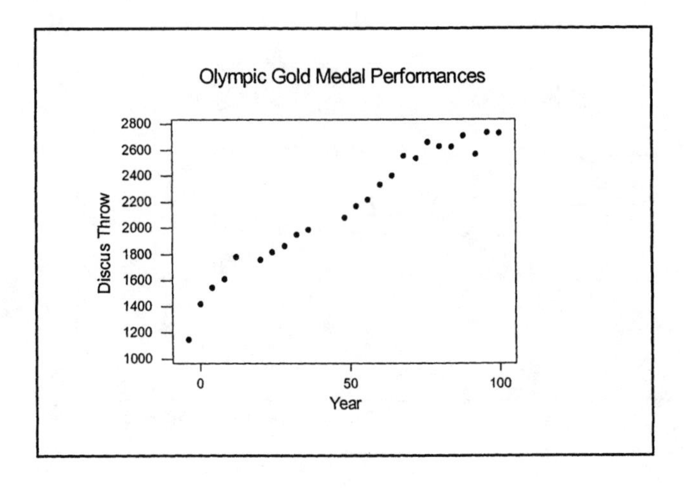

Olympic Gold Medal Performances
Discus Throw
2800
2600
2400
2200
2000
1800
1600
1400
1200
1000
0
50
100
Year

b. There is a very strong increasing pattern in all three, and even though it is not a perfectly straight line, they all seem to follow a fairly straight pattern.

c. These three scatter diagrams indicate that our Olympic athletes, at least the gold medal winning ones, jump higher and longer and throw the discus further. All of these are strength skills and indicate that these athletes are stronger today. It is reasonable to anticipate that the general population will follow a similar pattern.

d.

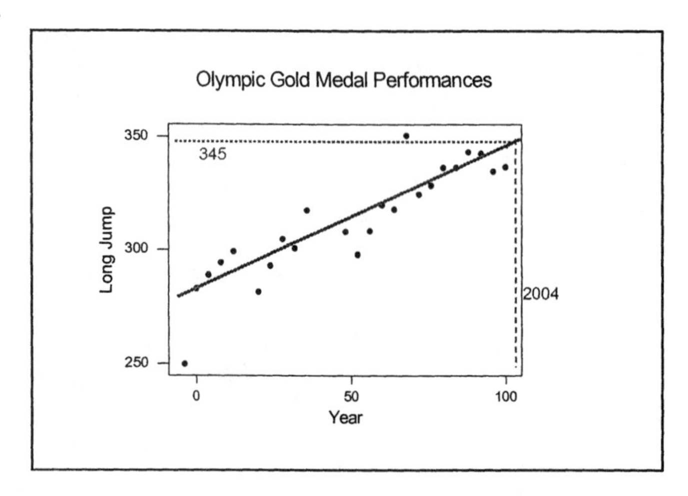

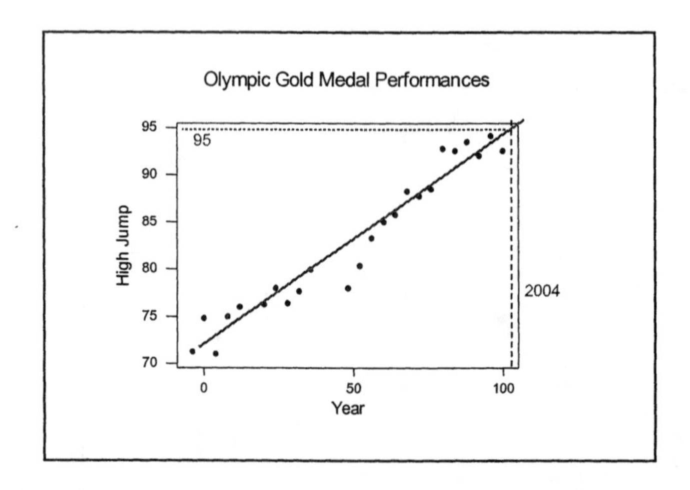

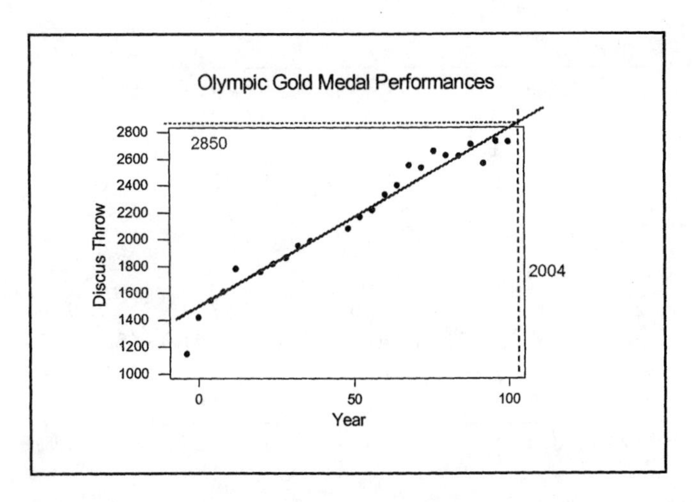

Predictions based on the scatter diagrams (in inches):
Long jump - 345, high jump - 95, discus throw - 2850.

e.

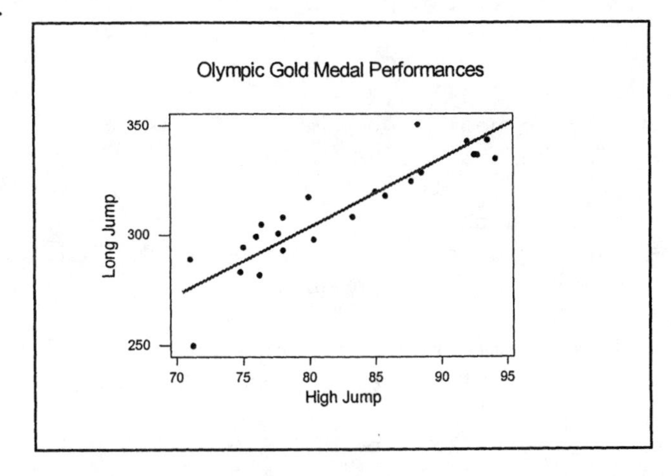

They seem to display a linear relationship. This should not be surprising.

SECTION 3.2 ANSWER NOW EXERCISES

3.27 a. Become closer to a straight line with a positive slope;
b. Become closer to a straight line with a negative slope

3.28 a.

x	y	x^2	xy	y^2
2	80	4	160	6400
5	80	25	400	6400
1	70	1	70	4900
4	90	16	360	8100
2	60	4	120	3600
14	380	50	1110	29,400

$SS(x) = \sum x^2 - ((\sum x)^2/n) = 50 - (14^2/5) = \underline{10.8}$
$SS(y) = \sum y^2 - ((\sum y)^2/n) = 29{,}400 - (380^2/5) = \underline{520}$
$SS(xy) = \sum xy - ((\sum x \cdot \sum y)/n) = 1110 - (14 \cdot 380/5) = \underline{46}$

b. $r = SS(xy)/\sqrt{SS(x) \cdot SS(y)} = 46/\sqrt{10.8 \cdot 520} = 0.6138 = \underline{0.61}$

3.29 Answers will vary: positive vs. negative; nearness to straight line, etc.

3.30 a. $r = 1$ or $r = -1$, two points make a straight line
b. Answers will vary.

3.31 Estimate r to be near -0.75
Estimate r to be near 0.00
Estimate r to be near +0.75

3.32 a. Manatees, powerboats
b. Number of registrations, manatee deaths
c. As one increases other does also

3.33 a. Summations from extensions tables: $n = 7$, $\sum x = 315$, $\sum y = 294.7$, $\sum x^2 = 14875$, $\sum xy = 15274$, $\sum y^2 = 19177$

$SS(x) = \sum x^2 - ((\sum x)^2/n) = 14875 - (315^2/7) = 700$
$SS(y) = \sum y^2 - ((\sum y)^2/n) = 19177 - (294.7^2/7) = 6770.13$
$SS(xy) = \sum xy - ((\sum x \cdot \sum y)/n) = 15274 - (315 \cdot 294.7/7) = 2012.5$

$r = SS(xy)/\sqrt{SS(x) \cdot SS(y)} = 2012.5/\sqrt{700} \cdot 6770.13 = \underline{0.924} = \underline{0.92}$

b.

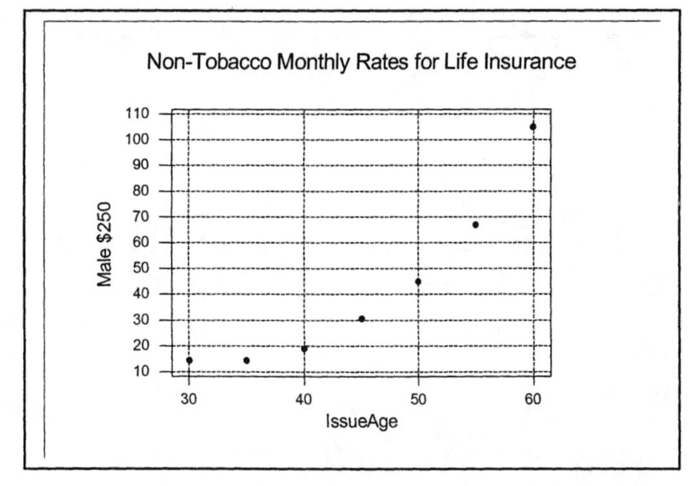

c. The relationship between age and monthly rate is not linear, it does however show a very definite and well-defined increasing pattern. It is this elongated pattern that causes the calculated r to be so large.

d. The pattern shown on the scatter diagrams showing age versus monthly rate should not surprise us, after all we have always been told the rate for insurance increases (accelerates) as the insured person's issue age increases, thus the "upward-bending" pattern.

3.34 a. Number of drownings, ice cream sales
b. No, both increase during hot weather

SECTION 3.2 EXERCISES

Based on the scatter diagram, we might suspect whether or not a linear relationship exists. If the y's increase linearly as the x's increase, there exists a relationship called positive correlation. If the y's decrease linearly as the x's increase, there exists a relationship called negative correlation. The measure of the strength of this linear relationship is denoted by *r*, the coefficient of linear correlation.

r = linear correlation coefficient

1. r has a value between -1 and +1, i.e. $-1 \le r \le +1$
2. r = -1 specifies perfect negative correlation. All of the data points would fall on a straight line slanted downward.
3. r = +1 specifies perfect positive correlation. All of the data points would fall on a straight line slanted upward.
4. $r \approx 0$ indicates little or no consistent linear pattern or a horizontal pattern.

The Student Suite CD has a video clip: "Linear Correlation".

3.35 a. Impossible. The correlation coefficient must be a numerical value between -1 and +1. There must be a calculation or typographical error.

b. Coefficient values near zero indicate that there is very little or no linear correlation.

Calculating r - the linear correlation coefficient

Preliminary Calculations:

1. Set up a table with the column headings: x, y, x^2, xy and y^2.
2. Insert the bivariate data into corresponding x and y columns. Perform the various algebraic functions to fill in the remaining columns.
3. Sum all columns, that is, find $\sum x$, $\sum y$, $\sum x^2$, $\sum xy$, $\sum y^2$.
4. Double check calculations and summations.
5. Calculate: SS(x) - the sum of squares of x
   SS(y) - the sum of squares of y
   SS(xy) - the sum of squares of xy

   where:

$$SS(x) = \sum x^2 - ((\sum x)^2/n)$$
$$SS(y) = \sum y^2 - ((\sum y)^2/n)$$
$$SS(xy) = \sum xy - ((\sum x \cdot \sum y)/n)$$

Final Calculation:

6. Calculate r:

$$r = \frac{SS(xy)}{\sqrt{SS(x)\,SS(y)}}$$

(round to the nearest hundredth) . . .

7. Retain the summations and the sums of squares, as they will be needed for later calculations.

NOTE: Remember $SS(x) \neq \sum x^2$, $SS(y) \neq \sum y^2$ and $SS(xy) \neq \sum xy$.

The computer and/or calculator command to calculate the correlation coefficient can be found in ES9-p148.

3.37 $n = 13$, $\sum x = 465$, $\sum y = 787$, $\sum x^2 = 19453$, $\sum xy = 27218$, $\sum y^2 = 48849$

a. $SS(x) = 19453 - (465^2/13) = 2820.307692$
b. $SS(y) = 48849 - (787^2/13) = 1205.23$
c. $SS(xy) = 27218 - (465 \cdot 787/13) = -932.385$

d. $r = SS(xy)/\sqrt{SS(x) \cdot SS(y)} = -932.385/\sqrt{2820.307692 \cdot 1205.23}$
$= \underline{-0.506} = \underline{-0.51}$

Estimating r - the linear correlation coefficient

1. Draw as small a rectangle as possible that encompasses all of the data on the scatter diagram. (Diagram should cover a "square window" - same length and width)
2. Measure the width.
3. Let k = the number of times the width fits along the length or in other words: length/width.
4. $r \approx \pm(1 - \frac{1}{k})$
5. Use +, if the rectangle is slanted positively or upward.
   Use -, if the rectangle is slanted negatively or downward.

3.39 a. Estimate r to be near 2/3 or 0.7.

b.

Data	x	y	x^2	xy	y^2
1	2	5	4	10	25
2	3	5	9	15	25
3	3	7	9	21	49
4	4	5	16	20	25
5	4	7	16	28	49
6	5	7	25	35	49
7	5	8	25	40	64
8	6	6	36	36	36
9	6	9	36	54	81
10	6	8	36	48	64
11	7	7	49	49	49
12	7	9	49	63	81
13	7	10	49	70	100
14	8	8	64	64	64
15	8	9	64	72	81
Σ	81	110	487	625	842

$$SS(x) = \Sigma x^2 - ((\Sigma x)^2/n) = 487 - (81^2/15) = 49.6$$
$$SS(y) = \Sigma y^2 - ((\Sigma y)^2/n) = 842 - (110^2/15) = 35.333$$
$$SS(xy) = \Sigma xy - ((\Sigma x \cdot \Sigma y)/n) = 625 - (81 \cdot 110/15) = 31.0$$

$$r = SS(xy)/\sqrt{SS(x) \cdot SS(y)} = 31.0/\sqrt{49.6 \cdot 35.333} = \underline{0.741} = \underline{0.74}$$

> * What is the r value of 0.74 in exercise 3.39 telling you? How do your estimated r and calculated r compare? They should be relatively close. (See the bottom of the next page for answer.)

3.41 a.

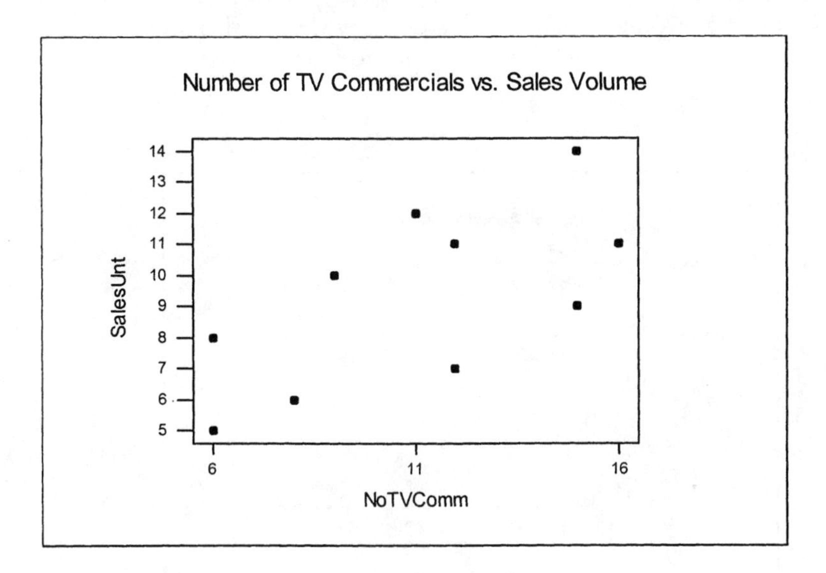

b. From 1/2 to 2/3

c. Summations from extensions table: $n = 10$, $\sum x = 110$, $\sum y = 93$, $\sum x^2 = 1332$, $\sum xy = 1085$, $\sum y^2 = 937$

$$SS(x) = \sum x^2 - ((\sum x)^2/n) = 1332 - (110^2/10) = 122.0$$
$$SS(y) = \sum y^2 - ((\sum y)^2/n) = 937 - (93^2/10) = 72.10$$
$$SS(xy) = \sum xy - ((\sum x \cdot \sum y)/n) = 1085 - (110 \cdot 93/10) = 62.00$$

$$r = SS(xy)/\sqrt{SS(x) \cdot SS(y)} = 62.0/\sqrt{122 \cdot 72.1} = \underline{0.66}$$

** What is the r value in exercise 3.41 telling you? (See the top of the next page for answer.)

*(3.39) There is a positive relationship between the variables, that is, as x increases, y increases. In this case, as the hours studied increased, the grade on the exam increased. What can one deduce from this?

**(3.41) As the number of television commercials increased, so did the amount of sales. This would be valuable information for the sales department as well as the advertising department.

3.43

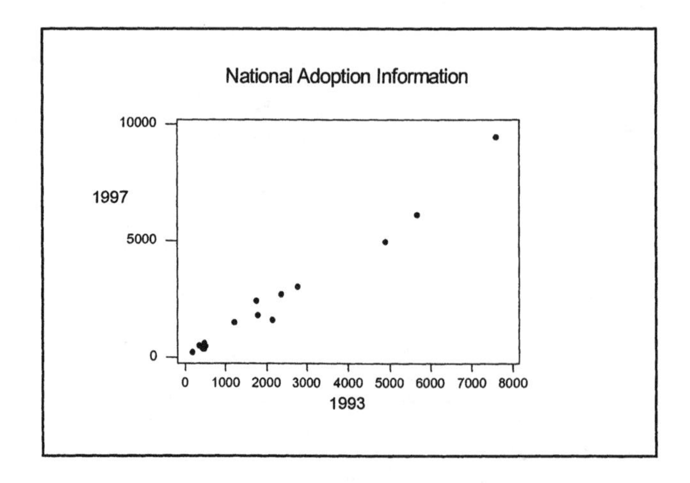

$n = 16$, $\sum x = 33019.0$, $\sum y = 36413.0$, $\sum x^2 = 140541807$,
$\sum xy = 159275750$, $\sum y^2 = 182712391$

$SS(x) = \sum x^2 - ((\sum x)^2/n) = 140541807 - (33019^2/16) = 72400909.44$
$SS(y) = \sum y^2 - ((\sum y)^2/n) = 182712391 - (36413^2/16) = 99843230.44$
$SS(xy) = \sum xy - ((\sum x \cdot \sum y)/n) = 159275750 - (33019 \cdot 36413/16)$
$= 84130697$

$r = SS(xy)/\sqrt{SS(x) \cdot SS(y)}$
$= 4130697/\sqrt{72400909.44 \cdot 99843230.44}$
$= \underline{0.990}$

There definitely appears to be a linear relationship between the two variables.

3.45 No. Both amount of ice cream sales and the number of drownings increase during months of warmer weather and decrease during months of cooler weather. Thus they would be expected to have a positive correlation.

SECTION 3.3 ANSWER NOW EXERCISES

3.47 $b_0 = \frac{\sum y - (b_1 \cdot \sum x)}{n} = \frac{\sum y}{n} - \frac{(b_1 \cdot \sum x)}{n} = \bar{y} - b_1\bar{x}$

3.48 a. $\sum x^2 = 13{,}717$, $SS(x) = 1396.9$
$\sum y^2 = 15{,}298$, $SS(y) = 858.0$
$\sum xy = 14{,}257$, $SS(xy) = 919.0$

b. Summations (Σ's) are sums of data values. Sum of squares (SS( )) are parts of complex formulas; they are calculated separately as preliminary values.

3.49 a. $\hat{y} = 14.9 + 0.66(20) = \underline{28.1}$

$\hat{y} = 14.9 + 0.66(50) = \underline{47.9}$

b. Yes, the line of best fit is made up of all points that satisfy its equation.

3.50 a. Summations from extensions table: $n = 5$, $\sum x = 14$, $\sum y = 380$, $\sum x^2 = 50$, $\sum xy = 1110$, $\sum y^2 = 29{,}400$

$SS(x) = \sum x^2 - ((\sum x)^2/n) = 50 - (14^2/5) = 10.8$
$SS(xy) = \sum xy - ((\sum x \cdot \sum y)/n) = 1110 - (14 \cdot 380/5) = 46$

$b_1 = SS(xy)/SS(x) = 46/10.8 = 4.259$

$b_0 = [\sum y - b_1 \cdot \sum x]/n = [380 - (4.259 \cdot 14)]/5 = 64.0748$

$\underline{\hat{y} = 64.1 + 4.26x}$

b. At $x = 1$, $\hat{y} = 64.1 + 4.26(1) = 68.36$; thus (1,68.4)
At $x = 3$, $\hat{y} = 64.1 + 4.26(3) = 76.9$; thus (3,76.9)
Points (1,68.4) and (3,76.9) are used to locate the line.

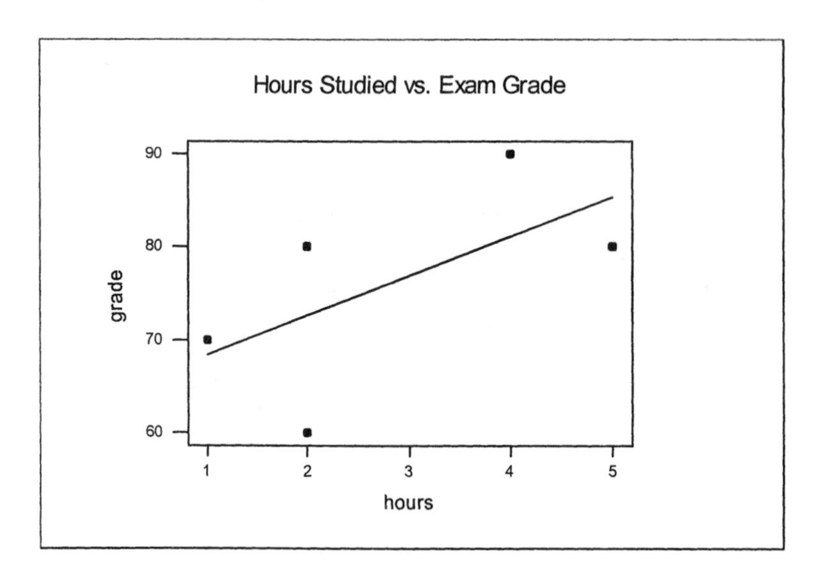

c. Yes, as the hours studied increased, the exam grades appear to increase, also.

3.51 a. The slope of 4.71 indicates that for each increase in height of one inch, college women's weight increased by 4.71 pounds.

b. The scale for the y-axis starts at y = 95 and the scale for the x-axis starts at x = 60. The y-intercept of -186.5 occurs when x = 0, so the x-axis would have to include x = 0 and the the y-axis would have to be extended down.

3.52 a. $\hat{y} = 14.9 + 0.66(40) = 41.3 = \underline{41}$

b. No

c. 41 is the average number of sit-ups expected for students who do 40 push-ups.

3.53 The vertical scale shown on figure 3.25 is located at x = 58 and therefore is not the y-axis; the y = 80 occurs at x = 58.Remember, the x-axis is the vertical line located at x= 0.

3.54 (61, 95) (67, 130)

$$b_1 \approx \frac{y_2 - y_1}{x_2 - x_1} = \frac{130 - 95}{67 - 61} = \frac{35}{6} = 5.83$$

$$b_0 \approx y - b_1 x = 130 - 5.83(67) = -260.61$$

3.55 a.

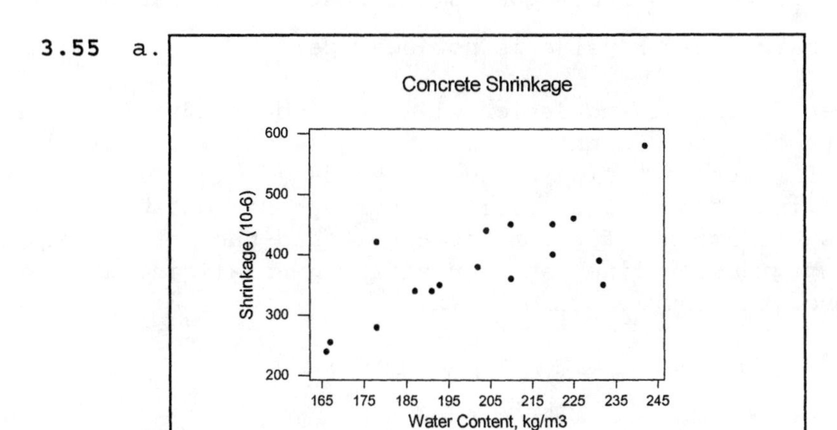

b. Yes; there is an elongated pattern from lower left to upper right corners of the scatter diagram.

c. Summations from extensions table: $n = 17$, $\sum x = 3456$, $\sum y = 6485$, $\sum x^2 = 711306$, $\sum xy = 1341865$, $\sum y^2 = 2586325$

$SS(x) = \sum x^2 - ((\sum x)^2/n) = 711306 - (3456^2/17) = 8721.529412$

$SS(xy) = \sum xy - ((\sum x \cdot \sum y)/n) = 1341865 - (3456 \cdot 6485/17)$
$= 23502.64706$

$b_1 = SS(xy)/SS(x) = 23502.64706/8721.529412 = 2.694785$

$b_0 = [\sum y - b_1 \cdot \sum x]/n = [6485-(2.694785 \cdot 3456)]/17 = -166.363$

$\hat{y} = -166.4 + 2.69x$

SECTION 3.3 EXERCISES

If a linear relationship exists between two variables, that is,

1. its scatter diagram suggests a linear relationship
2. its calculated r value is not near zero

the techniques of linear regression will take the study of bivariate data one step further. Linear regression will calculate an equation of a straight line based on the data. This line, also known as the line of best fit, will fit through the data with the smallest possible amount of error between it and the actual data points. The regression line can be used for generalizing and predicting over the sampled range of x.

FORM OF A LINEAR REGRESSION LINE

$$\hat{y} = b_0 + b_1x$$

where $\hat{y}$ (y hat) = predicted y
b_0 (b sub zero) = y intercept
b_1 (b sub one) = slope of the line
x = independent data value.

The Student Suite CD has a video clip: "Linear Regression".

3.57 a. The y-intercept of $23.65 is the amount of the total monthly telephone cost when x, the number of long distance calls, is equal to zero. That is, when no long distance calls are made there is still the monthly phone charge of $23.65.

b. The slope of $1.28 is the rate at which the total phone bill will increase for each additional long distance call; it is related to average cost of the long distance calls.

3.59 a. $\hat{y} = 185.7 - 21.52x$, when x = 3
$\hat{y} = 185.7 - 21.52(3) = \underline{121.14}$ or $\underline{\$12,114.}$

b. $\hat{y} = 185.7 - 21.52x$, when x = 6
$\hat{y} = 185.7 - 21.52(6) = \underline{56.58}$ or $\underline{\$5,658.}$

c. 21.52($100) = <u>$2,152</u> [the "slope" in dollars]

CALCULATING $\hat{y} = b_0 + b_1x$ - THE EQUATION OF THE LINE OF BEST FIT

1. Retrieve preliminary calculations from previous *r* calculations.

2. Calculate b_1 where $b_1 = \frac{SS(xy)}{SS(x)}$

3. Calculate b_0 where $b_0 = \frac{1}{n}(\sum y - b_1 \sum x)$

$\hat{y}$ = predicted value of y (based on the regression line)

NOTE: See Review Lessons for additional information about the concepts of slope and intercept of a straight line.

DRAWING THE LINE OF BEST FIT ON THE SCATTER DIAGRAM

1. Pick two *x*-values that are within the interval of the data *x*-values. (one value near either end of the domain)
2. Substitute these values into the calculated $\hat{y} = b_0 + b_1x$ equation and find the corresponding $\hat{y}$ values.
3. Plot these points on the scatter diagram in such a manner that they are distinguishable from the actual data points.
4. Draw a straight line connecting these two points. This line is a graph of the line of best fit.
5. Plot a third point, the ordered pair $(\bar{x}, \bar{y})$ as an additional check. It should be a point on the line of best fit.

OR:
Computer and/or calculator commands to find the equation of the line of best fit and also draw it on a scatter diagram can be found in ES9-pp161&162.

3.61 a.

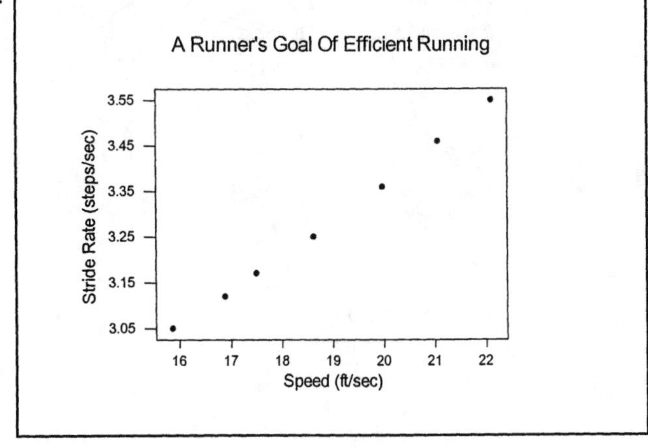

b. Yes the scatter diagram appears to show a linear pattern.

c. Summations from extensions table: $n = 7$, $\sum x = 132.00$, $\sum y = 22.96$, $\sum x^2 = 2520.61$, $\sum xy = 435.486$

$$SS(x) = \sum x^2 - ((\sum x)^2/n) = 2520.61 - (132^2/7) = 31.4621$$
$$SS(xy) = \sum xy - ((\sum x \cdot \sum y)/n) = 435.486 - (132 \cdot 22.96/7) = 2.52590$$

$$b_1 = SS(xy)/SS(x) = 2.52590/31.4621 = 0.0803$$
$$b_0 = [\sum y - b_1 \cdot \sum x]/n = [22.96 - (0.0803 \cdot 132)]/7 = 1.76577$$

$$\hat{y} = 1.77 + 0.0803x$$

d. For each one foot per second increase in speed, the stride will increase by 0.0803, or approximately (1/12)th, of a step per second.

e.

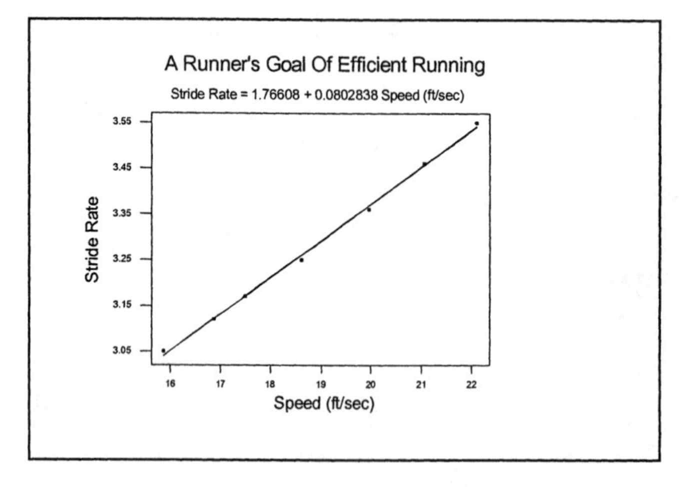

f.

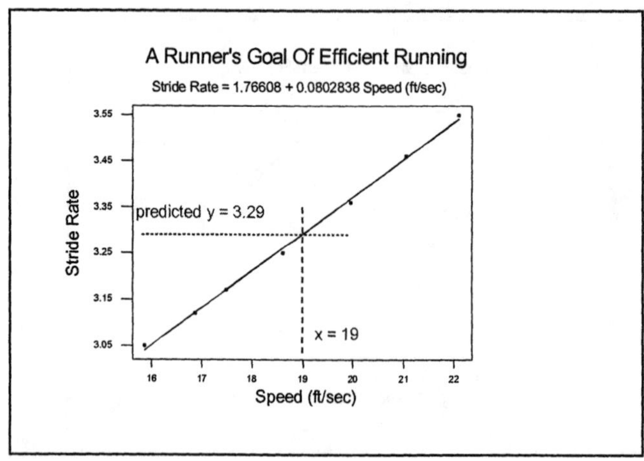

g. The stride rate at speed 0 using the line of best fit would be the y-intercept 1.77. However for this application, it seems only reasonable that the stride rate would be zero if the runner is standing still (speed zero). The domain of the study is speed 16 to 22 feet per second, a running speed and the regression methods do not apply outside the domain.

3.63 a.

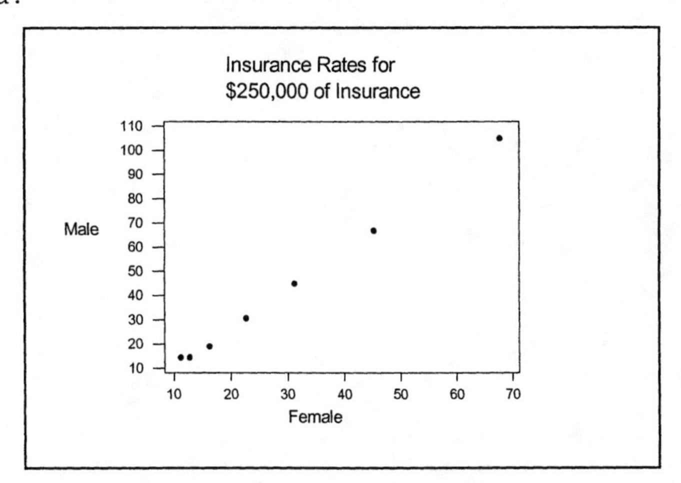

Yes, the scatter diagram shows a linear relationship. The ordered pairs follow very closely a straight line from lower left to upper right.

Summations from extensions table: $n = 7$, $\sum x = 206.47$, $\sum y = 294.7$, $\sum x^2 = 8640.98$, $\sum xy = 12845.6$, $\sum y^2 = 19176.8$

b. $SS(x) = \sum x^2 - ((\sum x)^2/n) = 8640.98 - (206.47^2/7) = 2551.00$

$SS(y) = \sum y^2 - ((\sum y)^2/n) = 19176.8 - (294.7^2/7) = 6769.93$

$SS(xy) = \sum xy - ((\sum x \cdot \sum y)/n) = 12845.6 - (206.47 \cdot 294.7/7) = 4153.21$

$r = SS(xy)/\sqrt{SS(x) \cdot SS(y)} = 4153.21/\sqrt{2551.0 \cdot 6769.93} = \underline{0.999}$

There is a strong linear relationship because the points all lay very close to a straight line.

c. $b_1 = SS(xy)/SS(x) = 4153.21/2551.0 = 1.628$

$b_0 = [\sum y - b_1 \cdot \sum x]/n = [294.7 - (1.628 \cdot 206.47)]/7 = -5.919$

$\underline{\hat{y} = -5.92 + 1.63x}$

d. male monthly rate = - 5.92 + (1.63·15.00) = -5.92 + 24.45 = $18.53

e. Males paid a higher rate. The slope of 1.63 means that males pay $1.63 for every $1.00 that females pay.

3.65 Multiplying 0.6 times an x value, even an x = 1, results in a 0.6 which is 5 times as much as the constant 0.12. Also, with a correlation coefficient of 0.99, it seems likely that the pattern shown on the scatter diagram must be linear and very strong. If that is the case, then the number of viewers must be quite effective in predicting the Nielsen Rating.

RETURN TO CHAPTER CASE STUDY

3.67 a. A relationship seems to exist, as the number of personal fouls committed increases, so does the number of points scored per game.

b. It seems contrary to common sense, but if there is a predictable pattern, then it might be possible to make such a prediction.

CHAPTER EXERCISES

3.71 a. Population: Workers ages 25-64 that have savings set aside for emergencies

Variables: Gender: Men or Women; nominal

Savings set aside: Less than a month's income, 1 to less than 3 months, 3 to less than 6 months, 6 or more months income or Don't know; ordinal

b.

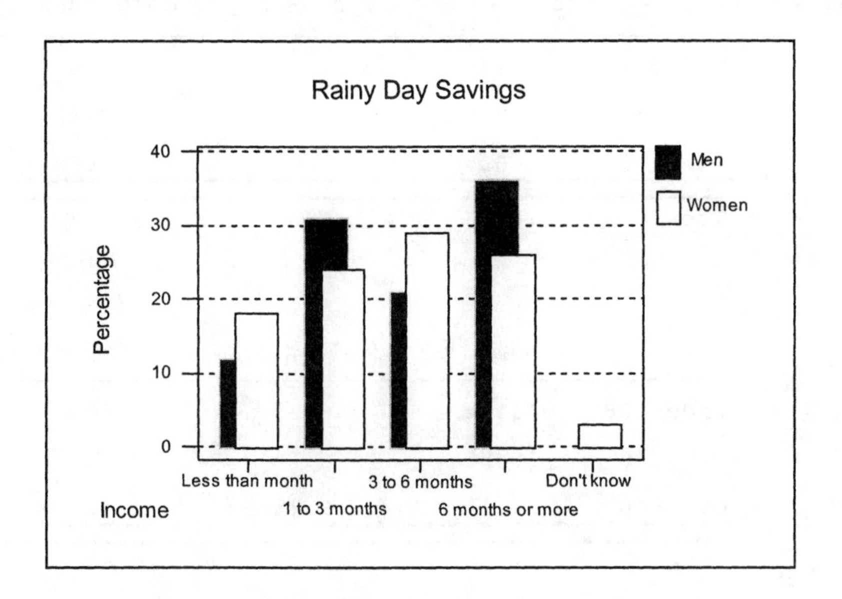

c. There does not seem to be a great difference between the genders and neither gender's distribution is consistently higher or lower than the other's gender's.

Exercises 3.72-3.74

<u>To find percentages based on the grand total</u>

- divide each count by the grand total

<u>To find percentages based on the row totals</u>

- divide each count by its corresponding row total
- each row should add up to 100%

<u>To find percentages based on the column totals</u>

- divide each count by its corresponding column total
- each column should add up to 100%

3.73 a.

	Less than 6mo	6 mo - 1 yr	More than 1yr	Total
Under 28	413	192	295	900
28 - 40	574	208	218	1000
Over 40	653	288	259	1200
Total	1640	688	772	3100

b.

	Less than 6mo	6 mo - 1 yr	More than 1yr	Total
Under 28	13.3%	6.2%	9.5%	29.0%
28 - 40	18.5%	6.7%	7.0%	32.2%
Over 40	21.1%	9.3%	8.4%	38.8%
Total	52.9%	22.2%	24.9%	100%

c.

	Less than 6mo	6 mo - 1 yr	More than 1yr	Total
Under 28	45.9%	21.3%	32.8%	100%
28 - 40	57.4%	20.8%	21.8%	100%
Over 40	54.4%	24.0%	21.6%	100%
Total	52.9%	22.2%	24.9%	100%

d.

	Less than 6mo	6 mo - 1 yr	More than 1yr	Total
Under 28	25.2%	27.9%	38.2%	29.0%
28 - 40	35.0%	30.2%	28.2%	32.3%
Over 40	39.8%	41.9%	33.6%	38.7%
Total	100%	100%	100%	100%

e.

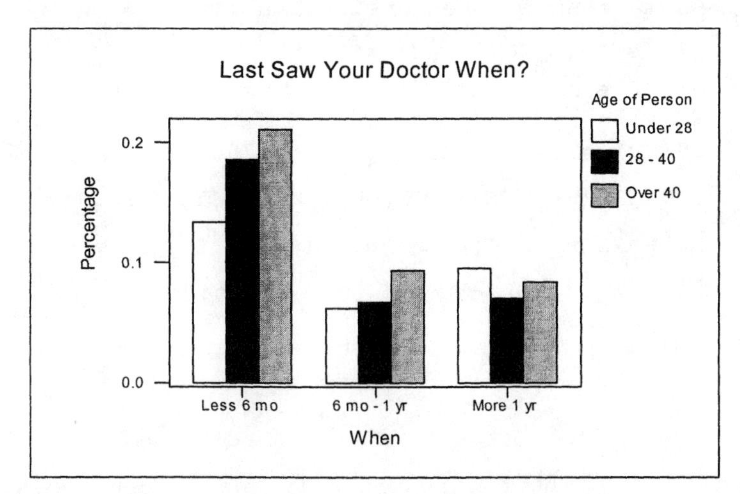

3.75 a. correlation b. regression c. correlation
d. regression e. correlation

3.77

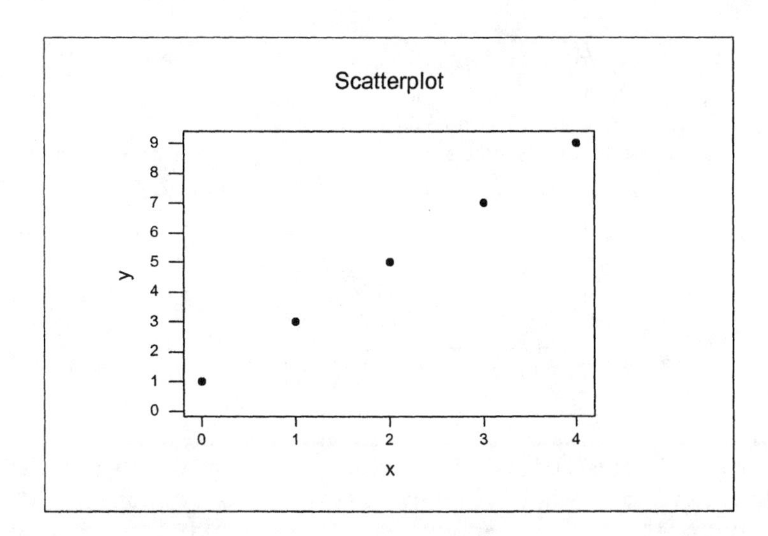

*What would you expect for the correlation coefficient in exercise 3.77 if all of these points fall exactly on a straight line? (See the bottom of the next page for answer.)

Summations from extensions table: $n = 5$, $\sum x = 10$, $\sum y = 25$, $\sum x^2 = 30$, $\sum xy = 70$, $\sum y^2 = 165$

$SS(x) = \sum x^2 - ((\sum x)^2/n) = 30 - (10^2/5) = 10.0$
$SS(y) = \sum y^2 - ((\sum y)^2/n) = 165 - (25^2/5) = 40.0$
$SS(xy) = \sum xy - ((\sum x \cdot \sum y)/n) = 70 - (10 \cdot 25/5) = 20.0$

$r = SS(xy)/\sqrt{SS(x) \cdot SS(y)} = 20.0/\sqrt{10.0 \cdot 40.0} = \underline{1.00}$

$b_1 = SS(xy)/SS(x) = 20.0/10.0 = 2.0$

$b_0 = [\sum y - b_1 \cdot \sum x]/n = [25 - (2.0 \cdot 10)]/5 = 1.0$

$\underline{\hat{y} = 1.0 + 2.0x}$

3.79 The answers shown here are only one of many possibilities.

a. (1,5), (2,5), (0,4), (0,6) [(5,5)]
Summations from extensions table: $n = 5$, $\sum x = 8$, $\sum y = 25$, $\sum x^2 = 30$, $\sum xy = 40$, $\sum y^2 = 127$

$SS(x) = \sum x^2 - ((\sum x)^2/n) = 30 - (8^2/5) = 17.2$
$SS(y) = \sum y^2 - ((\sum y)^2/n) = 127 - (25^2/5) = 2.0$
$SS(xy) = \sum xy - ((\sum x \cdot \sum y)/n) = 40 - (8 \cdot 25/5) = 0.0$

$r = SS(xy)/\sqrt{SS(x) \cdot SS(y)} = 0.0/\sqrt{17.2 \cdot 2.0} = \underline{0.00}$

b. (1,1), (2,2), (3,3), (4,4), [(5,5)]
Summations from extensions table: $n = 5$, $\sum x = 15$, $\sum y = 15$, $\sum x^2 = 55$, $\sum xy = 55$, $\sum y^2 = 55$

$SS(x) = \sum x^2 - ((\sum x)^2/n) = 55 - (15^2/5) = 10.0$
$SS(y) = \sum y^2 - ((\sum y)^2/n) = 55 - (15^2/5) = 10.0$
$SS(xy) = \sum xy - ((\sum x \cdot \sum y)/n) = 55 - (15 \cdot 15/5) = 10.0$

$r = SS(xy)/\sqrt{SS(x) \cdot SS(y)} = 10.0/\sqrt{10.0 \cdot 10.0} = \underline{1.00}$

*(3.77) If all of the points fall exactly on a straight line, perfect positive or negative correlation has occurred. The r value will be either +1 or -1, depending on the upward or downward trend of the data points.

c. (6,4), (7,3), (4,6), (3,7), [(5,5)]

Summations from extensions table: $n = 5$, $\sum x = 25$, $\sum y = 25$, $\sum x^2 = 135$, $\sum xy = 115$, $\sum y^2 = 135$

$SS(x) = \sum x^2 - ((\sum x)^2/n) = 135 - (25^2/5) = 10.0$
$SS(y) = \sum y^2 - ((\sum y)^2/n) = 135 - (25^2/5) = 10.0$
$SS(xy) = \sum xy - ((\sum x \cdot \sum y)/n) = 115 - (25 \cdot 25/5) = -10.0$

$r = SS(xy)/\sqrt{SS(x) \cdot SS(y)} = -10.0/\sqrt{10.0 \cdot 10.0} = \underline{-1.00}$

d. (1,5), (5,3), (7,3), (9,5) [(5,5)]

Summations from extensions table: $n = 5$, $\sum x = 27$, $\sum y = 21$, $\sum x^2 = 181$, $\sum xy = 111$, $\sum y^2 = 93$

$SS(x) = \sum x^2 - ((\sum x)^2/n) = 181 - (27^2/5) = 35.2$
$SS(y) = \sum y^2 - ((\sum y)^2/n) = 93 - (21^2/5) = 4.8$
$SS(xy) = \sum xy - ((\sum x \cdot \sum y)/n) = 111 - (27 \cdot 21/5) = -2.4$

$r = SS(xy)/\sqrt{SS(x) \cdot SS(y)} = -2.4/\sqrt{35.2 \cdot 4.8} = \underline{-0.185}$

e. (1,2), (3,3), (1,4), (3,5), [(5,5)]

Summations from extensions table: $n = 5$, $\sum x = 13$, $\sum y = 19$, $\sum x^2 = 45$, $\sum xy = 55$, $\sum y^2 = 79$

$SS(x) = \sum x^2 - ((\sum x)^2/n) = 45 - (13^2/5) = 11.2$
$SS(y) = \sum y^2 - ((\sum y)^2/n) = 79 - (19^2/5) = 6.8$
$SS(xy) = \sum xy - ((\sum x \cdot \sum y)/n) = 55 - (13 \cdot 19/5) = 5.6$

$r = SS(xy)/\sqrt{SS(x) \cdot SS(y)} = 5.6/\sqrt{11.2 \cdot 6.8} = \underline{0.642}$

3.81 The answers shown here are only one of many possibilities.

a. (1,3), (3,3), (3,5), (7,6), (9,7) [(5,5)]

Summations from extensions table: $n = 6$, $\sum x = 28$, $\sum y = 29$, $\sum x^2 = 174$, $\sum xy = 157$, $\sum y^2 = 153$

$SS(x) = \sum x^2 - ((\sum x)^2/n) = 174 - (28^2/6) = 43.33$
$SS(y) = \sum y^2 - ((\sum y)^2/n) = 153 - (29^2/6) = 12.83$
$SS(xy) = \sum xy - ((\sum x \cdot \sum y)/n) = 157 - (28 \cdot 29/6) = 21.67$

$r = SS(xy)/\sqrt{SS(x) \cdot SS(y)} = 21.67/\sqrt{43.33 \cdot 12.83} = \underline{0.919}$

$b_1 = SS(xy)/SS(x) = 21.67/43.33 = \underline{0.5}$

b. (1,2), (3,3), (1,4), (3,5), [(5,5)]

Summations from extensions table: $n = 5$, $\sum x = 13$, $\sum y = 19$, $\sum x^2 = 45$, $\sum xy = 55$, $\sum y^2 = 79$

$SS(x) = \sum x^2 - ((\sum x)^2/n) = 45 - (13^2/5) = 11.2$
$SS(y) = \sum y^2 - ((\sum y)^2/n) = 79 - (19^2/5) = 6.8$
$SS(xy) = \sum xy - ((\sum x \cdot \sum y)/n) = 55 - (13 \cdot 19/5) = 5.6$

$r = SS(xy)/\sqrt{SS(x) \cdot SS(y)} = 5.6/\sqrt{11.2 \cdot 6.8} = \underline{0.642}$

$b_1 = SS(xy)/SS(x) = 5.6/11.2 = \underline{0.5}$

c. (3,7), (3,5), (7,3), (7,5) [(5,5)]

Summations from extensions table: $n = 5$, $\sum x = 25$, $\sum y = 25$, $\sum x^2 = 141$, $\sum xy = 117$, $\sum y^2 = 133$

$SS(x) = \sum x^2 - ((\sum x)^2/n) = 141 - (25^2/5) = 16$
$SS(y) = \sum y^2 - ((\sum y)^2/n) = 133 - (25^2/5) = 8$
$SS(xy) = \sum xy - ((\sum x \cdot \sum y)/n) = 117 - (25 \cdot 25/5) = -8$

$r = SS(xy)/\sqrt{SS(x) \cdot SS(y)} = -8/\sqrt{16 \cdot 8} = \underline{-0.707}$

$b_1 = SS(xy)/SS(x) = -8/16 = \underline{-0.5}$

d. Impossible, no set of ordered pairs can have both a positive correlation coefficient and a negative slope.

3.83 a.

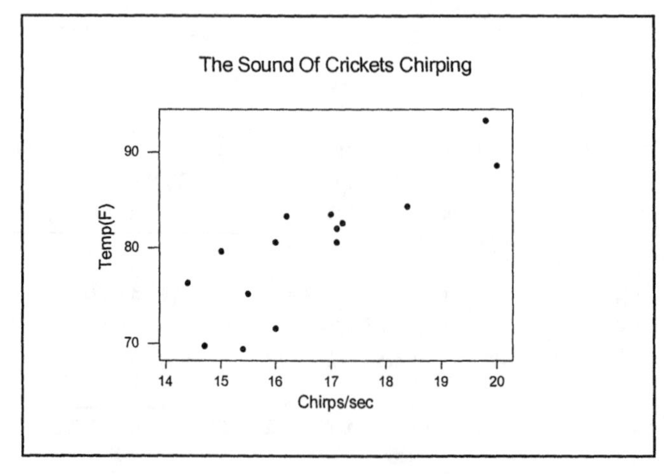

b. There is a strong linearly increasing pattern.

c. Summations from extensions table: $n = 15$, $\sum x = 249.8$, $\sum y = 1200.60$, $\sum x^2 = 4200.56$, $\sum xy = 20127.5$

(slight variation may occur due to number of decimal places used)

$SS(x) = \sum x^2 - ((\sum x)^2/n) = 4200.56 - (249.8^2/15) = 40.5573$
$SS(xy) = \sum xy - ((\sum x \cdot \sum y)/n) = 20127.5 - (249.8 \cdot 1200.6/15)$
$= 133.508$

$b_1 = SS(xy)/SS(x) = 133.508/40.5573 = 3.2918$

$b_0 = [\sum y - b_1 \cdot \sum x]/n = [1200.6 - 3.2918 \cdot 249.8)]/15 = 25.22$

$\underline{\hat{y} = 25.2 + 3.29x}$

d. $x = 14$, temperature(F) = $25.2 + 3.29(14) = 71.26$ or 71F
$x = 20$, temperature(F) = $25.2 + 3.29(20) = 91.0$ or 91F

e. It seems reasonable, temperatures do range from 70 to 90 degrees on summer nights.

f. $x = 16$, temperature(F) = $25.2 + 3.29(16) = 77.84$ or 78F

3.85 a.

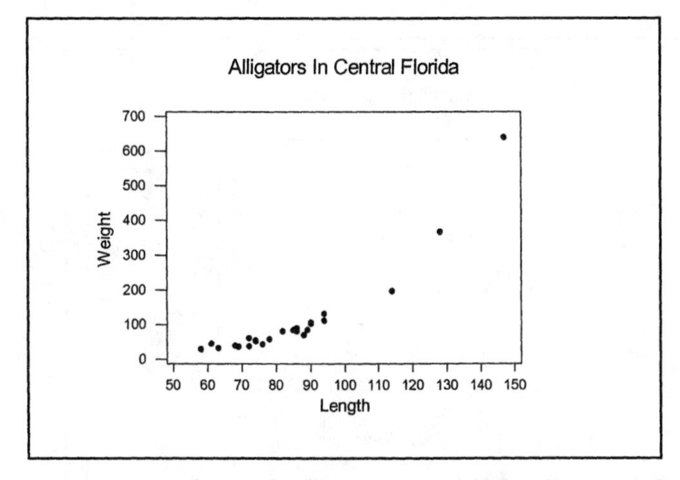

b. Yes, it looks like the weight of an alligator is predictable based on its length -the data very definitely follows a tight pattern along a line. Just not a straight line.

c. No, the pattern is curved. It looks like one side of a parabola.

d. The techniques described in this chapter are for straight line relationships, and this definitely is not a straight line.

e. Summations from extensions table: $n = 25$, $\sum x = 2124$, $\sum y = 2705$, $\sum x^2 = 190282$, $\sum xy = 287819$, $\sum y^2 = 702127$

$$SS(x) = \sum x^2 - ((\sum x)^2/n) = 190282 - (2124^2/25) = 9826.96$$
$$SS(y) = \sum y^2 - ((\sum y)^2/n) = 702127 - (2705^2/25) = 409446$$
$$SS(xy) = \sum xy - ((\sum x \cdot \sum y)/n) = 287819 - (2124 \cdot 2705/25) = 58002.2$$

$$r = SS(xy)/\sqrt{SS(x) \cdot SS(y)}$$
$$= 58002.2/\sqrt{9826.96 \cdot 409446} = \underline{0.914}$$

f. The pattern of the data is very elongated and shows an increasing relationship. This overall pattern dominates and the nonlinear nature of the relationship does not effect the calculation of r.

3.87 a.

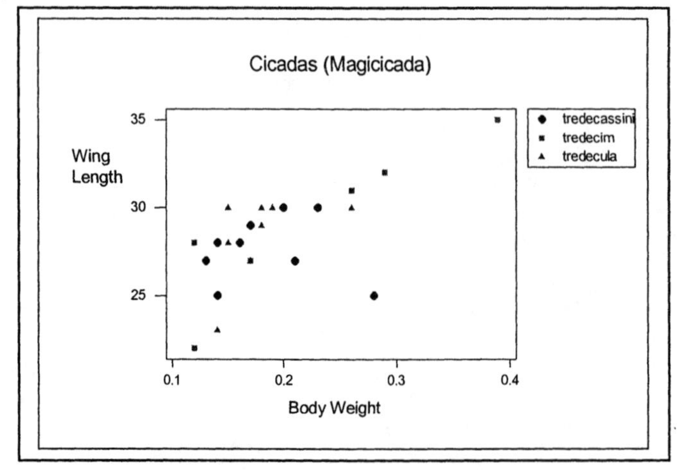

b. The overall pattern is linear, moderately strong and shows an increasing relationship. All three species are intermingled to a certain extent. The exceptions are: all the points in the upper right portion of the diagram are tredecim, and the tredecassini are all in the central part of the pattern.

c. Summations from extensions table: $n = 24$, $\sum x = 4.6$, $\sum y = 680$, $\sum x^2 = 0.978400$, $\sum xy = 133.050$, $\sum y^2 = 19448.0$

$$SS(x) = \sum x^2 - ((\sum x)^2/n) = 0.978400 - (4.6^2/24) = 0.0967333$$
$$SS(y) = \sum y^2 - ((\sum y)^2/n) = 19448.0 - (680^2/24) = 181.333$$
$$SS(xy) = \sum xy - ((\sum x \cdot \sum y)/n) = 133.050 - (4.6 \cdot 680/24) = 2.71667$$

$$r = SS(xy)/\sqrt{SS(x) \cdot SS(y)}$$
$$= 2.71667/\sqrt{0.0967333 \cdot 181.333} = \underline{0.649}$$

d. $b_1 = SS(xy)/SS(x) = 2.71667/0.0967333 = 28.084$

$$b_0 = [\sum y - b_1 \cdot \sum x]/n = [680 - (28.084 \cdot 4.6)]/24 = 22.95$$

$$\underline{\hat{y} = 23.0 + 28.1x}$$

e. Wing length = 23.0 + 28.1(body weight) = 23.0 + 28.1(0.2) = 28.62 mm. Probably the tredecassini species.

3.89 a. Numerator of formula (3-1):

$$\text{Numerator} = \sum(x-\bar{x})(y-\bar{y})$$
$$= \sum[xy - \bar{x}y - x\bar{y} + \bar{x}\,\bar{y}]$$
$$= \sum xy - \bar{x} \cdot \sum y - \bar{y} \cdot \sum x + n\bar{x}\,\bar{y}$$
$$= \sum xy - [(\sum x/n) \cdot \sum y] - [(\sum y/n) \cdot \sum x] + [n \cdot (\sum x/n)(\sum y/n)]$$
$$= \sum xy - [(\sum x \cdot \sum y/n) - (\sum x \cdot \sum y/n) + (\sum x \cdot \sum y/n)]$$
$$= \sum xy - [(\sum x \cdot \sum y)/n]$$
$$= SS(xy)$$

Denominator of formula (3-1):

$$\text{Denominator} = (n-1)s_x s_y$$
$$= (n-1) \cdot \sqrt{SS(x)/(n-1)} \cdot \sqrt{SS(y)/(n-1)}$$
$$= \sqrt{SS(x) \cdot SS(y)}$$

Therefore, formula (3-1) is equivalent to formula (3-2).

b. The numerators of formula (3-5) and (3-7) were shown to be equal in (a) above. (See numerator.)

   The denominators are equal by definition (formula 2-9).

CHAPTER 4 ∇ PROBABILITY

Chapter Preview

Chapter 4 deals with the basic theory and concepts of probability. Probability, in combination with the descriptive techniques in the previous chapters, allows us to proceed into inferential statistics in later chapters.

The results of the Global Color Vote conducted for M&M's Milk Chocolate Candies are presented in this chapter's case study.

CHAPTER 4 CASE STUDY

4.1 a. Most: red, yellow and purple
Least: blue, brown, green and orange
b. Not exactly, but similar

SECTION 4.1 EXERCISES

4.3 Each student will get different results. The four answers will each be a fraction with a denominator of 25.

> Computer and/or calculator commands to generate random integers and tally the findings can be found in ES9-pp183&184.
>
> Variations in the quantity of random integers and the interval are necessary for exercises 4.5 and 4.6.

4.5 Note: Each will get different results. MINITAB results on one run were:
a. Relative frequency for: 1 - 0.22, 2 - 0.16, 3 - 0.14, 4 - 0.22, 5 - 0.16, 6 - 0.10

b. Relative frequency for: H - 0.58, T - 0.42

SECTION 4.2 ANSWER NOW EXERCISES

4.7 P'(5) = 9/40 = 0.225

4.8 All three are calculated by dividing the experimental count by the sample size.

4.9 a. Answers will vary.
b. Yes, ½ = 0.5, P(red) = ½, P(black) = ½
c. Probability is substantially greater than 0.5

SECTION 4.2 EXERCISES

4.11 a. 197/365 = 0.5397
b. 31/365 = 0.0849

4.13 Each student will get different results. These are the results I obtained: [Note: 12 is an ordered pair (1,2)]

12	65	15	32	54	12	52	63	64	62
66	44	42	45	42	35	54	66	54	32
31	12	23	33	26	33	32	23	46	64
63	63	35	54	52	55	56	26	11	44
11	61	46	11	45	55	15	33	43	11

a. P'(white die is odd) = 27/50 = 0.54

b. P'(sum is 6) = 7/50 = 0.14

c. P'(both dice show odd number) = 14/50 = 0.28

d. P'(number on color die is larger) = 16/50 = 0.32

4.15 Each student will get different results. These are the results I obtained:

n(heads)/10	P'(head)/set of 10	Cum.P'(head)
6	0.6	6/10 = 0.60
3	0.3	9/20 = 0.45
5	0.5	14/30 = 0.47
5	0.5	19/40 = 0.48
7	0.7	26/50 = 0.52
4	0.4	30/60 = 0.50
6	0.6	36/70 = 0.51
6	0.6	42/80 = 0.52
6	0.6	48/90 = 0.53
5	0.5	53/100 = 0.53
3	0.3	56/110 = 0.51
4	0.4	60/120 = 0.50
7	0.7	67/130 = 0.52
3	0.3	70/140 = 0.50
6	0.6	76/150 = 0.51
3	0.3	79/160 = 0.49
7	0.7	86/170 = 0.51
7	0.7	93/180 = 0.52
4	0.4	97/190 = 0.51
6	0.6	103/200 = 0.52

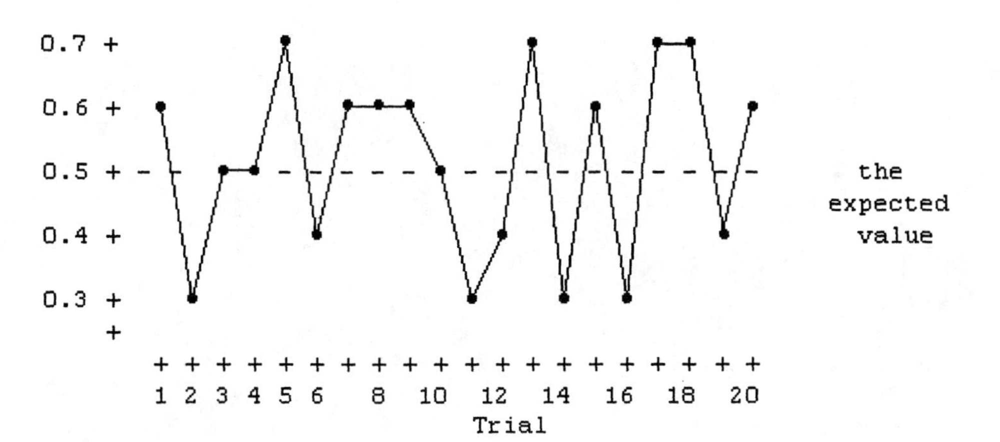

The observed probability varies above and below 0.5, but seems to average approximately 0.5.

Cumulative Observed Probability of Heads from Sets of Ten

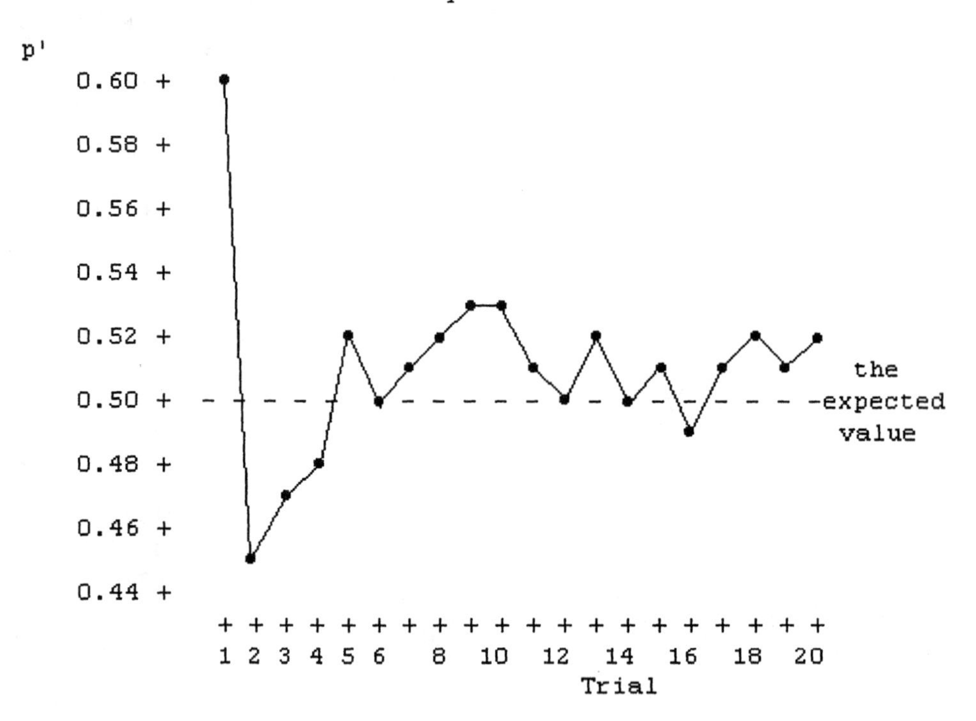

SECTION 4.3 ANSWER NOW EXERCISES

4.17 {0, 1, 2, 3, 4, 5, 6, 7, 8, 9}

4.18

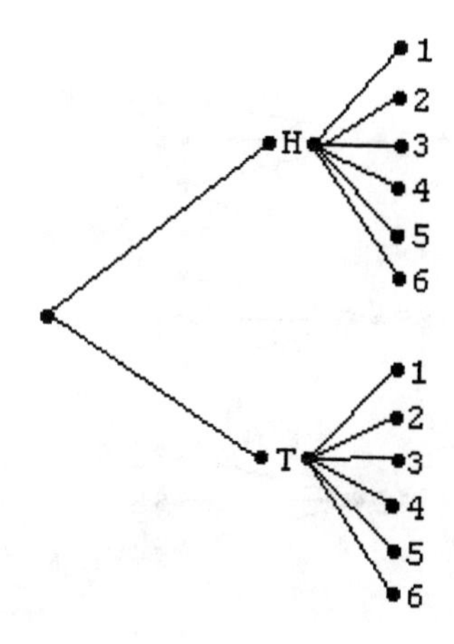

4.19

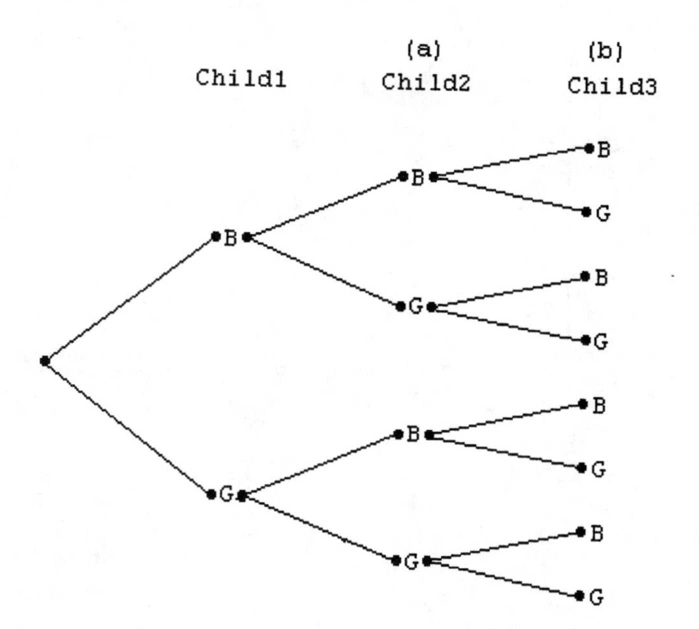

4.20 a. $S = \{\$1, \$5, \$10, \$20\}$

b.

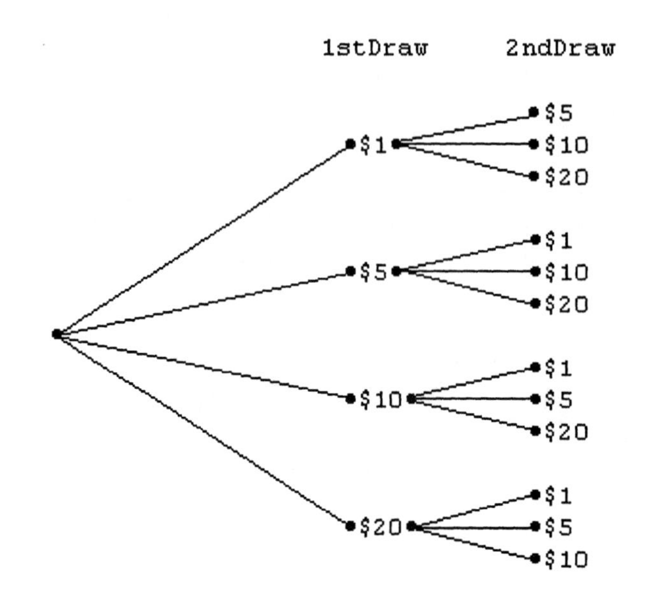
1stDraw
2ndDraw
$1
$5
$10
$20
$5
$10
$20
$1
$10
$20
$1
$5
$20
$1
$5
$10

c.

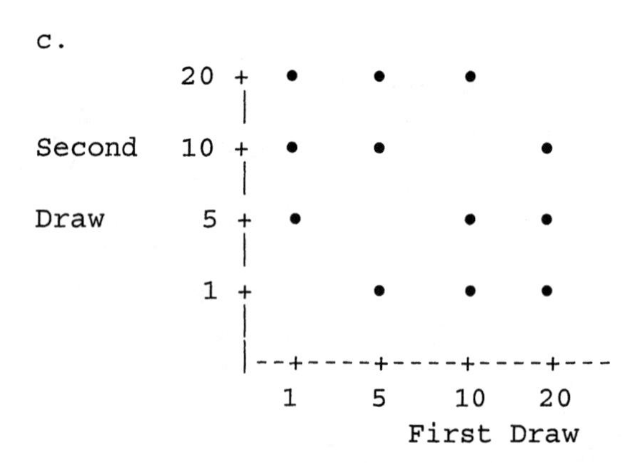
Second
Draw
20
10
5
1
1
5
10
20
First Draw

SECTION 4.3 EXERCISES

A REGULAR (BRIDGE) DECK OF PLAYING CARDS

52 cards 26 red, 26 black 4 suits (diamonds, hearts, clubs, spades)

each suit - 13 cards - 2,3,4,5,6,7,8,9,10,Jack,Queen,King,Ace

face cards = Jack, Queen, and King

4.21 Let J = jack, Q = queen, K = king, H = heart, C = club, D = diamond, S = spade.
S = {JH, JC, JD, JS, QH, QC, QD, QS, KH, KC, KD, KS}

A tree diagram may be helpful for exercise 4.23.

NOTE: See additional information about tree diagrams in Review Lessons, Tree Diagrams, in the Statistical Tutor.

4.23 Let: y = yes, n = no

Sample space = S = {yyy, yyn, yny, ynn, nyy, nyn, nny, nnn}

N(S) = 8

4.25 a. S = {HH, HT, TH, TT} and equally likely.

b. S = {HH, HT, TH, TT} and not equally likely; the possibilities are the same however.

A tree diagram is helpful for exercise 4.27.

4.27 S = {H1, H2, H3, H4, H5, H6, T1, T2, T3, T4, T5, T6}

NOTE: P' is the notation for an experimental or empirical probability.

4.29 Everyone's results will be different. One example of observed probabilities are:

a. $P'(H) = 90/200 = \underline{0.45}$

b. $P'(3) = 30/200 = \underline{0.15}$

c. $P'(H,3) = 12/200 = \underline{0.06}$

SECTION 4.4 ANSWER NOW EXERCISES

4.31 $P(5) = 4/36$; $P(6) = 5/36$; $P(7) = 6/36$; $P(8) = 5/36$; $P(9) = 4/36$; $P(10) = 3/36$; $P(11) = 2/36$; $P(12) = 1/36$

4.32
$P(P) = 4 \cdot P(F)$
$P(P) + P(F) = 1$
$4 \cdot P(F) + P(F) = 1$
$5 \cdot P(F) = 1$
$P(F) = 1/5$
$P(P) = 4 \cdot P(F) = 4 \cdot (1/5) = \underline{4/5}$

4.33 a. 1/13 b. 12:1

4.34 A = face card

$$P(\overline{A}) = 1 - P(A) = 1 - \frac{12}{52} = \frac{40}{52}$$

4.35 a. $\dfrac{64}{150,000} = \dfrac{1}{2343.75} = 0.0004267$

b. (senior:nonsenior) 2400:1400 (3800-2400=1400) reduces to $\underline{12:7}$

c. (no play:play) 2336:64 (2400-64=2336) reduces to $\underline{37:1}$ (rounded)

SECTION 4.4 EXERCISES

The sum of all the probabilities over a sample space is equal to one. $\sum P(x) = 1$

The probability of choosing a red marble plus the probability of choosing a yellow marble plus the probability of choosing a green marble should be equal to the probability of the sample space, which is equal to one.

$$P(\text{red}) + P(\text{yellow}) + P(\text{green}) = 1$$

4.37 Let $P(R) = X$, $P(Y) = X$, $P(G) = 2X$

$X + X + 2X = 1$
$4X = 1$
$X = 1/4$

therefore: $P(R) = \underline{1/4}$, $P(Y) = \underline{1/4}$, $P(G) = 2/4 = \underline{1/2}$.

4.39 a. 1/6 b. 3/6 c. 4/6 d. 3/6

The $\sum P(x) = 1$, therefore $P(A) + P(B) + P(C) = 1$

4.41 Let $P(A) = X$, $P(B) = 2X$, $P(C) = 4X$

$X + 2X + 4X = 1$
$7X = 1$
$X = 1/7$

therefore: $P(A) = \underline{1/7}$, $P(B) = \underline{2/7}$, $P(C) = \underline{4/7}$.

PROPERTIES OF PROBABILITY

1. $0 \le P(A) \le 1$ The probability of an event must be a value between 0 and 1, inclusive.

2. $\sum P(A) = 1$ The sum of all the probabilities for each event in the sample space equals 1.

4.43 The three success ratings (highly successful, successful, and not successful) appear to be non intersecting, and their union appears to be the entire sample space. If this is true, none of the three sets of probabilities are appropriate.

Judge A has a total probability of 1.2. The total must be exactly 1.0.

Judge B has a negative probability of -0.1 for one of the events. All probability numbers are between 0.0 and 1.0.

Judge C has a total probability of 0.9. The total must be exactly 1.0.

In exercise 4.45, add the rows and columns first, to find marginal totals.

4.45 a. (35+20)/100 = <u>0.55</u>

b. (20+20)/100 = <u>0.40</u>

4.47 a. 1 to 232, 232 to 1, 0.00429
b. 1 to 3699, 3699 to 1, 0.00027
c. 1 to 3999, 3999 to 1, 0.00025
d. 1 to 15, 15 to 1, 0.0625
e. 1 to 64, 64 to 1, 0.0154
f. 1 to 129, 129 to 1, 0.0077

<u>COMPLEMENT</u> - Probability of A complement = $P(\bar{A})$

$$P(\bar{A}) = P(\text{not } A) = 1 - P(A)$$

4.49 P(AIDS and 50 or older) = 0.10
P(AIDS and younger than 50) = 1 - 0.10 = <u>0.90</u>

SECTION 4.5 ANSWER NOW EXERCISES

4.51 a. Yes, they can not occur at the same time; i.e., a student can not be both male and female.

b. No, they can occur at the same time; i.e., a student can be both male and registered for statistics.

c. No, they can occur at the same time; i.e., a student can be both female and registered for statistics.

d. Yes, the probability of being female at this college plus the probability of being male at this college equals one.

e. No, the two events do not include all of the students.

f. Yes, in both situations there are no common elements shared by the two events.

g. No, two complementary events comprise the sample sample; two mutually exclusive events do not necessarily make up the whole sample space.

4.52 a. A & C and A & E are mutually exclusive because they cannot occur at the same time.

b. P(A or C) = P(A) + P(C) = 6/36 + 6/36 = <u>12/36</u>
P(A or E) = P(A) + P(E) = 6/36 + 5/36 = <u>11/36</u>
P(C or E) = P(C) + P(E) - P(C and E)
= 6/36 + 5/36 - 1/36 = <u>10/36</u>

SECTION 4.5 EXERCISES

<u>Mutually Exclusive Events</u> - events that cannot occur at the same time (they have no sample points in common).

<u>Not Mutually Exclusive Events</u> - events that can occur at the same time (they have sample points in common).

4.53 a. <u>Not mutually exclusive</u>. *One head* belongs to both events, therefore the two event intersect.

b. <u>Not mutually exclusive.</u> All sales that *exceed $1000* also *exceed $100*, therefore $1200 belongs to both events and the two events have an intersection.

c. Not mutually exclusive. The student selected could be both *male* and *over 21*, therefore the two events have an intersection.

d. Mutually exclusive. The total cannot be both *less than 7* and *more than 9* at the same time, therefore there is no intersection between these two events.

PROBABILITY - THE ADDITION RULE

P(A or B) = P(A) + P(B) - P(A and B) if A and B are not

The probability of event A or event B is equal to the probability of event A plus the probability of event B, minus the probability of events A and B occurring at the same time(otherwise that common probability is counted twice).

P(A or B) = P(A) + P(B) if A and B are mutually exclusive

The probability of event A or event B is equal to the probability of event A plus the probability of event B, if A and B have nothing in common (i.e., they cannot occur at the same time).

P(A and B) = 0 if A and B are mutually exclusive

The probability of events A and B occurring at the same time is impossible if A and B are mutually exclusive.

4.55 If two events are mutually exclusive, then there is no intersection. The event, *A and B*, is the intersection. If no intersection, then P(A and B) = 0.0.

4.57 a. $P(\overline{A}) = 1 - 0.3 = \underline{0.7}$

b. $P(\overline{B}) = 1 - 0.4 = \underline{0.6}$

c. P(A or B) = 0.3 + 0.4 = 0.7

d. P(A and B) = 0.0 (Mutually exclusive events have no intersection.)

4.59 No. *Female* students can be *working* students. Further, if the probabilities are correct, there must be an intersection otherwise the total probability would be more than 1.0.

4.61 a. P(A) = (28 + 17)/102 = 0.4412
b. P(B) = (17 + 34)/102 = 0.5000
c. P(A and B) = 17/102 = 0.1667
d. P(A or B) = (28 + 17 + 17 + 34 − 17)/102 = 0.7745;
P(not (A or B)) = 23/102 = 0.2255

e. Venn diagram:

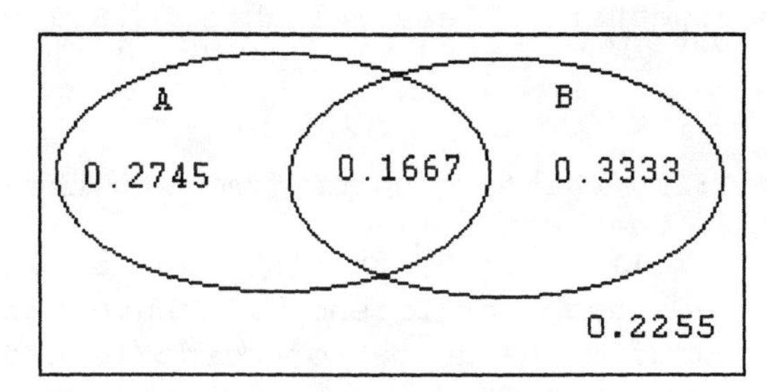

The events are not mutually exclusive. P(A or B) ≠ P(A)+P(B)

4.63 Let U = used part and D = defective part

Given info: P(U) = 0.60, P(U or D) = 0.61, P(D) = 0.05

P(U or D) = P(U) + P(D) - P(U and D)
0.61 = 0.60 + 0.05 - P(U and D)

P(U and D) = 0.60 + 0.05 - 0.61

P(U and D) = <u>0.04</u>; 4% are both used and defective

SECTION 4.6 ANSWER NOW EXERCISES

4.65 a. P(S) = 135/300 = <u>0.45</u>
b. P(S|viewer was female) = 80/200 = <u>0.40</u>
c. P(S|viewer was male) = 55/100 = <u>0.55</u>
d. No, P(S) ≠ P(S|F) ≠ P(S|M)

4.66 $P(R \text{ and } H) = P(R) \cdot P(H|R) = (0.6)(0.25) = 0.15$

4.67 $P(A \text{ and } B) = P(A) \cdot P(B) = 0.70 \cdot 0.40 = \underline{0.28}$

4.68 a. $P(A \text{ and } C) = 80/200 = 0.40$
b. Yes

4.69 a. The joints are working separately.
b. 0.869
c. Working together is less reliable, 0.869 versus 0.9777

4.70 a. 4:44 or 1:11, that's 4 cards complete the full house against 44 that won't.
b. $4/48 = 1/(1+11) = 1/12$, that's 4 out of 48
c. No, both you and your friend will have both drawn some cards, changing the number of cards left as possibilities for both the numerator and the denominator of the fraction forming the probability.

SECTION 4.6 EXERCISES

<u>Independent Events</u> - when there are independent events, the occurrence of one event **has no effect** on the probability of the other event.

4.71 a. Two events are mutually exclusive if they cannot occur at the same time or they have no elements in common.

b. Two events are independent if the occurrence of one has no effect on the probability of the other.

c. Mutually exclusive has to do with whether or not the events share common elements; while independence has to do with the effect one event has on the other event's probability.

4.73 a. independent b. not independent

c. independent d. independent

e. not independent (If you do not own a car, how can your car have a flat tire?)

f. not independent

PROBABILITY - THE MULTIPLICATION RULE & THE CONDITIONAL

The probability of event A given event B has occurred is a conditional probability, written as $P(A|B)$.

For any two events, the probability of events A and B occurring simultaneously is equal to:

1. the probability of event A times the probability of event B, given event A has already occurred: that is:

$$P(A \text{ and } B) = P(A) \cdot P(B|A)$$

OR

2. the probability of event B times the probability of event A, given event B has already occurred: that is:

$$P(A \text{ and } B) = P(B) \cdot P(A|B)$$

For two independent events:

1. the probability of events A and B occurring simultaneously is equal to the probability of event A times the probability of event B

$$P(A \text{ and } B) = P(A) \cdot P(B)$$

2. the conditional probabilities are equal to the single event probabilities

$$P(A|B) = P(A) \quad \text{AND} \quad P(B|A) = P(B)$$

FORMULAS FOR CONDITIONAL PROBABILITIES

$$P(A|B) = \frac{P(A \text{ and } B)}{P(B)} \quad \text{and} \quad P(B|A) = \frac{P(A \text{ and } B)}{P(A)}$$

Conditionals can also be computed without the formulas on the previous page. Suppose $P(A|B)$ is desired. The word *given*, $(|)$, in the conditional tells what the newly reduced sample space is. The number of elements in the reduced sample space, $n(B)$, becomes the denominator in the probability fraction. The numerator is the number of elements in the reduced sample space that satisfy the first event, $n(A \text{ and } B)$. Therefore:

$$P(A|B) = \frac{n(A \text{ and } B)}{n(B)}$$

The Student Suite CD has a video clip: "Probability".

4.75 a. $P(A \text{ and } B) = P(A) \cdot P(B) = 0.3 \cdot 0.4 = \underline{0.12}$

b. $P(B|A) = P(B) = \underline{0.4}$

c. $P(A|B) = P(A) = \underline{0.3}$

4.77 a. $P(A \text{ and } B) = P(B) \cdot P(A|B)$
$0.20 = 0.4 \cdot P(A|B)$; therefore, $P(A|B) = \underline{0.5}$

b. $P(A \text{ and } B) = P(A) \cdot P(B|A)$
$0.20 = 0.3 \cdot P(B|A)$; therefore, $P(B|A) = \underline{0.667}$

c. No, A and B are not independent events.

Note: A independent of B and A independent of C <u>does not</u> imply that B and C are independent.

4.79 a. $P(A) = 12/52 = 3/13$ and $P(A|B) = 6/26 = 3/13$
Therefore, A and B are <u>independent</u> events.

b. $P(A) = 12/52 = 3/13$ and $P(A|C) = 3/13$
Therefore, A and C are <u>independent</u> events.

c. $P(B) = 26/52 = 1/2$ and $P(B|C) = 13/13 = 1$
Therefore, B and C are <u>dependent</u> events.

4.81 a. P(no life insurance) = 1 - 0.49 = 0.51

b. P(18-24 purchase life insurance) = 0.15

c. P(no life insurance and 25-34 will purchase life insurance)

= (0.51)(0.26) = 0.1326

4.83 a. P(2 households have 3 or more vehicles)
= (0.17)(0.17) = 0.0289

b. P(neither household has 3 or more vehicles)
= (0.83)(0.83) = 0.6889

c. P(4 households have 3 or more vehicles)
= (0.17)(0.17)(0.17)(0.17) = 0.0008

4.85 P(3 grandparents are primary caregivers)

= (0.42)(0.42)(0.42) = 0.0741

4.87 Let C = correct decision, I = incorrect decision

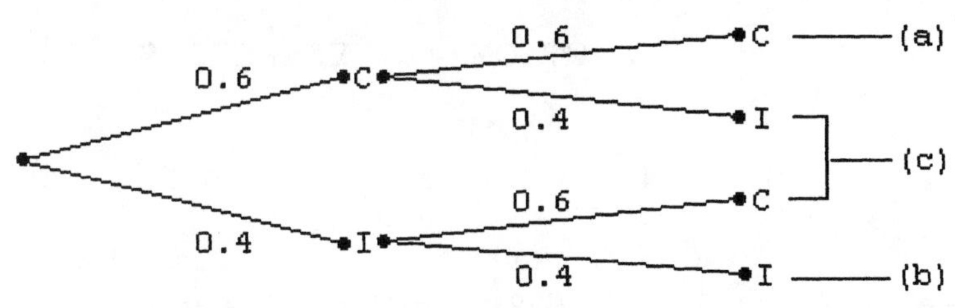

a. P(right decision) = P(C1 and C2) = 0.6·0.6 = 0.36

b. P(wrong decision) = P(I1 and I2) = 0.4·0.4 = 0.16

c. P(delay) = P[(C1 and I2) or (I1 and C2)]
= 0.6 · 0.4 + 0.4 · 0.6 = 0.24 + 0.24 = 0.48

4.89 a. P(odd) = 3/5

b. P(neither odd) = (2/5)·(2/5) = 4/25 = 0.16

P(exactly one odd) = [(2/5)·(3/5)] + [(3/5)·(2/5)]
= 12/25 = 0.48

P(both odd) = (3/5)·(3/5) = 9/25 = 0.36

4.91 a. Whether or not the events part-time and graduate within five years are independent.

b. No. Whether a student is part-time or full-time will make a difference in how soon he/she will graduate.

c. P(PT and G) = P(PT)·P(G|PR) + P(PT)·P(G|PU)
= (0.42)(0.419) + (0.42)(0.551)
= 0.17598 + 0.23142 = 0.4074

SECTION 4.7 ANSWER NOW EXERCISES

4.93 Knowing P(owner|married) and P($200+k income|married owner) would allow the question to be answered.

SECTION 4.7 EXERCISES

4.95 P(A or B) = 0.4 + 0.5 - (0.4×0.5) = 0.4 + 0.5 - 0.2 = 0.7

4.97 a. P(R and S) = P(R)·P(S) = 0.5·0.3 = 0.15

b. P(R or S) = P(R) + P(S) - P(R and S)
= 0.50 + 0.30 - 0.15 = 0.65

c. P($\overline{S}$) = 1 - P(S) = 1 - 0.3 = 0.7

d. P(R|S) = P(R and S)/P(S) = 0.15/0.30 = 0.5

e. $P(S \text{ and } R) + P(\overline{S} \text{ and } R) = P(R)$;

$0.15 + P(\overline{S} \text{ and } R) = 0.5$;
$P(\overline{S} \text{ and } R) = 0.35$

$P(\overline{S}|R) = P(\overline{S} \text{ and } R)/P(R) = 0.35/0.50 = \underline{0.7}$

f. No. Independent events can intersect, therefore R and S are not mutually exclusive events.

4.99 a. $P(\text{both red}) = (5/8)\cdot(4/7) = \underline{20/56}$

b. $P(\text{one of each color}) = (5/8)\cdot(3/7) + (3/8)\cdot(5/7) = \underline{30/56}$

c. $P(\text{both white}) = (3/8)\cdot(2/7) = \underline{6/56}$

4.101 a.

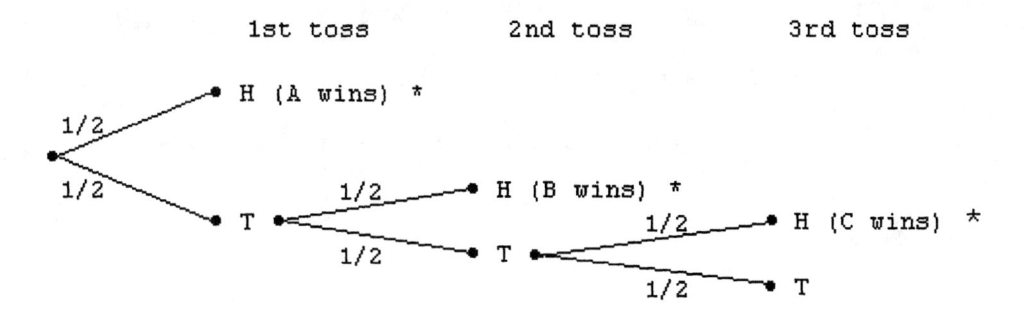

$P(\text{A wins on 1st turn}) = \underline{1/2}$

$P(\text{B wins on 1st turn}) = P(\text{A does not})\cdot P(\text{B wins}) = (1/2)\cdot(1/2) = \underline{1/4}$

$P(\text{C wins on } 1^{st} \text{ turn} = P(\text{A does not})\cdot P(\text{B does not})\cdot P(\text{C wins}) = (1/2)\cdot(1/2)\cdot(1/2) = \underline{1/8}$

b.

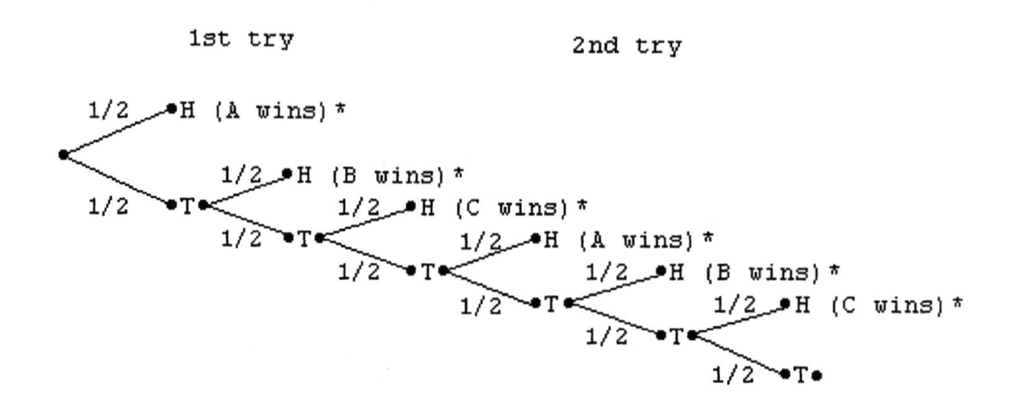

P(A wins on 2nd turn)
= P(A not on 1st)·P(B not)·P(C not)·P(A wins on 2nd)
= (1/2)·(1/2)·(1/2)·(1/2) = 1/16

P(A wins on 1st try or 2nd try) = 1/2 + 1/16 = 9/16

P(B wins on 1st try or 2nd try) = 1/4 + 1/32 = 9/32

P(C wins on 1st try or 2nd try) = 1/8 + 1/64 = 9/64

4.103

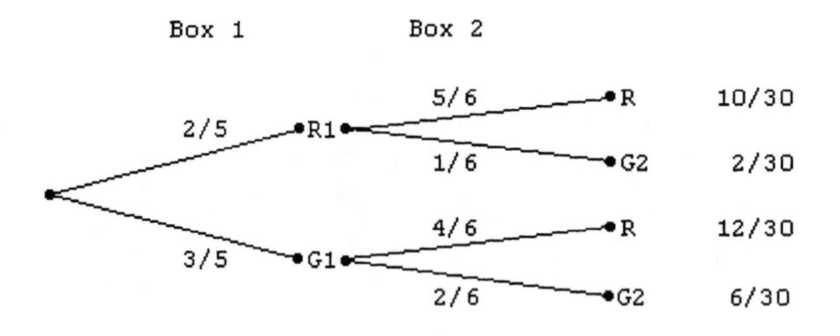

Let G2 = green ball is selected from Box 2 and R1 = red ball is selected from box 1.

P(G2) = P[(R1 and G2) or (G1 and G2)]
= P(R1)·P(G2|R1) + P(G1)·P(G2|G1)
= (2/5)·(1/6) + (3/5)·(2/6) = 2/30 + 6/30 = 8/30

4.105 P(2-wk and S) = 3/13 => 12 weeks of the 52 are 2-wk sculptor's, and that is 6 S showings - leaving 14 1-wk S showings - using 26 of the weeks.
Therefore the 22 painters must have 18 1-wk and 4 2-wk showings.

a. (18x1 + 4x2)/52 = 26/52 = 0.50
b. (14x1 + 6x2)/52 = 26/52 = 0.50
c. (18+14)/52 = 32/52 = 0.615
d. (4x2 + 6x2)/52 = 20/52 = 0.385

4.107 a. P(satisfied|unskilled) = (150+100)/(250+150) = 0.625
b. P(satisfied|skilled woman) = 25/100 = 0.25
c. Compare P(satisfied|skilled woman) to P(satisfied|unskilled woman)

P(satisfied|skilled woman) = 25/100 = 0.25

P(satisfied|unskilled woman) = 100/150 = 0.667

Since these two probabilities are not equal, therefore the events are not independent.

RETURN TO CHAPTER CASE STUDY

4.108

a.

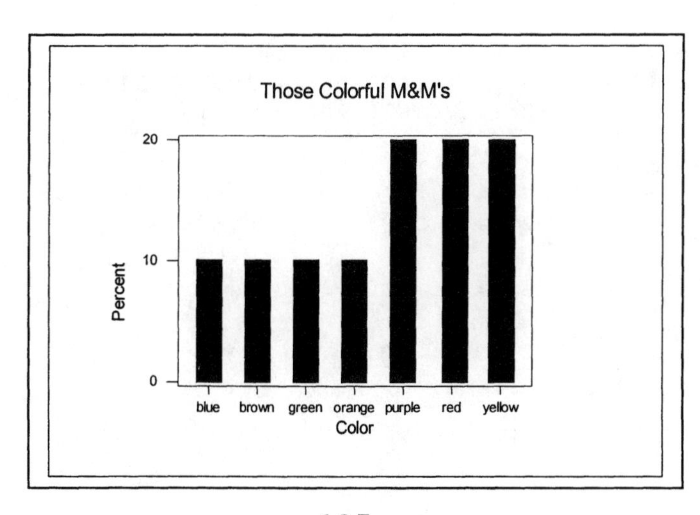

b. In general: 40 times the percentage for each color: blue - 4, brown - 4, green - 4, orange - 4, red - 8, yellow - 8, purple - 8.

c. No, one should expect the numbers to vary from those listed above, but on the average over many bags, the numbers in (b) should be close to the mean number for each color.

d. Very surprised! To observe any predetermined set of numbers for each color would be a fairly rare event.

CHAPTER EXERCISES

4.111 a. P(Male, over 59)
= (4,530,005 + 4,248,092 + 3,582,678 + 2,465,550 + 2,094,581)/176,628,482 = 0.0958

b. P(Female, under 30)
= (4,362,558 + 7,508,844 + 8,822,290)/176,628,482
= 0.1172

c. P(under 25)
= (4,761,567 +8,016,601 + 4,362,558 + 7,508,844)/176,628,482 = 0.1396

d. P(Female) = 87,414,115/176,628,482 = 0.4949

e. P(Male, between 35 and 49)
= (10,381,712 + 9,512,860 + 8,469,713)/176,628,482
= 0.1606

f. P(Female, between 25 and 44)
= (8,822,290 + 10,028,055 + 10,227,348 + 9,465,126)/176,628,482 = 0.2182

g. P(over 69)
= (3,582,678 + 2,465,550 + 2,094,581 + 3,702,020 + 2,577,527 + 2,149,589)/176,628,482 = 0.0938

4.113 a. P(blue eyes) = 90/300 = 0.30

b. P(yes) = 120/300 = 0.40

c. If independent; then P(A and B) = P(A)·P(B)

P(A and B) = 70/300 = 0.233, and

P(A)·P(B) = (90/300)·(120/300) = 0.12; <u>not independent</u>

d. Blue eyes and brown eyes are mutually exclusive events. They are not complementary since not everyone was classified as having brown or blue eyes. Since they are mutually exclusive, they cannot be independent events.

4.115 a. <u>False.</u> If mutually exclusive, P(R or S) is found by adding 0.2 and 0.5.

b. <u>True.</u> 0.2 + 0.5 - (0.2·0.5) = 0.6

c. <u>False.</u> If mutually exclusive, P(R and S) must be equal to zero; there is no intersection.

d. <u>False.</u> 0.2 + 0.5 = 0.7, not 0.6

4.117 a. S = {GGG, GGR, GRG, GRR, RGG, RGR, RRG, RRR}

b. P(exactly one R) = <u>3/8</u>

c. P(at least one R) = <u>7/8</u>

4.119 P(boy) approximately equal to 7/8

Rearrange probability formulas in order needed. Remember P(A and B) = P(A) · P(B|A) is the same as P(B and A) = P(B)· P(A|B) since A and B is the same as B and A.

4.121 a. P(A and B) = P(B)·P(A|B) = (0.36)·(0.88) = <u>0.3168</u>

b. P(B|A) = P(A and B)/P(A) = 0.3168/0.68 = <u>0.4659</u>

c. <u>No.</u> P(A) does not equal P(A|B)

d. <u>No.</u> P(A and B) does not equal 0.0

e. It would mean that the two events "candidate wants job" and "RJB wants candidate" could not both happen.

4.123 a. P(both damage free) = $(10/15)\cdot(9/14)$ = <u>0.429</u>

b. P(exactly one) = $(10/15)\cdot(5/14) + (5/15)\cdot(10/14)$
= <u>0.476</u>

c. P(at least one)= 0.429 + 0.476 = <u>0.905</u>

4.125 P[(med or sh) and (mod or sev)] =
= (90 + 121 + 35 + 54)/1000 = <u>0.300</u>

Note the wording: pink seedless denotes pink <u>and</u> seedless. Use formulas accordingly.

4.127 a. P(seedless) = (10+20)/100 = <u>0.30</u>

b. P(white) = (20+40)/100 = <u>0.60</u>

c. P(pink and seedless) = 10/100 = <u>0.10</u>

d. P(pink or seedless) = (10+20+30)/100 = <u>0.60</u>

e. P(pink|seedless) = 0.10/0.30 = <u>0.333</u>

f. P(seedless|pink) = 0.10/0.40 = <u>0.25</u>

4.129 P(satisfactory) = P(all good) = p^6

a. P(satisfactory|p=0.9) = 0.9^6 = <u>0.531</u>

b. P(satisfactory|p=0.8) = 0.8^6 = <u>0.262</u>

c. P(satisfactory|p=0.6) = 0.6^6 = <u>0.047</u>

Exercise 4.131 involves many possibilities and given conditionals. These are clues that a tree diagram should be used. Assign probabilities to the branches.

4.131

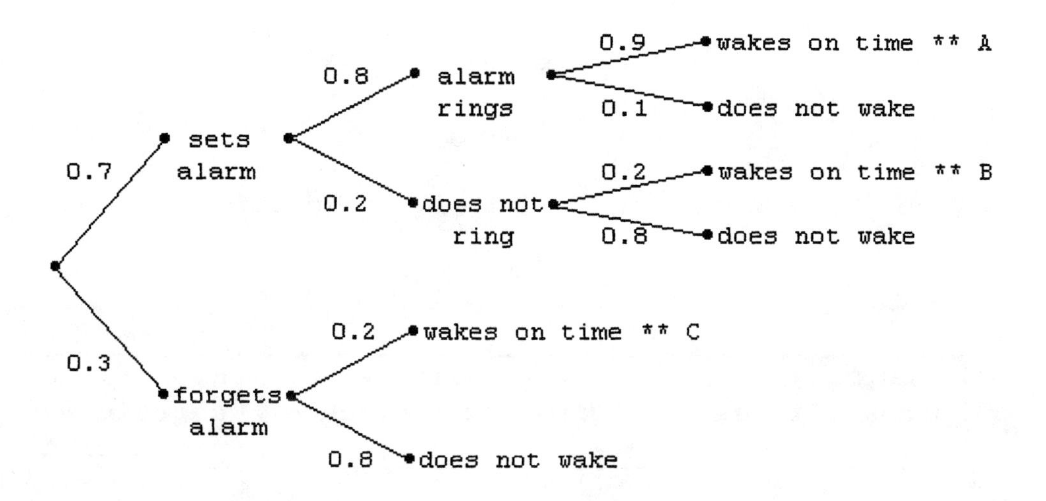

P(wakes on time) = P(A or B or C) = P(A) + P(B) + P(C)
= (0.7)(0.8)(0.9)+(0.7)(0.2)(0.2)+(0.3)(0.2)
= 0.504 + 0.028 + 0.060 = <u>0.592</u>

4.133 Let AW = Team A wins game, BW = Team B wins game

a. P(AW) = 0.60; P(A wins a one game series) = <u>0.60</u>

b. P(A wins best of 3 game series) =

= P[(AW1,AW2) or (AW1,BW2,AW3) or (BW1,AW2,AW3)]

= (0.6)(0.6) + (0.6)(0.4)(0.6) + (0.4)(0.6)(0.6)

= <u>0.648</u>

c. P(A wins best of 7 game series)
= P(A wins in 4 games) + P(A wins in 5 games) + P(A wins in 6 games) + P(A wins in 7 games)

$= 1\cdot(0.6)^4 + 4\cdot(0.6)^4(0.4)^1 + 10\cdot(0.6)^4(0.4)^2 + 20\cdot(0.6)^4(0.4)^3$

= 0.1296 + 0.20736 + 0.20736 + 0.16589 = <u>0.710</u>

d. (a) 0.70 (b) 0.784 (c) 0.874

e. (a) 0.90 (b) 0.972 (c) 0.997

f. The larger the number of games in the series, the greater the chance that the "best" team will win.
The greater the difference between the two teams individual chances, the more likely the "best" team wins.

PROBABILITY

BASIC PROPERTIES: $0 \le$ each probability ≤ 1 (1)

$$\Sigma P(A) = 1 \quad (2)$$

Finding probabilities	**From an equally likely sample space**	**By formula, given certain probabilities**
P(A), any event A	$P(A) = \frac{n(A)}{n(S)}$ (3)	-does not apply-
$P(\bar{A})$, complementary event	$P(\bar{A}) = \frac{n(\bar{A})}{n(S)}$ (4)	$P(\bar{A}) = 1.0 - P(A)$ (11)
Any 2 events, no special conditions or relations known: P(A|B), conditional Event	$P(A\|B) = \frac{n(A \text{ and } B)}{n(B)}$ (5)	$P(A\|B) = \frac{P(A \text{ and } B)}{P(B)}$ (12)
P(A or B), union of 2 events	$P(A \text{ or } B) = \frac{n(A \text{ or } B)}{n(S)}$ (6)	$P(A \text{ or } B) = P(A) + P(B) - P(A \text{ and } B)$ (13)
P(A and B), intersection of 2 events	$P(A \text{ and } B) = \frac{n(A \text{ and } B)}{n(S)}$ (7)	$P(A \text{ and } B) = P(A)\cdot P(B\|A)$ (14)
2 events, known to be mutually exclus. P(A or B)	$P(A \text{ or } B) = \frac{n(A) + n(B)}{n(S)}$ (8)	$P(A \text{ or } B) = P(A) + P(B)$ (15)
P(A and B)		$P(A \text{ and } B) = 0$ (16)
P(A|B)		$P(A\|B) = 0$ (17)
2 events, known to be independent P(A and B)	$P(A \text{ and } B) = \frac{n(A \text{ and } B)}{n(S)}$ (9	$P(A \text{ and } B) = P(A)\cdot P(B)$ (18)
P(A|B)	$P(A\|B) = \frac{n(A \text{ and } B)}{n(B)} = \frac{n(A}{n(S}$ (10)	$P(A\|B) = P(A)$ (19)

Resulting Properties:

(20) If $P(A) + P(B) = P(A \text{ or } B)$; then A and B are mutually exclusive.

(21) If $P(A) \cdot P(B) = P(A \text{ and } B)$; then A and B are independent.

(22) If $P(A|B) = P(A)$, then A and B are independent.

(23) If $P(A \text{ and } B) = 0$, then A and B are mutually exclusive.

(24) If $P(A \text{ and } B) \neq 0$, then A and B are not mutually exclusive.

The Relationship between Independence and Mutually Exclusive

(25) If events are independent, then they are NOT mutually exclusive.

(26) If events are mutually exclusive, then they are NOT independent.

CHAPTER 5 ∇ PROBABILITY DISTRIBUTIONS (DISCRETE VARIABLES)

Chapter Preview

Chapter 5 combines the "ideas" of a frequency distribution from Chapter 2 with probability from Chapter 4. This combination results in a discrete probability distribution. The main elements of this type distribution will be covered in this chapter. The elements include:

1. discrete random variables
2. discrete probability distributions
3. the mean and standard deviation of a discrete probability distribution
4. binomial probability distribution
5. the mean and standard deviation of a binomial distribution.

An article, published in the USA Today, showing the frequency of home cooked meals is presented in this chapter's case study.

CHAPTER 5 CASE STUDY

5.1 a. 19%, 8%
b. 5 nights
c. Number of evening meals American adults cook at home in an average week.
d. Histogram

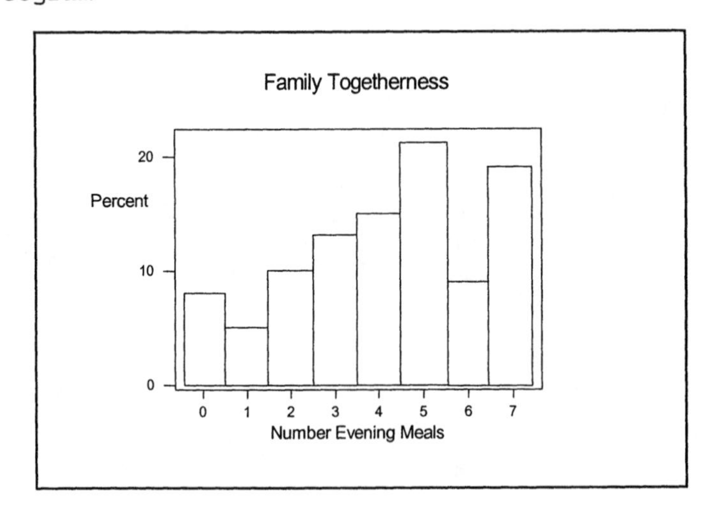

e. Relative frequency distribution, cumulative relative frequency distribution, ogive

SECTION 5.1 ANSWER NOW EXERCISES

5.2 One of the random variables is the number of siblings that a classmate has. The possible values for the random variable are x = 0, 1, 2, 3, ..., n.
The other random variable is the length of the last conversation a class mate had with their mother. The random variable will be a numerical value between 0 and 60 minutes for most classmates.

5.3 a. The random variables in Illustrations 5.1 and 5.2 are discrete because they can only assume a countable numerical value. A value of 1.73 for example would not make sense.
b. The random variables in Illustrations 5.3 and 5.4 are continuous because length and speed are measurements and can assume any value along a line interval including all possible fractions.

SECTION 5.1 EXERCISES

Random Variable - a numerical quantity whose value depends on the conditions and probabilities associated with an experiment.

Discrete Random Variable - a numerical quantity taking on or having a finite or countably infinite number of values.

Use x to denote a discrete random variable. (x is often a count of something; ex. the number of home runs in a baseball game)

Continuous Random Variable - a numerical quantity taking on or having an infinite number of values. (often a measurement of something; ex. a person's height)

5.5 The random variable is the *number of children per family.*

The possible values for the random variable are x = 0, 1, 2, 3, ... , n; where n is the maximum number of children for any family in the community. <u>The variable is discrete.</u>

5.7 The random variable is the *distance from center to arrow.* x = 0 to n, where n = radius of the target, measured in inches, including all possible fractions. The variable is continuous.

5.9 a. The random variable is the number of home runs hit by a player in one month.

b. The random variable is discrete because it represents a count.

SECTION 5.2 ANSWER NOW EXERCISES

5.11

x	0	1
P(x)	1/2	1/2

5.12

x	1	2	3	4	5	6
P(x)	1/6	1/6	1/6	1/6	1/6	1/6

5.13 The values of x in a probability distribution form a set of mutually exclusive events because they can never overlap. Each possible outcome is assigned a unique numerical value.

5.14 All possible outcomes (values of the random variable) are accounted for.

5.15 a. 1/6 = 0.1667 or 16.7%

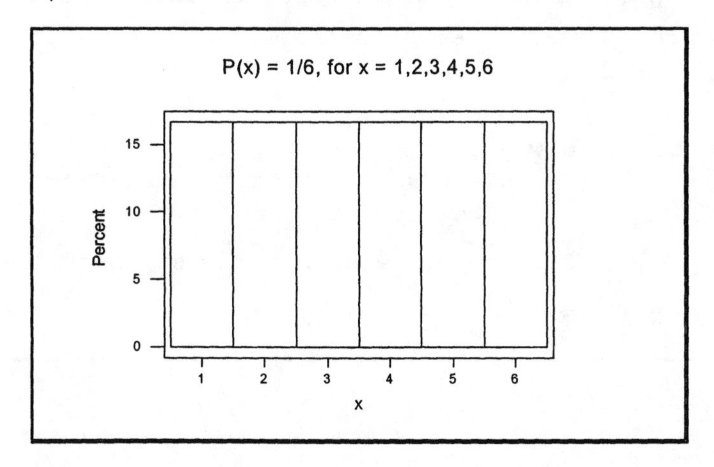

b. The distribution in (a) is uniform or rectangular.

5.16 a.

Number	Proportion
0	0.196
1	0.131
2	0.162
3	0.168
4	0.121
5	0.082
6	0.054
7 to 10	0.072
11 or more	0.014

b. Student now chooses among multiple acceptances

SECTION 5.2 EXERCISES

Probability distributions look very much like frequency distributions. The probability P(x) takes the place of frequency. The frequency *f* column contains integers (counts), whereas the probability *P(x)* column contains fractions or decimals between 0 and 1. The probability P(x) relates to relative frequency, the frequency divided by the size of the data set. . . .

The two main properties of a probability distribution are:

1. $0 \le$ each $P(x) \le 1$, each probability is a number between 0 and 1 inclusive.
2. $\sum P(x) = 1$, the sum of all the probabilities should be

Remember to always:

1. make sure each entry in the P(x) column is between 0 and 1, and
2. sum your P(x) column and check that it is

Both properties must exist.

5.17

x	0	1	2	3
P(x)	0.20	0.30	0.40	0.10

Notice that each P(x) is a value between 0.0 and 1.0, and the sum of all P(x) values is exactly 1.0.

Function Notation

P(x) ⇒ an equation with x as its variable, which assigns probabilities to the corresponding or given values.

P(0) ⇒ replace x on the right side of the equation with 0 and evaluate.

P(3) ⇒ replace x on the right side of the equation with 3 and evaluate.

ex.: $P(x) = \frac{x+1}{26}$ $\quad P(0) = \frac{0+1}{26} = \frac{1}{26}$ $\quad P(3) = \frac{3+1}{26} = \frac{4}{26}$

NOTE: Only evaluate P(x) for the x values in its domain, otherwise P(x) = 0. The domain of a variable is the specified set of replacements (x-values).

5.19 a.

x	P(x)
1	0.12
2	0.18
3	0.28
4	0.42
Σ	1.00

P(x) is a probability function:

1. Each P(x) is a value between 0 and 1.

2. The sum of the P(x)'s is 1.

b.

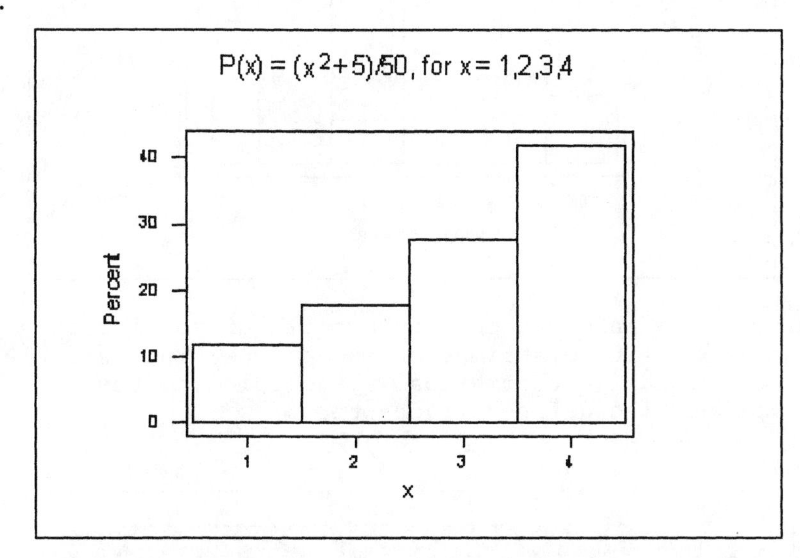

* What shape distribution does the histogram in exercise 5.19b depict? (See the bottom of the next page for answer.)

5.21 a. Yes. Each percentage in the table can be thought of as a probability multiplied by 100 (all are between 0 and 1), and the sum of the percentages is 100% (a probability of 1).

b.

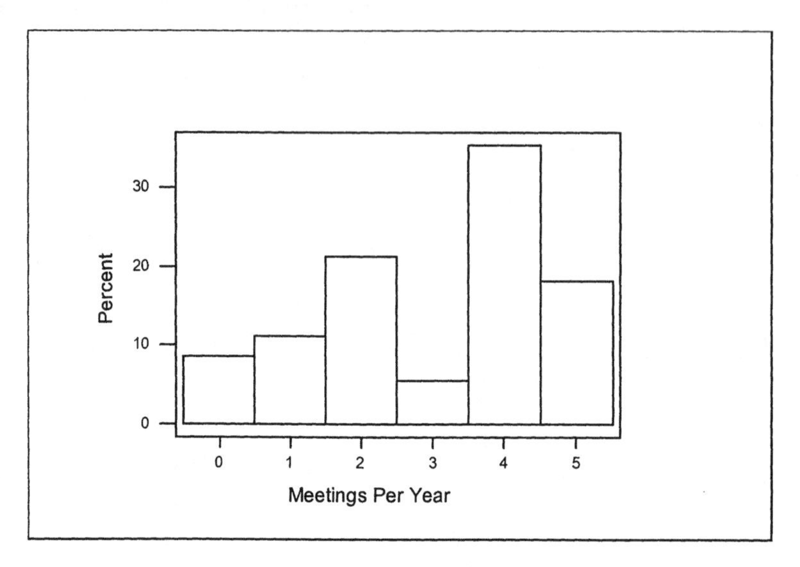

5.23 a. No. The information displays all properties of a probability distribution except one; the variable is attribute (not numerical - according to the properties defined in chapter 5).

b.

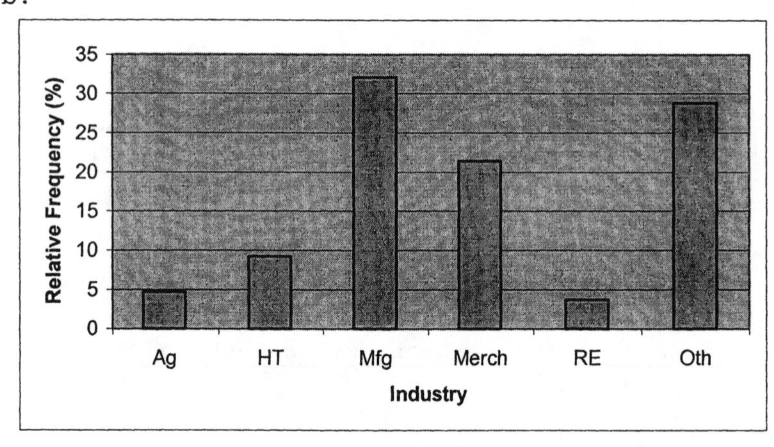

For more information on the computer commands to generate discrete data according to a probability distribution, see ES9-pp241&243. Note: The higher the probability, the more often the number will be generated.

*(5.19b) A J-shaped distribution.

5.25 a. Everyone's generated values will be different. Listed here is one such sample.

2 2 3 2 2 5 3 3 2 2 2 1 4 3 3 1
5 1 3 3 3 4 3 3 5

b. Sample obtained

x	rel.freq.
1	0.12
2	0.28
3	0.40
4	0.08
5	0.12
ALL	1.00

c. Given Distribution

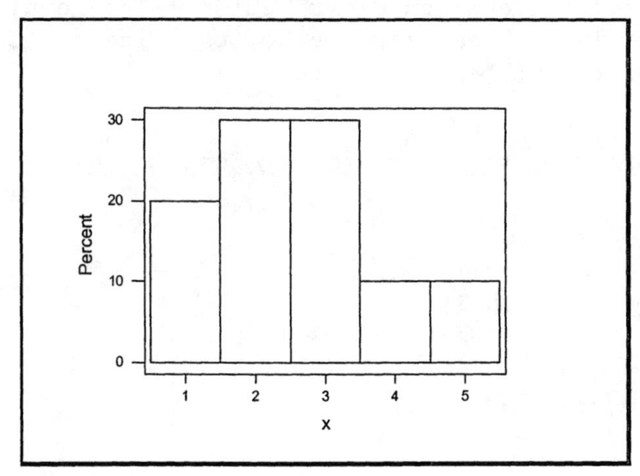

Sample Results

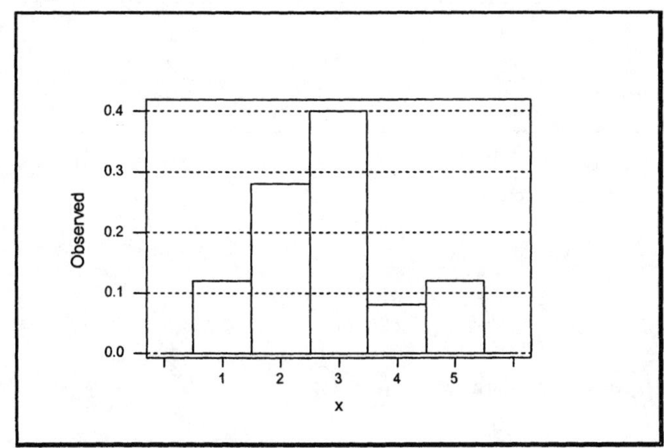

d. The distribution of the sample is somewhat similar to that of the given distribution. The two highest probabilities in the random data occurred at x = 2 and 3, matching the two highest probabilities for the given distribution.

e. Results will vary, but expect: occasionally a sample will have no 4 or no 5 in it, 2 or 3 is almost always the most frequent number, 4 or 5 is almost always the least frequent number, the histograms seem to vary but yet almost always look somewhat like the histogram of the given distribution.

f. Results will vary, but expect: little variability among the samples and the histograms, 4 or 5 occur nearly 10% most of the time, the histograms are quite similar to the histogram of the given distribution. The results indicate that the larger sample seems to stabilize the overall results (Law of Large Numbers).

SECTION 5.3 ANSWER NOW EXERCISES

5.27

$$\begin{aligned}\sigma^2 &= \sum[(x - \mu)^2 \cdot P(x)] \\ &= \sum[x^2 - 2x\mu + \mu^2) \cdot P(x)] \\ &= \sum[x^2 \cdot P(x) - 2x\mu \cdot P(x) + \mu^2 \cdot P(x)] \\ &= \sum[x^2 \cdot P(x)] - 2\mu \cdot \sum[x \cdot P(x)] + \mu^2 \cdot [\sum P(x)] \\ &= \sum[x^2 \cdot P(x)] - 2\mu \cdot [\mu] + \mu^2 \cdot [1] \\ &= \sum[x^2 \cdot P(x)] - 2\mu^2 + \mu^2 \\ &= \sum[x^2 \cdot P(x)] - \mu^2 \quad \text{or} \quad \sum[x^2 \cdot P(x)] - \{\sum[x \cdot P(x)]\}^2\end{aligned}$$

5.28 a.

x	P(x)	b) xP(x)	$x^2P(x)$
1	1/6	1/6	1/6
2	2/6	4/6	8/6
3	3/6	9/6	27/6
Σ	6/6 = 1.0 ck	c) 14/6 = 2.33	36/6 = 6.0

d. $\mu = \Sigma[xP(x)] = \underline{2.33}$

e. $\sigma^2 = \sum[x^2P(x)] - \{\sum[xP(x)]\}^2 = 6.0 - \{2.3333\}^2 = 0.55556$

f. $\sigma = \sqrt{\sigma^2} = \sqrt{0.55556} = \underline{0.745}$

5.29 The sum of the number values, once each. Nothing of any meaning.

5.30 a. vary, close to -$0.20
b. -$0.20;
c. close, no, need mean = 0 for a fair game

SECTION 5.3 EXERCISES

The mean and standard deviation of a probability distribution are μ and σ respectively. They are parameters since we are using theoretical probabilities.

$$\mu = \Sigma[xP(x)] \qquad \sigma^2 = \Sigma[x^2P(x)] - \{\Sigma[xP(x)]\}^2 \Rightarrow \sigma = \sqrt{\sigma^2}$$

OR

$$\sigma^2 = \Sigma[x^2P(x)] - \mu^2 \Rightarrow \sigma = \sqrt{\sigma^2} \quad \text{(easier formula)}$$

5.31

x	$P(x)$	$xP(x)$	$x^2P(x)$
1	4/10	4/10	4/10
2	3/10	6/10	12/10
3	2/10	6/10	18/10
4	1/10	4/10	16/10
Σ	10/10 = 1.0 ck	20/10 = 2.0	50/10 = 5.0

$\mu = \Sigma[xP(x)] = \underline{2.0}$

$\sigma^2 = \Sigma[x^2P(x)] - \{\Sigma[xP(x)]\}^2 = 5.0 - \{2.0\}^2 = 1.0$

$\sigma = \sqrt{\sigma^2} = \sqrt{1.0} = \underline{1.0}$

5.33 a.

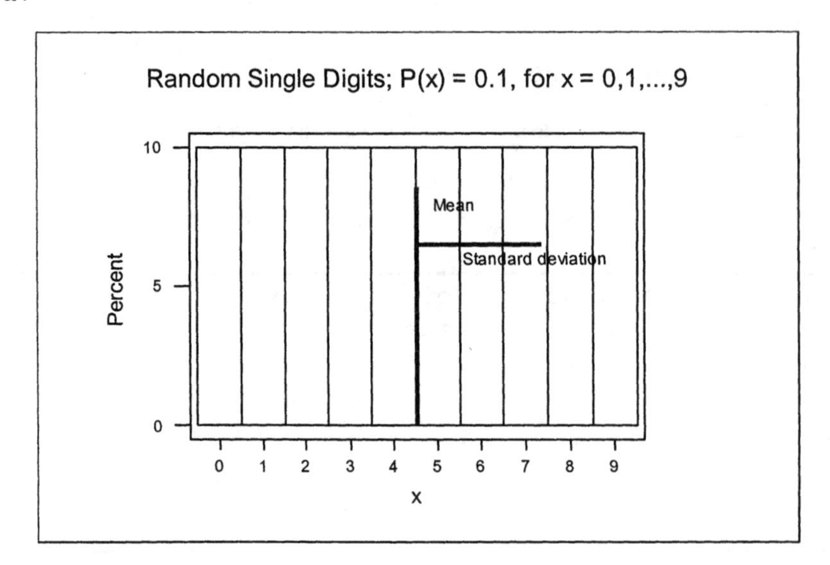

> * What shape distribution does the histogram in exercise 5.33a depict? Is the mean where you would expect it? (See the bottom of the next page for the answer.)

b.

x	P(x)	xP(x)	$x^2P(x)$
0	0.1	0.0	0.0
1	0.1	0.1	0.1
2	0.1	0.2	0.4
3	0.1	0.3	0.9
4	0.1	0.4	1.6
5	0.1	0.5	2.5
6	0.1	0.6	3.6
7	0.1	0.7	4.9
8	0.1	0.8	6.4
9	0.1	0.9	8.1
Σ	1.0 ck	4.5	28.5

$\mu = \Sigma[xP(x)] = \underline{4.5}$

$\sigma^2 = \Sigma[x^2P(x)] - \{\Sigma[xP(x)]\}^2 = 28.5 - \{4.5\}^2 = 8.25$

$\sigma = \sqrt{\sigma^2} = \sqrt{8.25} = \underline{2.87}$

c. See graph in part (a).

d. $\mu \pm 2\sigma = 4.5 \pm 2(2.87) = 4.5 \pm 5.74$ or -1.24 to 10.24

The interval from -1.24 to 10.24 contains all the x-values of this probability distribution; 100%

** How does the 100% from exercise 5.33d compare to Chebyshev's Theorem? (See bottom of the next page for answer.)

5.35 a.

x	P(x)	xP(x)	$x^2P(x)$
1	0.6	0.6	0.6
2	0.1	0.2	0.4
3	0.1	0.3	0.9
4	0.1	0.4	1.6
5	0.1	0.5	2.5
Σ	1.0 ck	2.0	6.0

$\mu = \Sigma[xP(x)] = 2.0$

$\sigma^2 = \Sigma[x^2P(x)] - \{\Sigma[xP(x)]\}^2 = 6.0 - \{2.0\}^2 = 2.0$

$\sigma = \sqrt{\sigma^2} = \sqrt{2.0} = 1.4$

b. $\mu - 2\sigma = 2.0 - 2(1.4) = -0.8$

$\mu + 2\sigma = 2.0 + 2(1.4) = 4.8$

The interval from -0.8 to 4.8 encompasses the numbers 1, 2, 3 and 4.

c. The total probability associated with these values of x is 0.9.

How does this value of 0.9 in exercise 5.35 compare with Chebyshev's theorem? (See the bottom of the next page for answer.)

* (5.33a) A uniform distribution. The mean is exactly in the center since all outcomes are equally likely.

5.37 a.

x	P(x)	xP(x)	$x^2P(x)$
0	0.103	0.000	0.000
1	0.342	0.342	0.342
2	0.384	0.768	1.536
3	0.171	0.513	1.539
Σ	1.000 ck	1.6230	3.417

$\mu = \Sigma[xP(x)] = \underline{1.623}$

$\sigma^2 = \Sigma[x^2P(x)] - \{\Sigma[xP(x)]\}^2 = 3.417 - \{1.623\}^2 = 0.78287$

$\sigma = \sqrt{\sigma^2} = \sqrt{0.78287} = 0.8848 = 0\underline{.9}$

b. It made both statistics slightly smaller in value.

SECTION 5.4 ANSWER NOW EXERCISES

5.39 Each question is in itself a separate trial with its own outcome having no effect on the outcomes of the other questions.

5.40 There are four different ways that one correct and three wrong answers can be obtained in four questions, each with the same probability. The sum of the 4 probabilities is the same value as 4 times one of them.

**(5.33d) Chebyshev's theorem states that for any shape distribution, at least 75% of the data is within 2 standard deviations of the mean. 100% is well over the minimum limit of 75%.

*(5.35) Chebyshev's theorem states that for any shape distribution, at least 75% of the data is within 2 standard deviations of the mean. 90% is well over the minimum limit of 75%.

5.41 The 1/3 is the probability of success for each question, i.e., the probability of choosing the right answer from the 3 choices. The 4 is the number of independent trials, i.e., the number of questions.

The expected average would be the sample size times the probability of success; if the probability of guessing one answer correctly is 1/3, then it seems reasonable that on the average one should be able to guess 1/3 of all questions correctly.

5.42 Property 1: One trial is the flip of <u>one coin</u>, repeated <u>n = 50</u> times. The trials are <u>independent</u> because the probability of a head on any one toss has no effect on the probabilities for the other tosses.

Property 2: Two outcomes on each trial:
<u>success = H</u>, heads
<u>failure = T</u>, tails

Property 3: <u>p = P(heads) = 1/2</u> and
<u>q = P(tails) = 1/2</u> [p+q=1]

Property 4: <u>x = the number of heads</u> for the experiment and can be any integer value from <u>0 to 50</u>.

5.43 a. $4! = 4 \cdot 3 \cdot 2 \cdot 1 = 24$

b. $\binom{4}{3} = \frac{4!}{3! \cdot 1!} = \frac{4 \cdot 3 \cdot 2 \cdot 1}{3 \cdot 2 \cdot 1 \cdot 1} = 4$

5.44 $P(x) = \binom{3}{x}(0.5)^x(0.5)^{3-x}$

$P(0) = \binom{3}{0}(0.5)^0(0.5)^3 = 1(1)(0.125) = 0.125$

$P(2) = \binom{3}{2}(0.5)^2(0.5)^1 = 3(0.25)(0.5) = 0.375$

$P(3) = \binom{3}{3}(0.5)^3(0.5)^0 = 1(0.125)(1) = 0.125$

5.45 a. $P(4) = \binom{5}{4}\left(\frac{1}{4}\right)^4\left(\frac{3}{4}\right)^1 = 5(0.0039)(0.75) = 0.0146$

$P(5) = \binom{5}{5}\left(\frac{1}{4}\right)^5\left(\frac{3}{4}\right)^0 = 1(0.0009766)(1) = 0.00098$

b.

x	P(x)
0	0.2373
1	0.3955
2	0.2637
3	0.0879
4	0.0146
5	0.00098
Σ	0.99998

$\Sigma P(x) = 0.99998 \approx 1$ (round-off error)

$0 \leq$ each $P(x) \leq 1$

5.46 P(replacement) = risk = 1 - (0.886 + 0.107) = 0.007 = 0.7%

5.47 a. 0.3585
b. 0.0159
c. 0.9245

5.48 a. $P(x = 3|B(12, 0.30)) = \underline{0.240}$
b. $\underline{0.240}$

5.49 a. Percentage of minorities is "less than would be reasonably expected"
b. Percentage of minorities is "not less than would be reasonably expected";
c. **0.96**, "not less than";
d. **0.035**, "less than"

SECTION 5.4 EXERCISES

FACTORIALS

$n! = n(n-1)(n-2)\ldots 1$ ex.: $3! = 3\cdot 2\cdot 1 = 6$

$0! = 1$ (this is defined this way so that the algebra of factorials will work)

$$\binom{n}{x} = \binom{n(\text{trials})}{n(\text{successes})\ \ n(\text{failures})} = \frac{n!}{x!\,(n-x)!}$$

ex.: $$\binom{8}{3} = \binom{8\text{trials}}{3\text{successes}\ \ 5\text{failures}} = \frac{8!}{3!5!} = \frac{8\cdot 7\cdot 6\cdot 5\cdot 4\cdot 3\cdot 2\cdot 1}{3\cdot 2\cdot 1\cdot 5\cdot 4\cdot 3\cdot 2\cdot 1}$$ or

EXPONENTS

$b^n = b\cdot b\cdot b\cdots$ (n times) ex.: $.2^3 = (.2)(.2)(.2) = .008$

NOTE: See additional information about factorial notation in Review Lessons.

5.51 a. $4! = 4\cdot 3\cdot 2\cdot 1 = \underline{24}$

b. $7! = 7\cdot 6\cdot 5\cdot 4\cdot 3\cdot 2\cdot 1 = \underline{5{,}040}$

c. $0! = \underline{1}$ (by definition)

d. $$\frac{6!}{2!} = \frac{6\cdot 5\cdot 4\cdot 3\cdot 2\cdot 1}{2\cdot 1} = 6\cdot 5\cdot 4\cdot 3 = \underline{360}$$

e. $$\frac{5!}{3!\cdot 2!} = \frac{5\cdot 4\cdot 3\cdot 2\cdot 1}{3\cdot 2\cdot 1\cdot 2\cdot 1} = \underline{10}$$

f. $$\frac{6\cdot 5\cdot 4\cdot 3\cdot 2\cdot 1}{4\cdot 3\cdot 2\cdot 1\cdot 2\cdot 1} = \underline{15}$$

g. $(0.3)^4 = (0.3)(0.3)(0.3)(0.3) = \underline{0.0081}$

h. $$\frac{7\cdot6\cdot5\cdot4\cdot3\cdot2\cdot1}{3\cdot2\cdot1\cdot4\cdot3\cdot2\cdot1} = \underline{35}$$

i. $$\frac{5!}{2!\cdot3!} = \frac{5\cdot4\cdot3\cdot2\cdot1}{2\cdot1\cdot3\cdot2\cdot1} = \underline{10}$$

j. $$\frac{3!}{0!\cdot3!} = \frac{3\cdot2\cdot1}{1\cdot3\cdot2\cdot1} = \underline{1}$$

> $\binom{4}{1}(0.2)^1(0.8)^3 = \binom{4}{1}\cdot(0.2)^1\cdot(0.8)^3$ The use of the multiplication dot is optional. They are sometimes used to emphasize that each of the three parts to a binomial must be evaluated separately first, then multiplication can take place.

k. $4\cdot(0.2)(0.8)(0.8)(0.8) = \underline{0.4096}$

l. $1\cdot1\cdot(0.7)^5 = \underline{0.16807}$

> <u>BINOMIAL EXPERIMENTS</u> must have:
>
> 1. n independent repeated trials
>   a) n - the number of times the trial is repeated
>   b) independent - the probabilities of the outcomes remain the same throughout the entire experiment
>
> 2. two possible outcomes for each trial
>   a) success - the outcome or group of outcomes that is the focus of the experiment
>   b) failure - the outcome or group of outcomes not included in success
>
> 3. p = probability of success on any one trial
>   q = probability of failure on any one trial (q = 1 - p)
>
> 4. x = number of successes when the experiment of all n trials is completed. x can range in value from 0 through n. However, when the experiment is completed, x will have exactly one value, that is, the number of successes that occurred.

5.53 Binomial properties:
n = 100 trials (shirts),
two outcomes (first quality or irregular),
p = P(irregular),
x = n(irregular); any integer value from 0 to 100.

5.55 a. x is not a binomial random variable because the trials are not independent. The probability of success (get an ace) changes from trial to trial. On the first trial it is 4/52. The probability of an ace on the second trial depends on the outcome of the first trial; it is 4/51 if an ace is not selected, and it is 3/51 if an ace was selected. The probability of an ace on any given trial continues to change when the experiment is completed without replacement.

b. x is a binomial random variable because the trials are independent. n = 4, the number of independent trials; two outcomes, success = ace and failure = not ace; p = P(ace) = 4/52 and q = P(not ace) = 48/52; x = n(aces drawn in 4 trials) and could be any number 0, 1, 2, 3 or 4. Further, the probability of success (get an ace) remains 4/52 for each trial throughout the experiment, as long as the card drawn on each trial is replaced before the next trial occurs.

5.57 a.

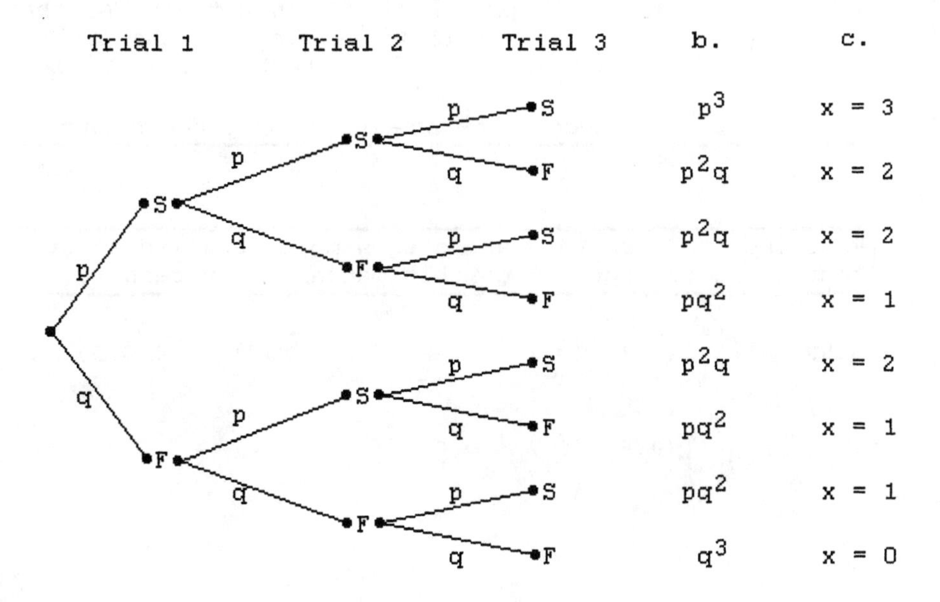

e. $P(x) = \binom{3}{x} p^x q^{3-x}$, for x = 0, 1, 2, 3

BINOMIAL PROBABILITY FUNCTION

$$P(x) = \binom{n}{x} p^x q^{n-x} \quad \text{for } x = 0,1,2,\ldots n$$

where: $P(x)$ = probability of x successes

n = the number of independent trials

$\binom{n}{x} = \frac{n!}{x!\,(n-x)!}$ = binomial coefficient = the number of combinations of successes and failures that result in exactly x successes in n trials.

p^x = probability of x successes, that is, p·p·p···, x times. Remember x is the number of successes, therefore every time a success occurs, the probability p is multiplied in.

q^{n-x} = probability of (n-x) failures, that is, q·q·q···, (n-x) times. This is the probability for "all of the rest of the trials."

Check: The sum of the exponents should equal n.

Exercise 5.59, parts a and c show more detailed solutions. Review factorials on page 157 of this manual, if necessary.

5.59 a. $\binom{4}{1}(0.3)^1(0.7)^3 = \frac{4!}{1!3!}(0.3)^1(0.7)^3 = 4(0.3)(0.343) = \underline{0.4116}$

b. $\binom{3}{2}(0.8)^2(0.2)^1 = \underline{0.384}$

c. $\binom{2}{0}(1/4)^0(3/4)^2 = \frac{2!}{0!2!}(1/4)^0(3/4)^2 = 1(1)(9/16) = \underline{0.5625}$

d. $\binom{5}{2}(1/3)^2(2/3)^3 = \underline{0.329218}$

e. $\binom{4}{2}(0.5)^2(0.5)^2 = \underline{0.375}$

f. $\binom{3}{3}(1/6)^3(5/6)^0 = \underline{0.0046296}$

5.61 By inspecting the function we see the binomial properties:

1. n = 5,
2. p = 1/2 and q = 1/2 (p + q = 1),
3. The two exponents x and 5-x add up to n = 5, and
4. x can take on any integer value from zero to n = 5;

therefore it is binomial.

By inspecting the probability distribution:

x	T(x)
0	1/32
1	5/32
2	10/32
3	10/32
4	5/32
5	1/32
Σ	32/32 = 1.0

It is a probability distribution.

1. Each T(x) is between 0 and 1.
2. ΣT(x) = 1.0

$$T(x) = \binom{5}{x}(1/2)^x(1/2)^{5-x} \text{ for } x = 0, 1, \ldots, 5$$

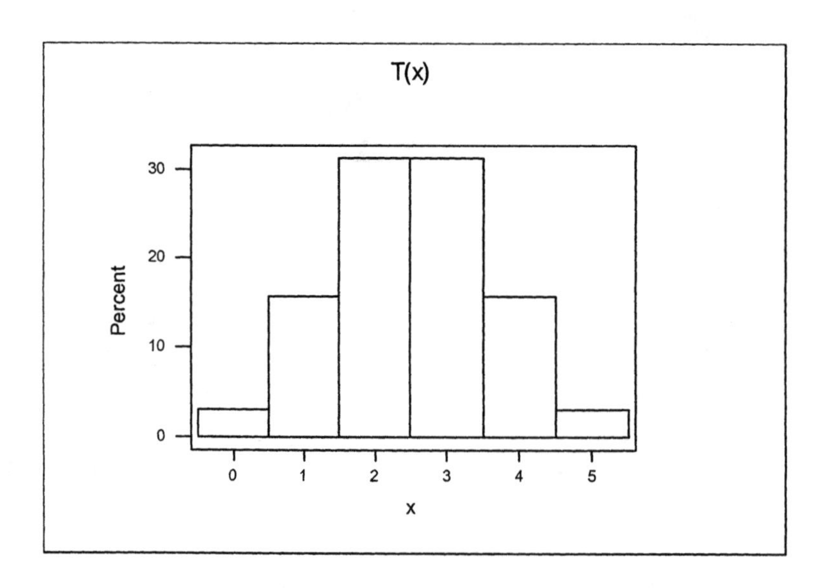

<u>Notation for a Binomial Probability Distribution</u>

P(x|B(n,p)) where x = whatever values (# of successes) are
B = binomial distribution
n = number of trials
p = probability of success

Use Table 2, (Appendix B, ES9-pp711-713), the Binomial Probabilities table, to find needed probabilities.

5.63 P(x = 8, 9, 10|B(n = 10,p = 0.90)) = P(8) + P(9) + P(10)

= 0.194 + 0.387 + 0.349 = <u>0.930</u>

5.65 a. P(x = 2|B(n = 3, p = 0.67)) = 0.444
b. P(x = 8|B(n = 12, p = 0.67)) = 0.238
c. P(x = 20|B(n = 30, p = 0.67)) = 0.153

Terminology in probability problems

For a binomial problem with $\mathbf{n = 10}$:

at least 5 successes $\Rightarrow x = 5,6,7,8,9$ or 10
at most 5 successes $\Rightarrow x = 0,1,2,3,4$ or 5

at most 9 successes $\Rightarrow x = 0,1,2,3,4,5,6,7,8$ or 9;
Since $\mathbf{P(0) + P(1) + \ldots + P(9)} + P(10) = 1$, then
$P(0) + P(1) + \ldots + P(9) = 1 - P(10)$.
Therefore use: P(at most 9) = 1 - P(10).

at least 2 successes $\Rightarrow x = 2,3,4,5,6,7,8,9,10$
Since $P(0) + P(1) + \mathbf{P(2) + P(3) + \ldots + P(9) + P(10)} = 1$, then
$P(2) + P(3) + \ldots + P(9) + P(10) = 1 - [P(0) + P(1)]$.
Therefore use: P(at least 2) = 1 - [P(0) + P(1)].

5.67 a. $P(x = 5|B(n = 5, p = 0.90)) = \underline{0.590}$

b. $P(x = 4,5|B(n = 5, p = 0.90)) = 0.328 + 0.590 = \underline{0.918}$

5.69 P(shut down) = $P(x \geq 2)$, where x represents the number defective in the sample of n = 10.

$P(x \geq 2) = 1.0 - [P(x = 0) + P(x = 1)]$

$$P(x = 0) = \binom{10}{0}(0.005)^0(0.995)^{10} = 0.9511$$

$$P(x = 1) = \binom{10}{1}(0.005)^1(0.995)^9 = 0.0478$$

$P(x \geq 2) = 1.0 - [0.9511 + 0.0478] = \underline{0.0011}$

5.71 $P(x = 1, 2, 3, 4, 5, 6|B(n = 6, p = 0.5))$
$= 1 - P(x = 0) = 1 - \binom{6}{0}(0.5)^0(0.5)^6 = \underline{0.984}$

5.73 a. $P(x < 2|B(n = 5, p = 0.373)) = 0.0969 + 0.2882 = 0.3851$
b. $P(x > 3|B(n = 5, p = 0.373)) = 0.0607 + 0.0072 = 0.0679$
c. $P(x = 5|B(n = 5, p = 0.373)) = 0.0072$

5.75 a. $P(x = 3|B(n = 5, p = 0.45)) = \binom{5}{3}(0.45)^3(0.55)^2 = \underline{0.2757}$

b. $P(x = 7|B(n = 15, p = 0.45)) = [(0.177 + 0.196)/2] = \underline{0.1865}$

c. $P(x \geq 7|B(n = 15, p = 0.45)) = [0.1865 + 0.157 + 0.107 + 0.058 + 0.0245 + 0.008 + 0.0015] = \underline{0.5425}$

d. $P(x \leq 7|B(n = 15, p = 0.45)) = [0.0025 + 0.0125 + 0.0385 + .0845 + 0.139 + 0.18 + 0.1865] = \underline{0.6435}$

Additional information on computer and/or calculator commands for binomial probabilities can be found in ES9-pp255-256.

5.77

x	P(x)	x	P(x)	x	P(x)
1*	0.0000	8	0.1009	15	0.0351
2	0.0003	9	0.1328	16	0.0177
3	0.0015	10	0.1502	17	0.0079
4	0.0056	11	0.1471	18	0.0031
5	0.0157	12	0.1254	19	0.0010
6	0.0353	13	0.0935	20	0.0003
7	0.0652	14	0.0611	21*	0.0001

* any other probabilities are each less than 0.00005

5.79 a. $P(x \leq 10|B(n = 25, p = 0.5)) = \underline{0.2122}$

b. These are independent events, so the simplified multiplication rule applies after the individual probabilities are calculated:

P(x ≤ 13 in San Antonio and x ≤ 13 in Salt Lake City)
= (0.6550)(0.9745) = <u>0.6383</u>

5.81 a. Using p = 0.555.

Row	xTwoShots	ShaqP2	ShaqCumP2
1	0	0.198025	0.19802
2	1	0.493950	0.69197
3	2	0.308025	1.00000

Row	x15Shots	ShaqP15	ShaqCumP15
1	0	0.000005	0.00001
2	1	0.000099	0.00010
3	2	0.000868	0.00097
4	3	0.004690	0.00566
5	4	0.017549	0.02321
6	5	0.048153	0.07137
7	6	0.100093	0.17146
8	7	0.160502	0.33196
9	8	0.200176	0.53214
10	9	0.194178	0.72631
11	10	0.145307	0.87162
12	11	0.082375	0.95400
13	12	0.034246	0.98824
14	13	0.009856	0.99810
15	14	0.001756	0.99985
16	15	0.000146	1.00000

Shaq had already made 9 of 13 free throws and he had two more attempts. That would make 15 free throw attempts for the game. The chances of him making 2 more seem to be against him. The binomial probability of him making 2 of 2 is only 0.308, while the probability that he makes more than 9 in a set of 15 is 0.274. Neither are strong probabilities.

b. Using p = 0.829.

Row	xTwoShots	KobeP2	KobeCmP2
1	0	0.029241	0.02924
2	1	0.283518	0.31276
3	2	0.687241	1.00000

Row	x10Shots	KobeP10	KobeCumP10
1	0	0.000000	0.00000
2	1	0.000001	0.00000
3	2	0.000023	0.00002
4	3	0.000292	0.00032
5	4	0.002480	0.00280
6	5	0.014426	0.01722
7	6	0.058282	0.07550
8	7	0.161455	0.23696
9	8	0.293522	0.53048
10	9	0.316218	0.84670
11	10	0.153301	1.00000

Kobe had made 6 of 8 previous attempts and these 2 would make a total of 10 for the game. The probability of him making 2 shots in a set of 2 is 0.687 and the probability that he makes 8 or more in a set of 10 is 0.763 - both are strong probabilities.

SECTION 5.5 ANSWER NOW EXERCISES

5.83 $\mu = np = 30 \cdot 0.6 = \underline{18}$

$\sigma = \sqrt{npq} = \sqrt{30 \cdot 0.6 \cdot 0.4} = \sqrt{7.2} = 2.68 = \underline{2.7}$

5.84 a. $\mu = np = 11 \cdot 0.05 = \underline{0.55}$

$\sigma = \sqrt{npq} = \sqrt{11 \cdot 0.05 \cdot 0.95} = \sqrt{0.5225} = 0.7228 = \underline{0.72}$

b.

x	P(x)	xP(x)	$x^2P(x)$
0	0.569	0	0
1	0.329	0.329	0.329
2	0.087	0.174	0.348
3	0.014	0.042	0.126
4	0.001	0.004	0.016
5	0+	0	0
Σ	1.0	0.549	0.819

The probabilities for x = 6 through 11 are all 0+.

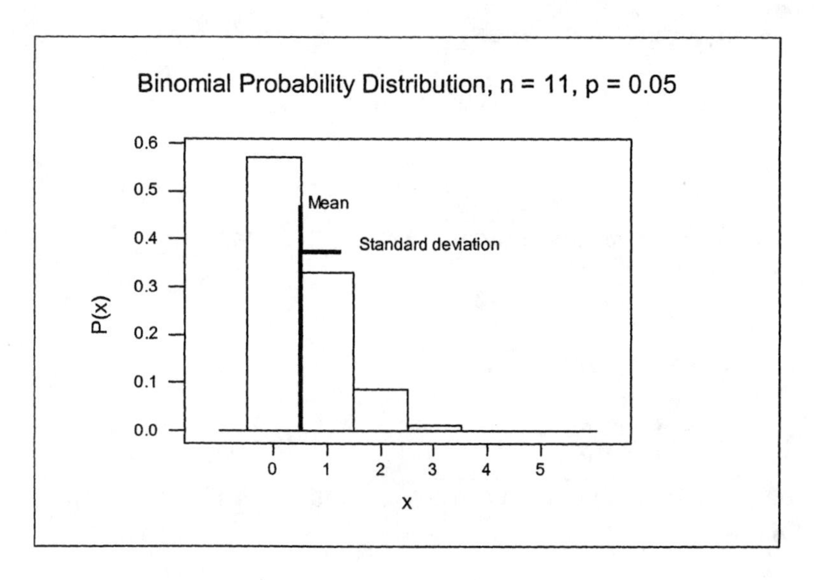

b. See histogram in (b).

SECTION 5.5 EXERCISES

5.85 a. Use extension table in exercise 5.84

$\mu = \Sigma[xP(x)] = 0.549 = \underline{0.55}$

$\sigma^2 = \Sigma[x^2P(x)] - \{\Sigma[xP(x)]\}^2 = 0.819 - \{0.549\}^2 = 0.5176$

$\sigma = \sqrt{\sigma^2} = \sqrt{0.5176} = 0.7194 = \underline{0.72}$

b. The mean and standard deviation of the probability distribution round to exactly the mean and standard deviation of the given binomial distribution.

For a binomial distribution the mean can be calculated using $\mu = np$ and the standard deviation by using $\sigma = \sqrt{npq}$.

Remember for any type discrete probability distribution:
$\mu = \Sigma[xP(x)]$ and $\sigma = \sqrt{\sigma^2}$ where $\sigma^2 = \Sigma[x^2P(x)] - [\Sigma[xP(x)]]^2$

Both formulas work for the binomial, however np and $\sqrt{npq}$ are quicker and less prone to computational error.

5.87 a. $\mu = np = 50 \cdot (0.5) = \underline{25.0}$

$$\sigma = \sqrt{npq} = \sqrt{50 \cdot (0.5) \cdot (0.5)} = 3.5355 = \underline{3.5}$$

b. $\mu = np = 100 \cdot (1/13) = 7.692 = \underline{7.7}$

$$\sigma = \sqrt{npq} = \sqrt{100 \cdot (1/13) \cdot (12/13)} = 2.665 = \underline{2.7}$$

c. $\mu = np = 400 \cdot 0.06 = \underline{24.0}$

$$\sigma = \sqrt{npq} = \sqrt{400 \cdot 0.06 \cdot 0.94} = 4.7497 = \underline{4.7}$$

d. $\mu = np = 50 \cdot 0.88 = \underline{44.0}$

$$\sigma = \sqrt{npq} = \sqrt{50 \cdot 0.88 \cdot 0.12} = 2.298 = \underline{2.3}$$

5.89 $\mu = np = 200$ and $\sigma = \sqrt{npq} = 10$, therefore:

$$npq = 100$$
$$200q = 100$$
$$q = 100/200 = \underline{0.5}$$
$$p = 1 - q = \underline{0.5}$$
$$n = 200/0.5 = \underline{400}$$

5.91 $\mu = np = 15 \cdot 0.3 = 4.5$

$$\sigma = \sqrt{npq} = \sqrt{15 \cdot 0.3 \cdot 0.7} = 1.77$$

$\mu \pm 2\sigma = 4.5 \pm 2(1.77) = 4.5 \pm 3.54$; 0.96 to 8.04

The variable is more than 2 standard deviations from the mean when x = 0, 9, 10, 11, 12, 13, 14, 15.

The probabilities are found on Table 2 (Appendix B; ES9-p711)

$P(x = 0, 9, 10, \ldots, 15) = 0.005 + 0.012 + 0.003 + 0.001 + 4(0+) = \underline{0.021}$

RETURN TO CHAPTER CASE STUDY

5.93 a. 19%
b. 8%
c. 5 nights
d. Number of evening meals American adults cook at home in an average week.
e. The circle graph represents 100% of the distribution and each sector of the circle is mutually exclusive of the others.
f. Histogram

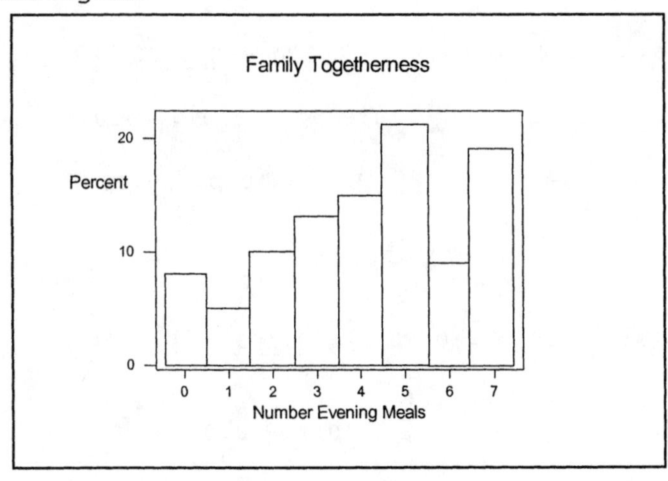

g. Relative frequency distribution, cumulative relative frequency distribution, ogive

CHAPTER EXERCISES

5.97 a. A probability function is generally thought to be the *algebraic formula*, and a probability distribution is generally thought to be the *chart* listing the pairs of x and P(x) values. They are equivalent, just two very different-looking ways to express this one concept.
b. A probability distribution relates to a population, whereas a frequency distribution relates to a sample. To compare, a frequency distribution needs to become a relative frequency distribution by dividing each frequency by the sample size.

5.99 a. P(exactly 14 arrive) = P(x = 14) = 0.1

b. P(at least 12 arrive) = P(x = 12, 13 or 14)
= P(12) + P(13) + P(14)
= 0.2 + 0.1 + 0.1 = 0.4

c. P(at most 11 arrive) = P(x = 10 or 11)
= P(10) + P(11)
= 0.4 + 0.2 = 0.6

5.101 a. $P(x = 0, 1, 2|B(n = 10, p = 0.10)) = P(0) + P(1) + P(2)$

$= 0.349 + 0.387 + 0.194 = 0.930$

b. $P(x = 2, 3, 4, \ldots, 10|B(n = 10, p = 0.10))$

$= 1 - P(x = 0, 1|B(n = 10, p = 0.10))$

$= 1.000 - [0.349 + 0.387] = 0.264$

5.103 $P(x = 5|B(n = 10, p = 0.70)) = 0.103$

5.105 a. $P(x > 4|B(n = 15, p = 0.10)) = 0.010 + 0.002 = 0.012$
b. $P(x = 2|B(n = 15, p = 0.10) = 0.267$
c. $P(x < 2|B(n = 15, p = 0.10)) = 0.206 + 0.343 = 0.549$

5.107 2/3(9) = 6 or more votes for a proposal to be accepted.

$P(x = 6, 7|B(n = 7, p = 0.5)) = 0.055 + 0.008 = 0.063$

5.109 If x is the random variable n(defective), then *success* is *defective*. On the first selection, P(defective) = 3/10. However, the P(defective) changes for the next selection: it is either 3/9 or 2/9 depending on whether or not the first selection resulted in a defective or not. Since the probability of defective changes, the trials are not independent. Thus, the experiment is not binomial.

5.111 $\sigma^2 = \sum x^2 P(x) - \mu^2$ (Formula 5.3b)
$100 = \sum x^2 P(x) - 50^2$ or $\sum x^2 P(x) = 2600$

5.113 a.

x	f	xf
39	1	39
40	2	80
41	3	123
42	4	168
43	6	258
44	7	308
45	8	360
46	4	184
47	3	141
48	1	48
49	1	49
	40	1758

Mean 43.95

Estimated probability of germination = 43.95/50
= 0.879 or 0.88

b. B(50, 0.88) partly listed below

x	B(50,0.88)	Cum B(50,0.88)
31	0.000002	0.00000
32	0.000008	0.00001
33	0.000032	0.00004
34	0.000118	0.00016
35	0.000395	0.00056
36	0.001208	0.00176
37	0.003352	0.00512
39	0.018974	0.03250
40	0.038264	0.07076
41	0.068439	0.13920
42	0.107547	0.24675
43	0.146731	0.39348
44	0.171186	0.56466
45	0.167382	0.73205
46	0.133420	0.86547
49	0.011424	0.99832
50	0.001675	1.00000

c.

x	B(50,0.88)	40(B(50,0.88))	Expected	Actual
39	0.018974	0.75895	1	1
40	0.038264	1.53055	2	2
41	0.068439	2.73756	3	3
42	0.107547	4.30188	4	4
43	0.146731	5.86923	6	6
44	0.171186	6.84744	7	7
45	0.167382	6.69527	7	8
46	0.133420	5.33681	5	4
47	0.083269	3.33078	3	3
48	0.038165	1.52661	2	1
49	0.011424	0.45694	0	1

d. The distribution for the expected results and the distribution for the actual occurrences are very similar.

5.115 Tool Shop: x = profit

x	P(x)	x·P(x)
100,000	0.10	10,000.0
50,000	0.30	15,000.0
20,000	0.30	6,000.0
-80,000	0.30	-24,000.0
Σ	1.00ck	7,000.0

mean profit = $\mu = \Sigma[x \cdot P(x)] = 7{,}000.0$

Book Store: x = profit

x	P(x)	x·P(x)
400,0000.20		80,000.0
90,000	0.10	9,000.0
-20,0000.40		-8,000.0
-250,000	0.30	-75,000.0
Σ	1.00ck	6,000.0

mean profit = $\mu = \Sigma[x \cdot P(x)] = 6{,}000.0$

The Tool Shop has a slightly higher mean profit.

5.117

$$\begin{aligned}
\mu &= \Sigma[x \cdot P(x)] \\
&= (1)\cdot(1/n) + (2)\cdot(1/n) + \ldots + (n)\cdot(1/n) \\
&= (1/n)\cdot[1 + 2 + 3 + \ldots + n] \\
&= (1/n)\cdot[(n)(n+1)/2] \\
&= (n + 1)/2
\end{aligned}$$

CHAPTER 6 ∇ NORMAL PROBABILITY DISTRIBUTIONS

Chapter Preview

Chapter 6 continues the presentation of probability distributions started in Chapter 5. In this chapter, the random variable is a continuous random variable (versus a discrete random variable in Chapter 5); therefore, the probability distribution is a continuous probability distribution. There are many types of continuous distributions, but this chapter will limit itself to the most common, namely, the normal distribution. The main elements of a normal probability distribution to be covered are:

1. how probabilities are found
2. how they are represented
3. how they are used.

Intelligence and aptitude test score measuring is presented in this chapter's case study.

CHAPTER 6 CASE STUDY

6.1
a. It's a quotient defined by
[100 x (Mental Age/Chronological Age)]
b. I.Q.: 100, 16; SAT: 500, 100; Standard score: 0, 1;
c. z = (I.Q. - 100)/16; z = (SAT - 500)/100
d. 2, 132, 700
e. The percentages are the same (other than for round-off)

SECTION 6.2 ANSWER NOW EXERCISES

6.2 0.4147

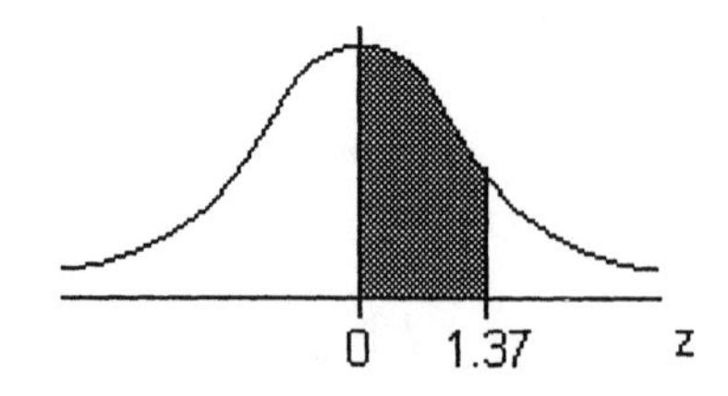

6.3 $P(z > 2.03) =$
$0.5000 - 0.4788 = \underline{0.0212}$

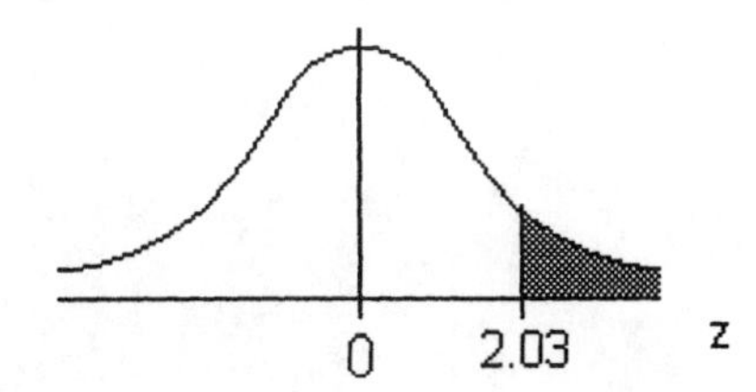

6.4 $P(z < 1.73) =$
$0.5000 + 0.4582 = \underline{0.9582}$

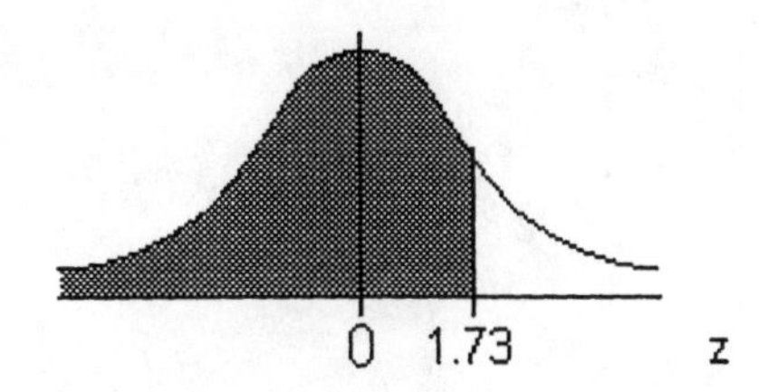

6.5 $P(-1.39 < z < 0.00) = \underline{0.4177}$

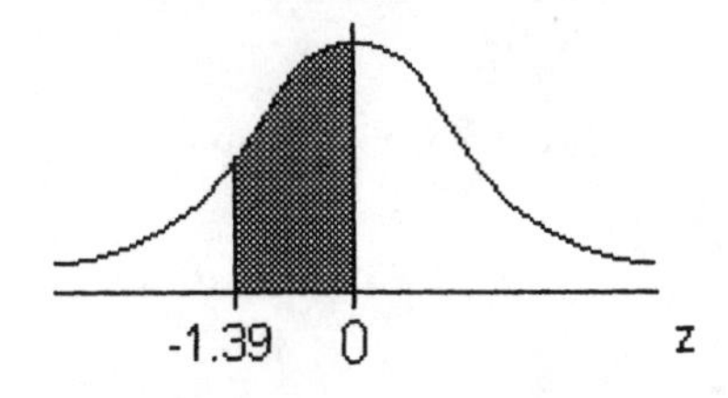

6.6 $P(z < -1.53) =$
$0.5000 - 0.4370 = \underline{0.0630}$

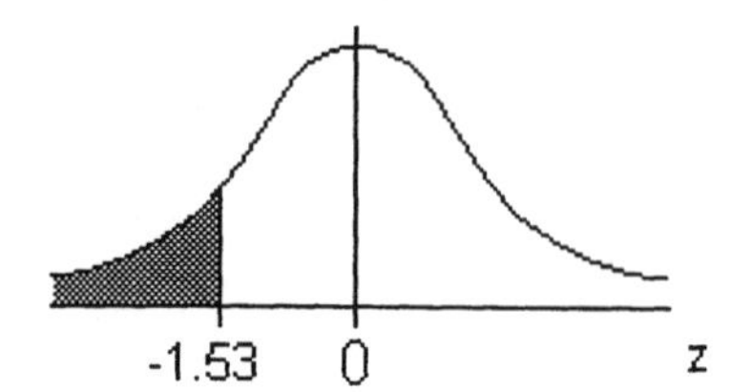

6.7 $P(-1.83 < z < 1.23) =$
$0.4664 + 0.3907 = \underline{0.8571}$

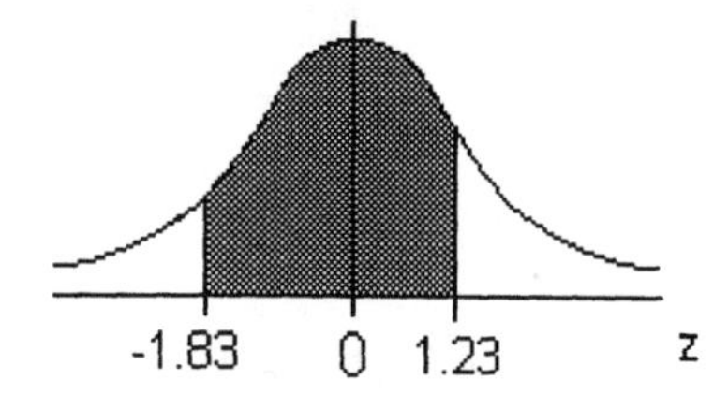

6.8 $P(0.75 < z < 2.25) =$
$0.4878 - 0.2734 = \underline{0.2144}$

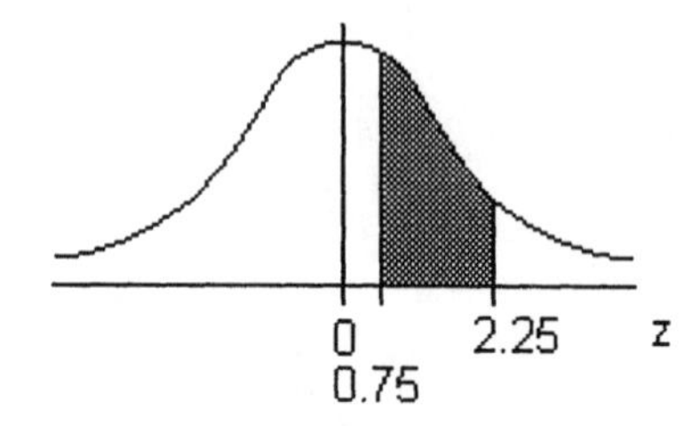

6.9 a. $z = \underline{0.84}$

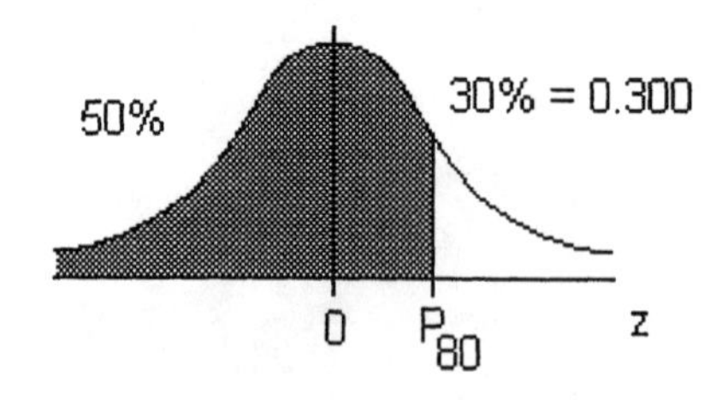

b. 0.75/2 = 0.375 -1.15 and +1.15

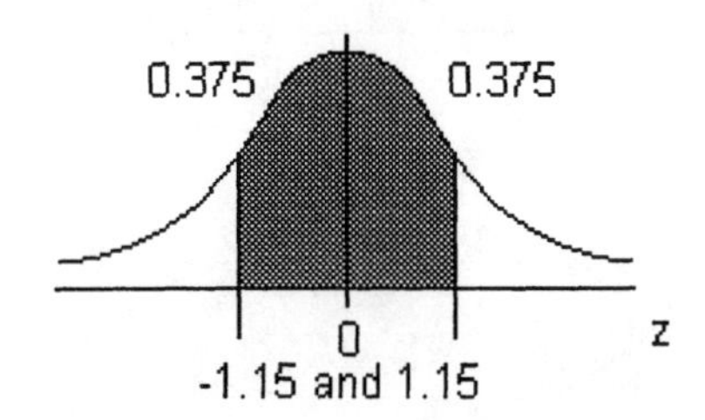

SECTION 6.2 EXERCISES

Continuous Random Variable
- a numerical quantity that can take on values on a certain interval

STANDARD NORMAL DISTRIBUTION

- bell shaped, symmetric curve
- $\mu = 0, \quad \sigma = 1$
- distribution for the standard normal score *z*

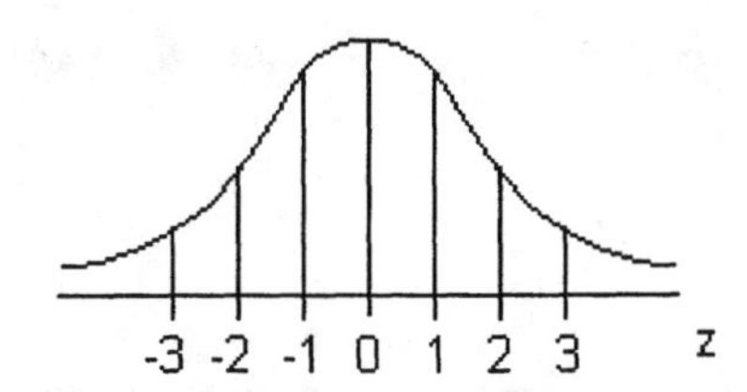

$z = 0 \Rightarrow \mu$

$z = 1 \Rightarrow \mu + 1\sigma$	$z = -1 \Rightarrow \mu - 1\sigma$
$z = 2 \Rightarrow \mu + 2\sigma$	$z = -2 \Rightarrow \mu - 2\sigma$
$z = 3 \Rightarrow \mu + 3\sigma$	$z = -3 \Rightarrow \mu - 3\sigma$

. . .

- area under the curve = 1

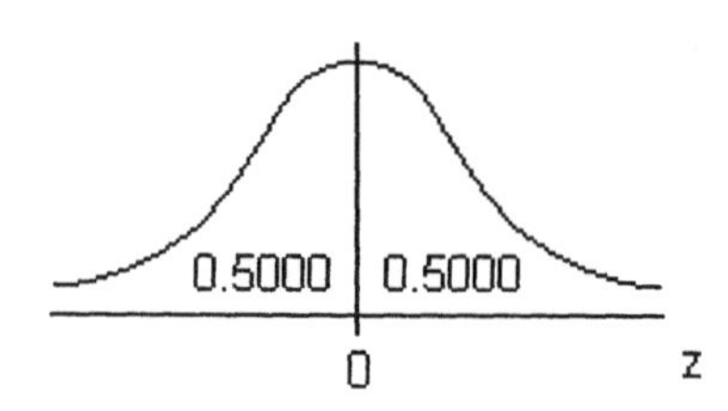

- symmetric, therefore

The Empirical Rule and Table 3

≈ 68% of the data lies within 1 standard deviation of the mean.
- Note z = 1, gives .3413. Since symmetric, from z = -1 to z = 1 would be 2(.3413) = .6828 ≈ 68%.

≈ 95% of the data lies within 2 standard deviations of the mean.
- Note z = 2, gives .4772. Since symmetric, from z = -2 to z = 2 would be 2(.4772) = .9544 ≈ 95%.

≈ 99.7% of the data lies within 3 standard deviations of the mean.
- Note z = 3, gives .4987. Since symmetric, from z = -3 to z = 3 would be 2(.4987) = .9974 ≈ 99.7%.

Draw a picture of a normal distribution and shade in the section representing the area desired. Remember .5000 or 50% of the distribution is on either side of μ or the z = 0.

6.11 a. 0.4032

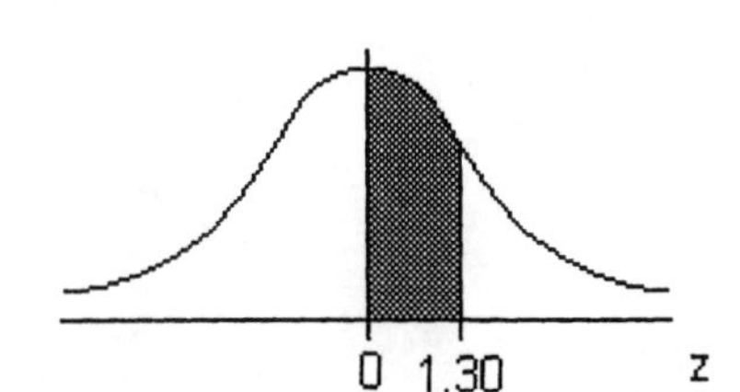

b. 0.3997

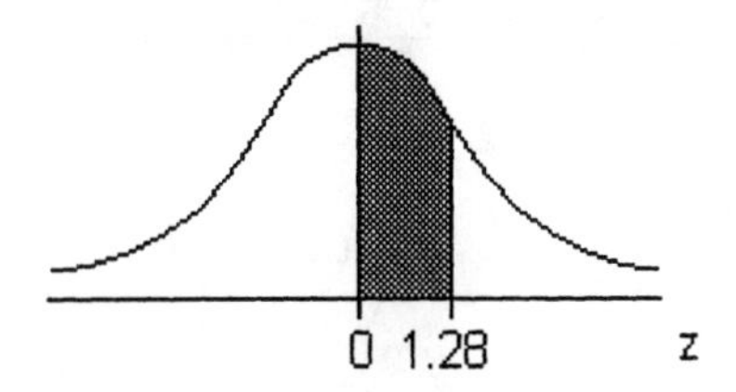

c. 0.4993

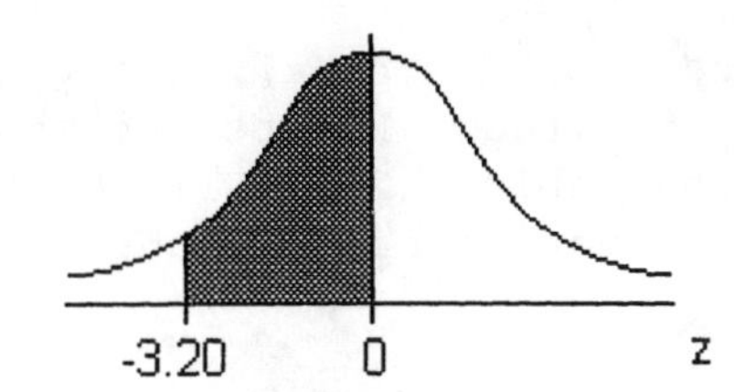

d. 0.4761

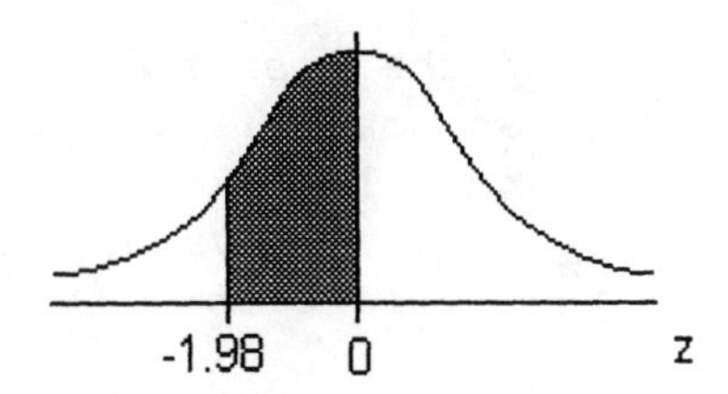

To find area in the right or left tail, subtract the Table 3 (Appendix B, ES9-p714) area (probability) for the given z value from 0.5000.

To find area that extends from one side of the mean (z = 0) through the tail of the other side, add the Table 3 area (probability) for the given z value to 0.5000.

6.13 a. 0.4394

b. 0.5000 - 0.4394 = 0.0606

c. 0.5000 + 0.4394 = 0.9394

d. 2(0.4394) = 0.8788

6.15 a. 0.3849 + 0.3888 = 0.7737

b. 0.4599 + 0.4382 = 0.8981

c. 0.4032 + 0.4951 = 0.8983

d. 0.4998 - 0.1368 = 0.3630

6.17 a. <u>0.5000</u>

b. 0.5000 - 0.3531= <u>0.1469</u>

c. 0.5000 + 0.4893 = <u>0.9893</u>

d. 0.4452 + 0.5000 = <u>0.9452</u>

e. 0.5000 - 0.4452 = <u>0.0548</u>

6.19 a. <u>0.4906</u>

b. 0.4821 + 0.4904 = <u>0.9725</u>

c. 0.5000 - 0.0517 = <u>0.4483</u>

d. 0.5000 + 0.4306 = <u>0.9306</u>

Look at the *inside* of Table 3 (Appendix B, ES9-p714), and get as close as possible to the probability desired. Locate the position (row and column) on the outside of the table. This will be the corresponding z value. Remember the negative sign if the z value is to the left of μ.

NOTE: NORMAL CURVE

TABLE 3
- inside ⇒ probabilities
- outside ⇒z-values

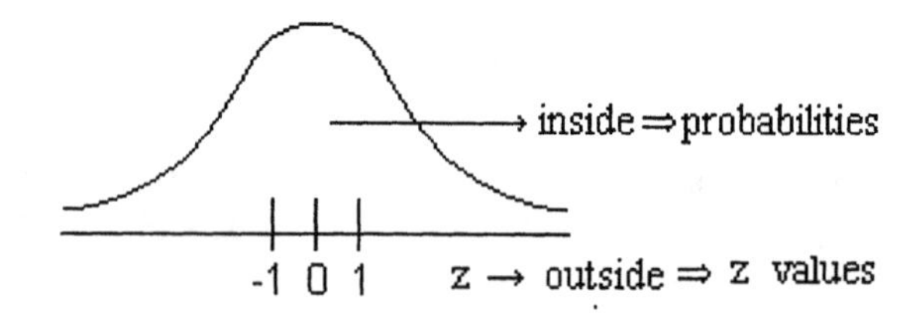

6.21 a. <u>1.14</u> b. <u>0.47</u> c. <u>1.66</u>

d. <u>0.86</u> e. <u>1.74</u> f. <u>2.23</u>

Subtract the area desired from 0.5000 first. Table 3 (Appendix B, ES9-p714) works from 0 to a *z* value.
Remember that one half of the distribution is .5000.

6.23 a. <u>1.65</u> b. <u>1.96</u> c. <u>2.33</u>

Draw a picture of a normal distribution with the desired area shaded in.
Fill in the needed probabilities.
Locate the appropriate probability in Table 3 (Appendix B, ES9-p714).
Locate the corresponding z value.

6.25 -1.28 or +1.28

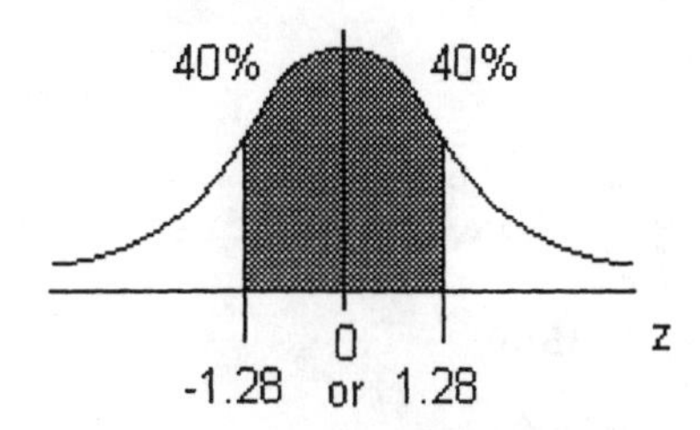

6.27 0.2500 0.2500

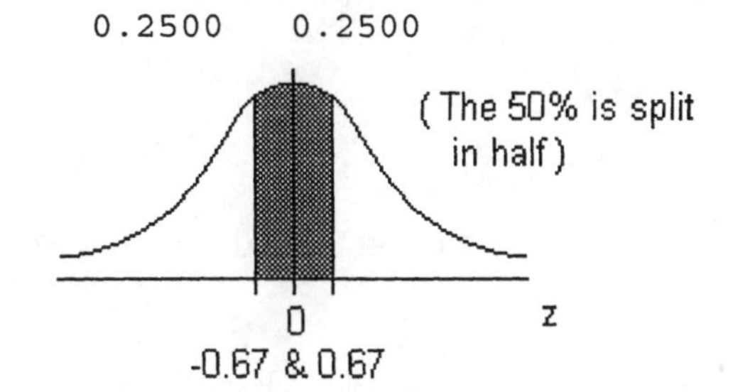

6.29

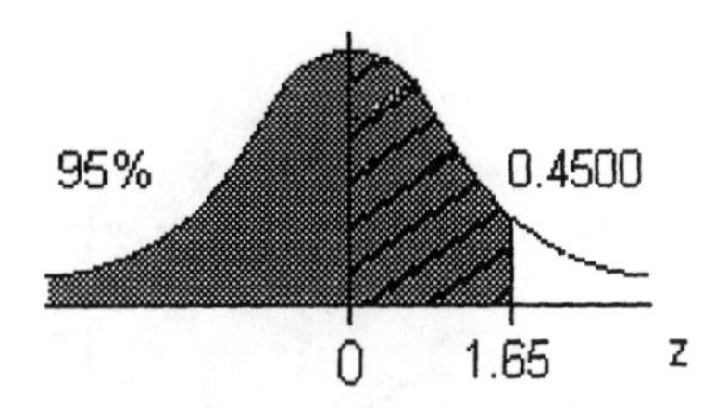

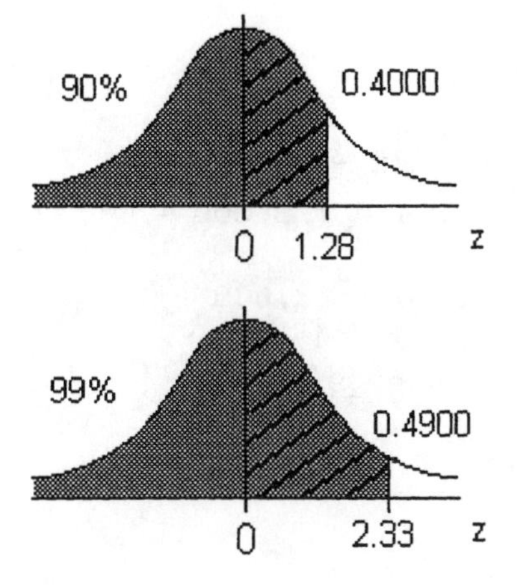

SECTION 6.3 ANSWER NOW EXERCISES

6.31 $z = (x - \mu)/\sigma = (58 - 43)/5.2 = \underline{2.88}$

6.32 a. $P(100 < x < 120) = P[(100 - 100)/16 < z < (120 - 100)/16]$
$= P[0.00 < z < 1.25] = \underline{0.3944}$

b. $P(x > 80) = P[z > (80 - 100)/16]$
$= P[z > - 1.25]$
$= 0.3944 + 0.5000 = \underline{0.8944}$

6.33 a. No; the amount varies from car to car.
b. The mean and standard deviation of the distribution.

6.34 a. 15% = 0.1500 = 0.5000 - 0.3500
z closest to 0.3500 = 1.04
$z = (x - \mu)/\sigma$
$1.04 = (x - 72)/13$
$13.52 = x - 72$
$x = 85.52 = \underline{86}$

b.

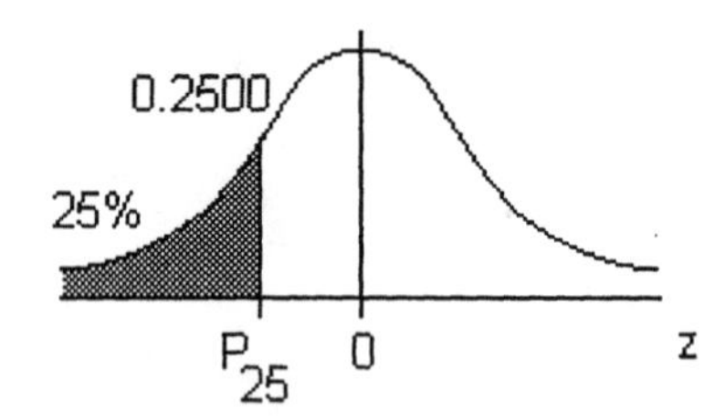

$P(a < z < 0) = 0.2500$
$a = -0.67$

$z = (x - \mu)/\sigma$
$-0.67 = (x - 100)/16$
$-10.72 = x - 100$
$x = 89.28 = 89$

c.

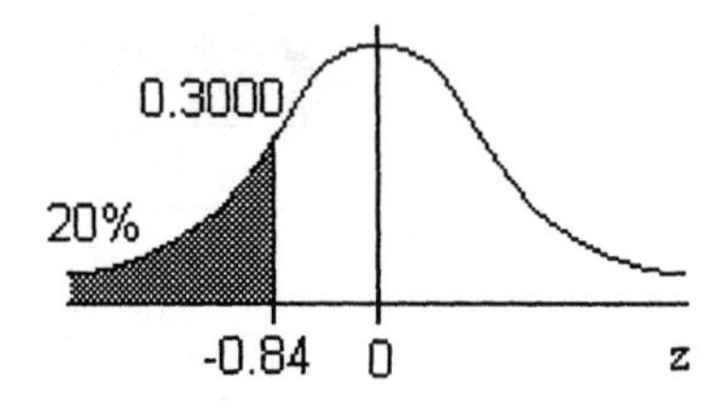

$z = (x - \mu)/\sigma$
$-0.84 = (28{,}000 - \mu)/1200$

$-1008 = 28{,}000 - \mu$

$\mu = \underline{29{,}008}$

6.35 a. 0.7606
b. 0.0386
c. 0.2689

6.36 a. shifts to the right 3 units
b. short and wider
c. tall and narrow

6.37 Everyone's results will be different. Computer and calculator commands can be found in ES9-pp285&286.

6.38 Everyone's results will be different. Computer and calculator commands can be found in ES9-pp285&286.

6.39 Everyone's results will be different. Computer and calculator commands can be found in ES9-pp286&287.

6.40 Everyone's results will be different. Each curve should be approximately normally distributed. Computer and calculator commands can be found in ES9-pp286&287.

6.41 With 55 and 65 as the data values, use the computer and/or calculator commands in ES9-p287&288.

$P(55 < x < 65) = 0.8944 - 0.6615 = 0.2329$

$z = (55 - 50)/12 = 0.4167 = 0.42$
$z = (65 - 50)/12 = 1.25$
$P(0.42 < z < 1.25) = 0.3944 - 0.1628 = \underline{0.2316}$

The difference is due to round-off error in the calculation of $z = 0.42$.

6.42 a. $z = (0.5505 - 0.5535)/0.0015 = -2.00$

b. $z = (0.5555 - 0.5535)/0.0015 = 1.33$

c. Using average value of 0.5530,

$z = (0.5505 - 0.5530)/0.0015 = -1.667$

$z = (0.5555 - 0.5530)/0.0015 = 1.667$

$z = 1.67$ is nearest table entry

$0.5000 - 0.4525 = 0.0475$ is the percentage associated with each tail

100,000(0.0475) = 4750, therefore anticipate 4750 items of scrap and another 4750 to be reworked

4750($5.25) = $24,937.50 for rework cost

4750($15.34) = $72,865.00 for scrap parts

These costs are approximately the same as reported. The difference was caused by the round-off error when we used $z = 1.67$ instead of 1.66667 and Table 3 instead of a more precise probability value.

SECTION 6.3 EXERCISES

<u>x</u> is still used to denote a <u>continuous random variable</u>, but *x* is now referred to on an interval, not a single value. (ex.: $a < x < b$) As before, any letter may be used; *x* is the most common.

<u>Applications of the Normal Distribution</u>

1. Draw a sketch of the desired area, noting given μ and σ.
2. Write the desired probability question in terms of the given variable - usually x (ex.: $P(x > 10)$).
3. Transform the given variable *x* into a *z* value using $z = (x - \mu)/\sigma$.
4. Rewrite the probability question using z (ex.: $P(z > *)$, where * = calculated z value).
5. Use Table 3 (Appendix B, ES9-p714) and find the probability.

Remember: Probabilities (areas) given in Table 3 are from $z = 0$ to $z = \#$.

If the interval starts at $z = 0$ and continues either to the right or to the left, the probability is given in the table.

If the interval contains $z = 0$, add the probabilities.

If the interval extends from one positive z to another positive z, subtract the smaller probability from the larger probability (works the same with 2 negative z's).

If the interval does not contain z = 0 but does contain either the right or left tail, subtract the probability given in the table from 0.5000.

Reminder: When calculating z → use 2 decimal places.

When calculating areas or probabilities → use 4 decimal places.

6.43 Use formula $z = (x - \mu)/\sigma$:

a. $P[x > 60] = P[z > (60 - 60)/10]$

$= P[z > 0.00] = \underline{0.5000}$

b. $P[60 < x < 72] = P[(60 - 60)/10 < z < (72 - 60)/10]$
$= P[0.00 < z < 1.20] = \underline{0.3849}$

c. $P[57 < x < 83] = P[(57 - 60)/10 < z < (83 - 60)/10]$
$= P[-0.30 < z < 2.30]$
$= 0.1179 + 0.4893 = \underline{0.6072}$

d. $P[65 < x < 82] = P[(65 - 60)/10 < z < (82 - 60)/10]$
$= P[0.50 < z < 2.20]$
$= 0.4861 - 0.1915 = \underline{0.2946}$

e. $P[38 < x < 78] = P[(38 - 60)/10 < z < (78 - 60)/10]$
$= P[-2.20 < z < 1.80]$
$= 0.4861 + 0.4641 = \underline{0.9502}$

f. $P[x < 38] = P[z < (38 - 60)/10]$
$= P[z < -2.20]$
$= 0.5000 - 0.4861 = \underline{0.0139}$

6.45 a. $P(x < 24{,}000) = P[z < (24{,}000 - 27{,}937)/1700]$
$= P[z < -2.32]$
$= 0.5000 - 0.4898 = \underline{0.0102} = \underline{1.02\%}$

b. $P(x > 30{,}000) = P[z > (30{,}000 - 27{,}937)/1700]$
$= P[z > 1.21]$
$= 0.5000 - 0.3869 = \underline{0.1131} = \underline{11.31\%}$

6.47 a. When x = \$11.00: $z = (\$11.00 - \$13.47)/\$4.75 = -0.52$
When x = \$15.00: $z = (\$15.00 - \$13.47)/\$4.75 = 0.32$
Therefore, $P(\$11.00 < x < \$15.00) = P(-0.52 < z < 0.32) = 0.1985 + 0.1255 = \underline{0.3240 \text{ or } 32.4\%}$

b. When x = \$8.00: $z = (\$8.00 - \$13.47)/\$4.75 = -1.15$
When x = \$19.00: $z = (\$19.00 - \$13.47)/\$4.75 = 1.16$
Therefore, $P(\$8.00 < x < \$19.00) = P(-1.15 < z < 1.16)$
$= 0.3749 + 0.3770 = \underline{0.7519 \text{ or } 75.19\%}$

c. If x = \$20.00: $z = (\$20.00 - \$13.47)/\$4.75 = 1.35$
Therefore, $P(x > \$20.00) = P(z > 1.35) = 0.5000 - 0.4115 = \underline{0.0885 \text{ or } 8.9\%}$

d. If x = \$6.00: $z = (\$6.00 - \$13.47)/\$4.75 = -1.57$
Therefore, $P(x < \$6.00) = P(z < -1.573) = 0.5000 - 0.4418$
$= \underline{0.0582 \text{ or } 5.8\%}$

6.49 a. $P(x < 20) = P[z < (20 - 21.3)/6.0]$
$= P[z < -0.22]$
$= 0.5000 - 0.0871 = \underline{0.4129} = \underline{41.29\%}$

b. When x = 18: $z = (18 - 21.3)/6.0 = -0.55$
When x = 24: $z = (24 - 21.3)/6.0 = 0.45$
Therefore, $P(18 < x < 24) = P(-0.55 < z < 0.45)$
$= 0.2088 + 0.1736 = \underline{0.3824 \text{ or } 38.24\%}$

c. $P(x > 30) = P[z > (30 - 21.3)/6.0]$
$= P[z > 1.45]$
$= 0.5000 - 0.4265 = \underline{0.0735} = \underline{7.4\%}$

> 75th percentile = $P_{75} \Rightarrow$ 75% of the data is less than this value, therefore shade in the left side of the normal curve (all 0.5000) plus part of the right side (0.2500 of it). Locate the probability 0.2500 in Table 3 and find the corresponding z value. Substitute known values into the z formula and solve for the unknown.

d.

0.2500

$\sigma = 6.0$

75%

21.3 S x

$z = +0.67$

$z = (x - \mu)/\sigma$

$0.67 = (S - 21.3)/6.0$

$S = (0.67)(6.0) + 21.3$

$S = 25.32$

Draw a picture first, filling in the given and calculated probabilities.
Based on the probability, determine the z value.
Use $z = (x - \mu)/\sigma$ to find x.

6.51 a.

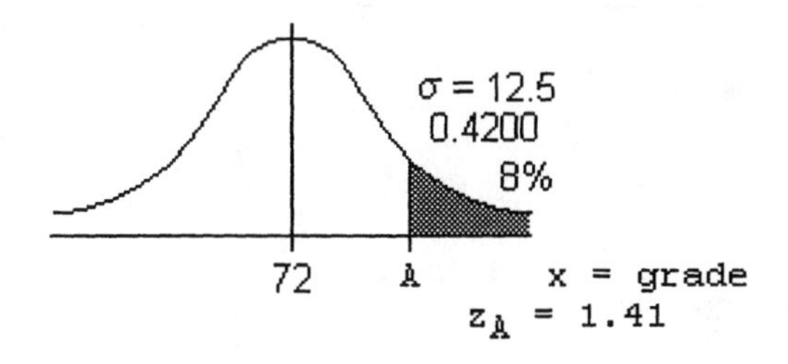

$z = (x - \mu)/\sigma$

$1.41 = (A - 72)/12.5$

$A = (1.41)(12.5) + 72$

$A = \underline{89.6}$

b.

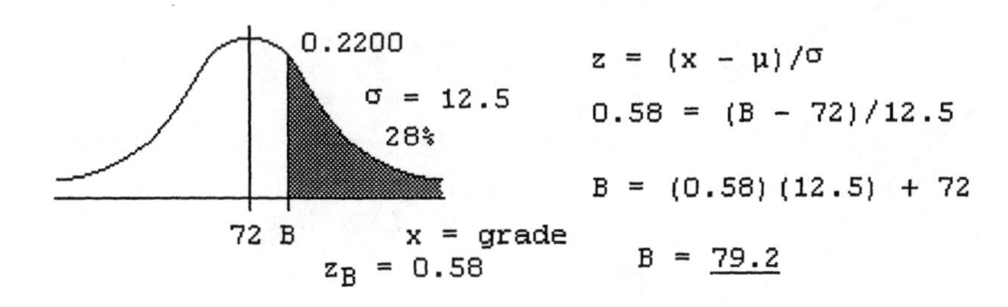

c.

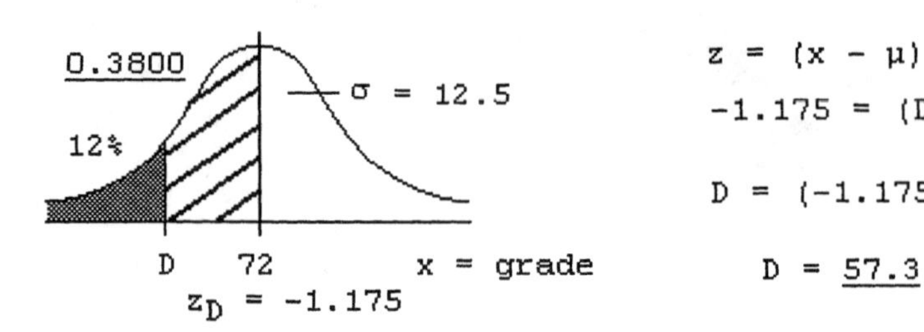

$z = (x - \mu)/\sigma$

$-1.175 = (D - 72)/12.5$

$D = (-1.175)(12.5) + 72$

$D = \underline{57.3}$

6.53

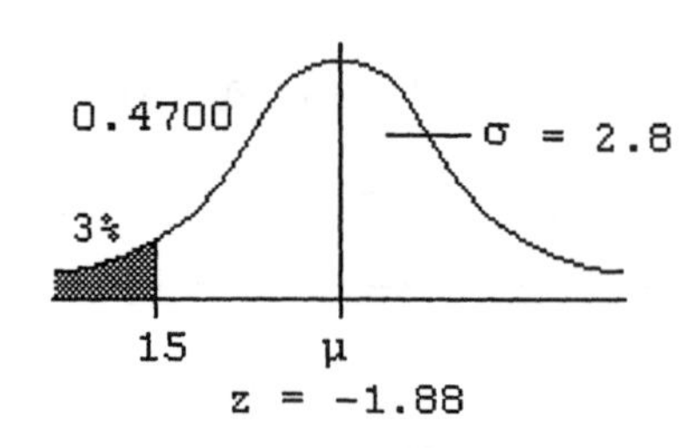

$z = (x - \mu)/\sigma$

$-1.88 = (15 - \mu)/2.8$

$\mu = 15 - (-1.88)(2.8)$

$\mu = \underline{20.26}$

6.55 a.

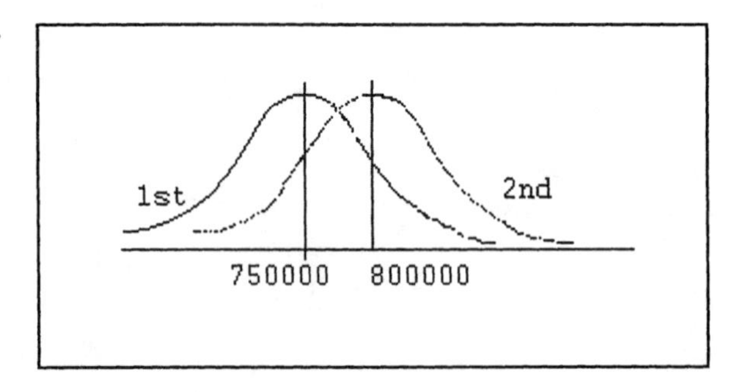

b. $P(x < 800{,}000) = P[z < (800{,}000 - 750{,}000)/60{,}000]$
$= P[z < 0.83]$
$= 0.5000 + 0.2967 = \underline{0.7967} = \underline{79.7\%}$

c. $P(x > 750{,}000) = P[z < (750{,}000 - 800{,}000)/60{,}000]$
$= P[z < -0.83]$
$= 0.5000 + 0.2967 = \underline{0.7967} = \underline{79.7\%}$

Additional information on computer and calculator commands regarding the normal distribution, can be found in ES9-pp285-288.

6.57 a. $P(x < 525) = \underline{0.056241}$

b. $P(525 < x < 590) = 0.561785 - 0.056241 = \underline{0.505544}$

c. $P(x \geq 590) = 1.000000 - 0.561785 = \underline{0.438215}$

d. $P(x < 525) = P[z < (525 - 584.2)/37.3]$
$= P[z < -1.59] = 0.5000 - 0.4441 = \underline{0.0559}$

$P(525 < x < 590) = P[-1.59 < z < (590 - 584.2)/37.3]$
$= P[-1.59 < z < 0.16]$
$= 0.4441 + 0.0636 = \underline{0.5077}$

$P(x > 590) = P(z > 0.16) = 0.5000 - 0.0636 = \underline{0.4364}$

e. Round-off errors; specifically in (d) when z is calculated to the nearest hundredth and the rounded z score is used with Table 3.

6.59 Everyone's generated values will be different, but should have a mean and standard deviation close to 100 and 16, respectively, and be approximately normally distributed.

SECTION 6.4 ANSWER NOW EXERCISES

6.61 a.

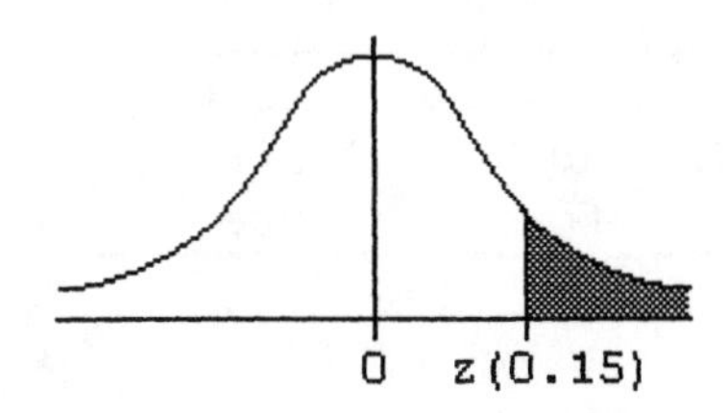

b.

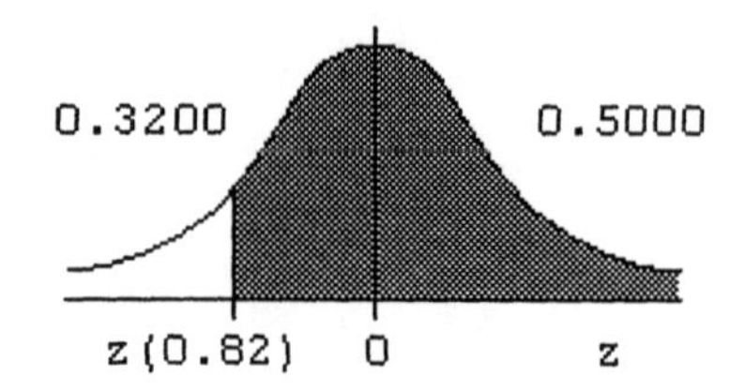

SECTION 6.4 EXERCISES

Z - NOTATION

$z(\alpha) = z_\alpha$ = the z value that has α area to the right of it

1. Draw a picture.
2. Shade in desired α area, starting from the far right tail.
3. Based on the diagram and location, determine the z value using Table 3 (Appendix B, ES9-p714).

6.63 a. z(0.03) b. z(0.14) c. z(0.75)

d. z(0.13) e. z(0.90) f. z(0.82)

6.65 a. 1.96 b. 1.65 c. 2.33

Solve exercise 6.67 using Table 3 (Appendix B, ES9-p714).

6.67 a. 1.28, 1.65, 1.96, 2.05, 2.33, 2.58

Note the z values. They are the most common occurring z values. For that reason, Table 4, Part A, has been included in Appendix B (ES9-p715). Note Table 4, Part B, for later use.

b. -2.58, -2.33, -2.05, -1.96, -1.65, -1.28

6.69 a. A is an area. z is 0.10 and the area to the right of z = 0.10 is 0.5000 - 0.0398 = 0.4602.

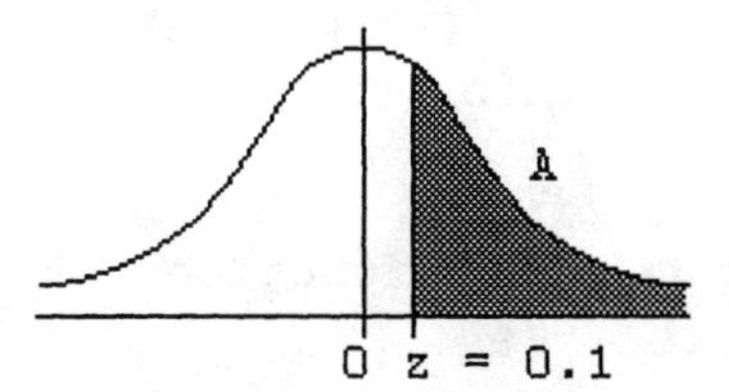

b. B is a z-score. 0.10 is the area to the right of z = B. Use 0.4000 [0.5000 - 0.1000] to look up the z-score on Table 3 (Appendix B, ES7-p760). z = B = 1.28.

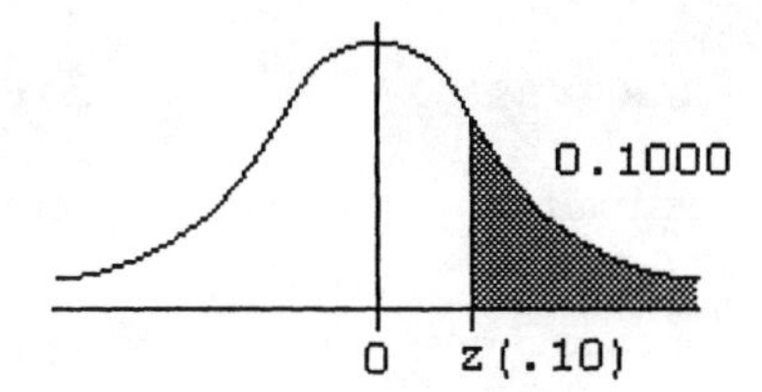

c. C is an area. z is -0.05 and the area to the right of z = -0.05 is 0.5000 + 0.0199 = 0.5199.

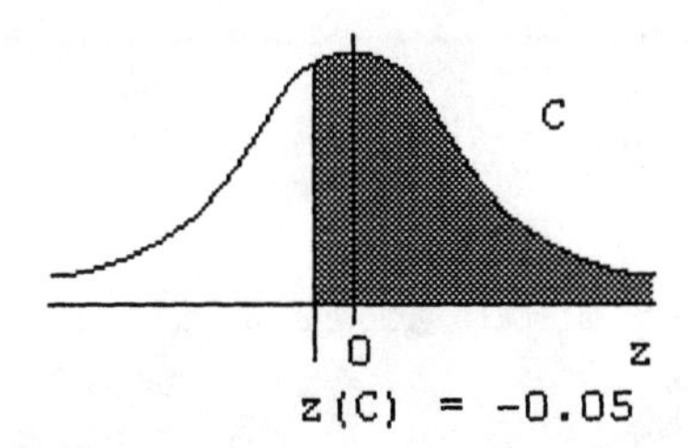

d. D is a z-score. D is to the left of zero [negative], use 0.4500 [0.5000 - 0.0500] to look it up; z = D = -1.65.

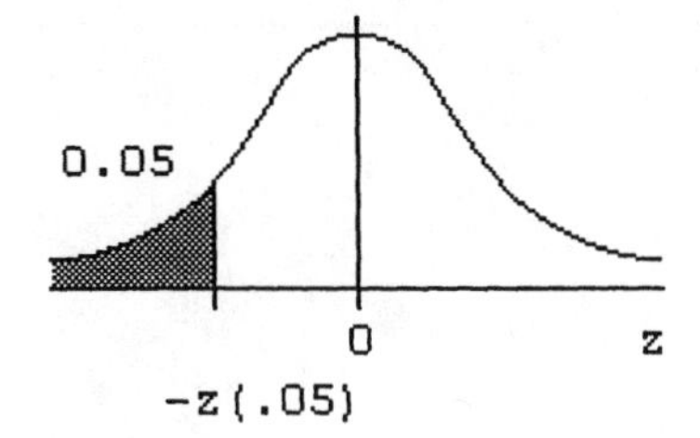

SECTION 6.5 ANSWER NOW EXERCISES

6.71 $np = (100)(0.02) = 2$ $nq = (100)(0.98) = 98$
No, both np and nq must be greater than or equal to 5.

6.72 a. Using the binomial probability function:

$\binom{25}{3}(1/3)^3(2/3)^{22} = 0.011386$

$\binom{25}{2}(1/3)^2(2/3)^{23} = 0.002970$

$\binom{25}{1}(1/3)^1(2/3)^{24} = 0.000495$

$\binom{25}{0}(1/3)^0(2/3)^{25} = 0.000040$

$P(x \le 3|B(25, 1/3)) = \underline{0.014891}$

Using the cumulative binomial probability function on a computer or calculator:
$P(x \le 3|B(25, 1/3)) = \underline{0.0149}$

b. Norm. approx. 0.0202 vs. 0.0149; only different by 0.0053

SECTION 6.5 EXERCISES

Criteria for the Normal Approximation of the Binomial

1. np and n(1-p) must both be ≥ 5.
2. 0.5 must be added to and/or subtracted from the x values to allow for an interval. ex.: $x = 5$ (discrete) $\Rightarrow$ $4.5 < x < 5.5$ (continuous)
3. $\mu = np$ and $\sigma = \sqrt{npq}$
4. $z = (x - \mu)/\sigma$ and Table 3 (Appendix B, ES9-p714) are used to find probabilities.

6.73 a. Rule of thumb: np = 3 and nq = 7, therefore the approximation is not appropriate, since np < 5. The distribution is slightly skewed, not symmetrical.

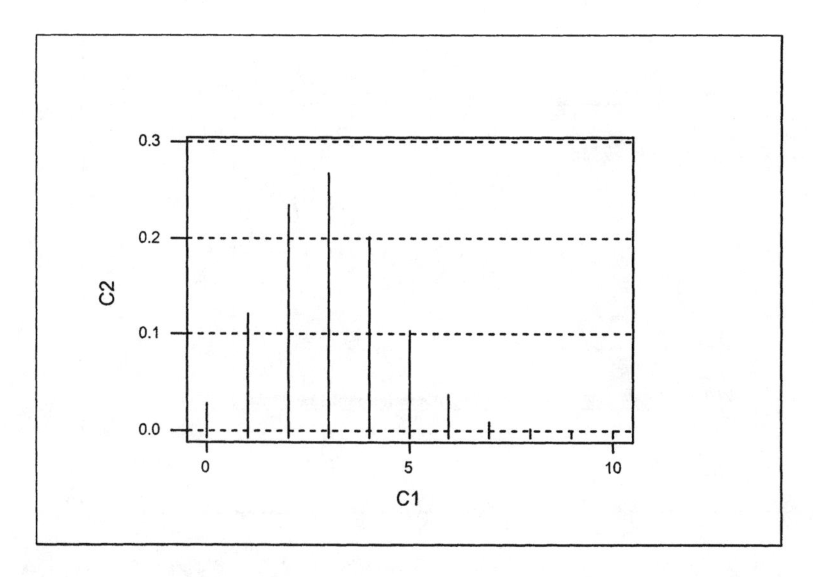

b. np = 0.5 and nq = 99.5, therefore the approximation is not appropriate, since np < 5.

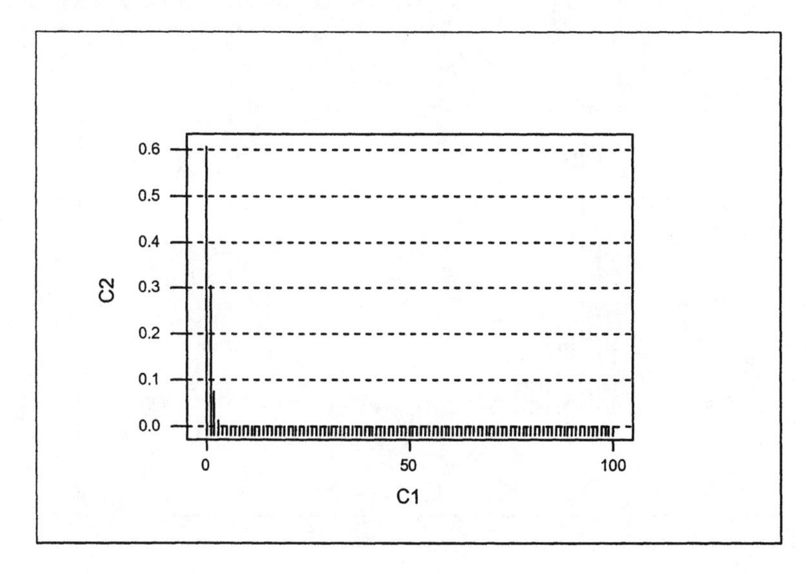

c. $np = 50$ and $nq = 450$, therefore the approximation is appropriate, since both $np > 5$ and $nq > 5$.

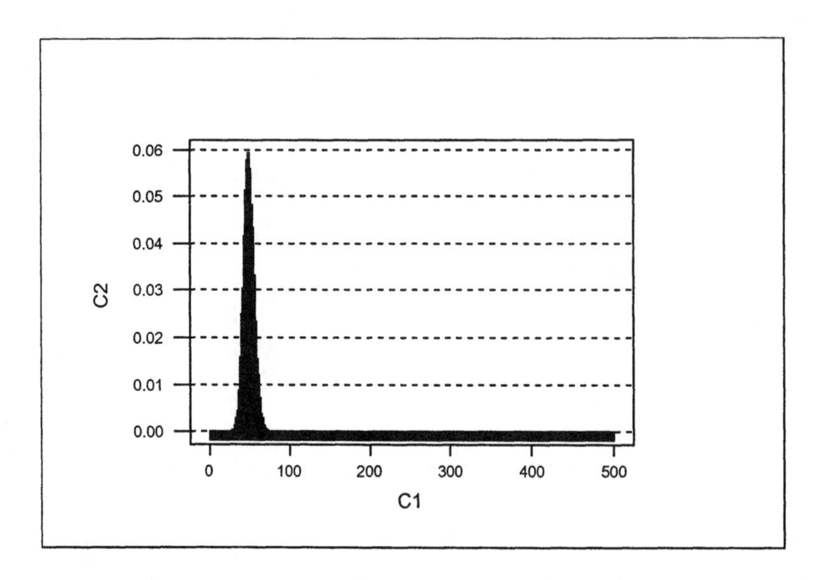

d. $np = 10$ and $nq = 40$, therefore the approximation is appropriate, since both $np > 5$ and $nq > 5$.

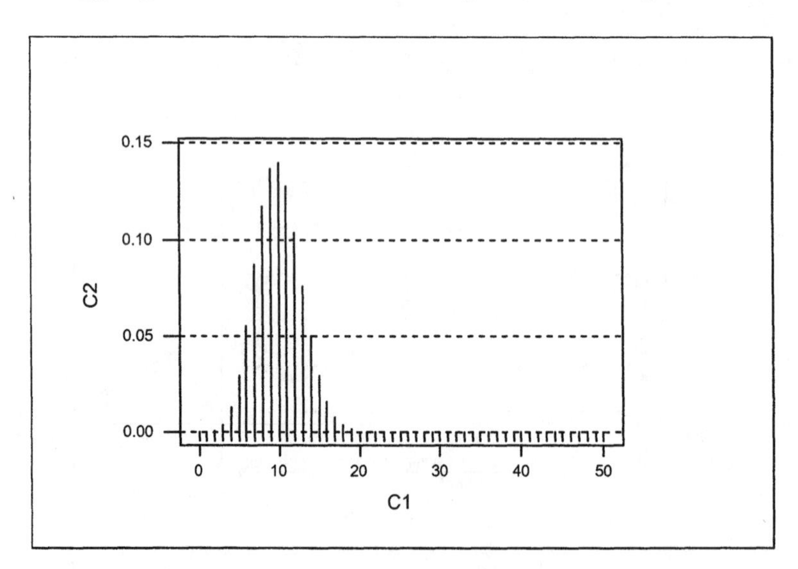

6.75 $P(x = 6) = P(5.5 < x < 6.5)$
$= P[(5.5 - 7.2)/\sqrt{2.88} < z < (6.5 - 7.2)/\sqrt{2.88}]$
$= P[-1.00 < z < -0.41]$
$= 0.3413 - 0.1591 = \underline{0.1822}$

$P[x = 6|B(n=12, p = 0.6)] = \underline{0.177}$ from Table 2 (Appendix B, ES9-p711)

6.77 $P(x \leq 8) = P(x < 8.5)$
$= P[z < (8.5 - 5.6)/\sqrt{3.36}]$
$= P[z < 1.58]$
$= 0.5000 + 0.4429 = \underline{0.9429}$

$P[x \leq 8|B(n = 14, p = 0.4)]$
$= P(0)+P(1)+P(2)+P(3)+P(4)+P(5)+P(6)+P(7)+P(8)$
$= 0.001+0.007+0.032+0.085+0.155+0.207+0.207+0.157+0.092$
$= \underline{0.943}$ from Table 2 (Appendix B, ES9-p711)

6.79 Let x represent the number of patients in the 250 who will survive melanoma.
x = n(survive)

$\mu = np = (250)(0.90) = \underline{225}$
$\sigma = \sqrt{npq} = \sqrt{(250)(0.90)(0.10)} = \underline{4.74}$

$P(x > 199.5) = P[z > (199.5 - 225.0)/4.74]$
$= P[z > -5.38]$
$= 0.5000 + 0.4999997 = \underline{0.9999997}$

6.81 Let x represent the number of dog walkers in the 50 who clean up after their dogs.

$\mu = np = (50)(0.59) = \underline{29.5}$
$\sigma = \sqrt{npq} = \sqrt{(50)(0.59)(0.41)} = \underline{3.48}$

a. $P(x < 25.5) = P[z < (25.5 - 29.5)/3.48]$
$= P[z < -1.15]$
$= 0.5000 - 0.3749 = \underline{0.1251}$

b. $P(x > 37.5) = P[z > (37.5 - 29.5)/3.48]$
$= P[z > 2.30]$
$= 0.5000 - 0.4893 = \underline{0.0107}$

6.83 $\mu = np = (60)[944/(944 + 1{,}106)] = (60)(0.4605) = 27.63$

$\sigma = \sqrt{npq} = \sqrt{(60)(0.4605)(0.5395)} = 3.86$

At x = 29.5: $z = (29.5 - 27.63)/3.86 = 0.48$

Therefore,

$P(x < 29.5) = P(z < 0.48) = 0.5000 + 0.1844 = \underline{0.6844}$

6.85 a. $\mu = np = (2500)(0.158) = \underline{395}$

$\sigma = \sqrt{npq} = \sqrt{(2500)(0.158)(0.842)} = \underline{18.237}$

$P(x \leq 450) = P(x < 450.5) = P[z < (450.5 - 395)/18.237]$
$= P[z < 3.04]$
$= 0.5000 + 0.4988 = \underline{0.9988}$

b. Use the data value of 450 and the cumulative normal computer or calculator commands found in ES9 pp.287&288. The answer is 0.9988

RETURN TO CHAPTER CASE STUDY

6.86 a. z = (I.Q. - 100) / 16

b. x = 90; z = (90 - 100) / 16 = -0.63
x = 110; z = (110 - 100) / 16 = 0.63
x = 120; z = (120 - 100) / 16 = 1.25

c. z = (SAT - 500) / 100
x = 465; z = (465 - 500) / 100 = -0.35
x = 575; z = (575 - 500) / 100 = 0.75
x = 650; z = (650 - 500) / 100 = 1.50

d. (i) x = 132; z = (132 - 100) / 16 = 2.00; 2%
(ii) x = 700; z = (700 - 500) / 100 = 2.00; 98%

CHAPTER EXERCISES

6.89 The range from z = -2 to z = +2 represents two standard deviations on either side of the mean. According to Chebyshev's theorem, there should be *at least 3/4* or *at least 0.75* of a distribution in this interval.

The area under a normal curve is 2(0.4772) or 0.9544.

6.91 a. 1.26 b. 2.16 c. 1.13

6.93 a. $P(|z| > 1.68) = P(z < -1.68) + P(z > +1.68)$
$= 2(0.5000 - 0.4535) = 0.0930$

b. $P(|z| < 2.15) = P(-2.15 < z < +2.15)$
$= 2(0.4842) = 0.9684$

6.95 a. 1.175 or 1.18 b. 0.58

c. -1.04 d. -2.33

6.97 a.

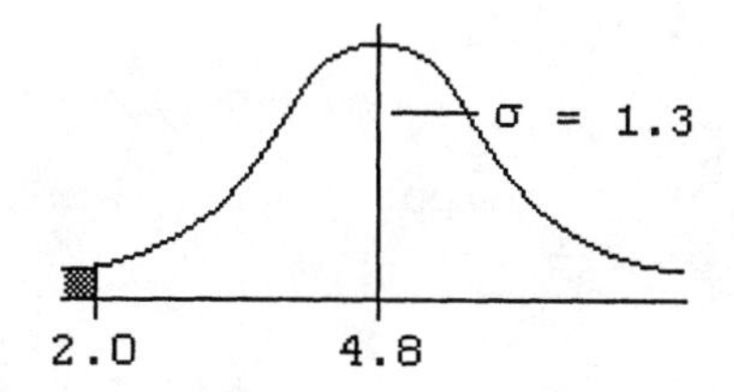

$P[x < 2.0] = P[z < (2.0-4.8)/1.3]$
$= P[z < -2.15]$
$= 0.5000 - 0.4842 = 0.0158$

b.

σ = 1.3
0.005
A
4.8
z = -2.58

$z = (x - \mu)/\sigma$

$-2.58 = (A - 4.8)/1.3$

$A = 4.8 + (-2.58)(1.3)$

$A = 1.446$ years

6.99

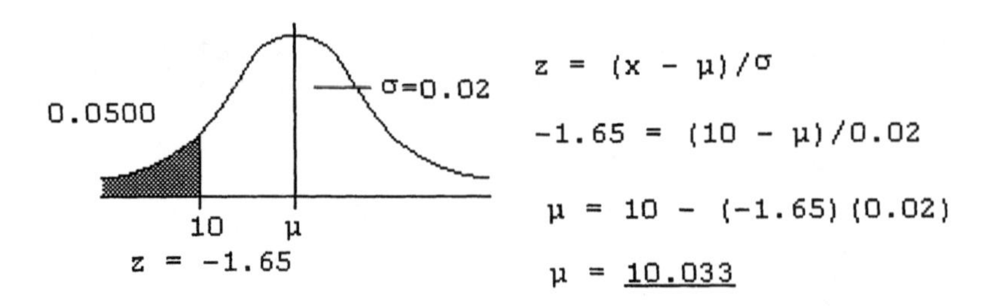

6.101 a. The normal approximation is reasonable since both $np = 7.5$ and $nq = 17.5$ are greater than 5.

b. $\mu = np = (25)(0.3) = \underline{7.5}$

$\sigma = \sqrt{npq} = \sqrt{(25)(0.3)(0.7)} = \underline{2.29}$

6.103 a. Use the binomial probablity computer or calculator commands found in ES9 pp.255-256.

b. $P(x \leq 6) = 0.005154 + 0.028632 + 0.077943 + 0.138565 + 0.180904 + 0.184925 + 0.154104 = \underline{0.77023}$

c. $\mu = np = (50)(0.1) = \underline{5}$
$\sigma = \sqrt{npq} = \sqrt{(50)(0.1)(0.9)} = \underline{2.12}$

Use -0.05 and 6.5 as the data values in the cumulative normal computer or calculator commands found in ES9 pp.287&288.

$P(x \leq 6) = 0.760387 - 0.008608 = \underline{0.751779}$

6.105 a. $P[x \leq 75 | B(n = 300, p = 0.2] = P(0) + P(1) + P(2) + P(3) + P(4) + \ldots + P(75)$

b. Use the cumulative binomial probablity computer or calculator commands found in ES9 pp.255&256. Result: $\underline{0.9856}$

c. $\mu = np = (300)(0.2) = \underline{60}$
$\sigma = \sqrt{npq} = \sqrt{(300)(0.2)(0.8)} = \underline{6.93}$

Use -0.05 and 75.5 as the data values in the cumulative normal computer or calculator commands found in ES9 pp.287&288.

$P(x \leq 75) = 0.987346 - 0.0000 = \underline{0.9873}$

d. (b) and (c) result in answers that are very close in value.

6.107 a.

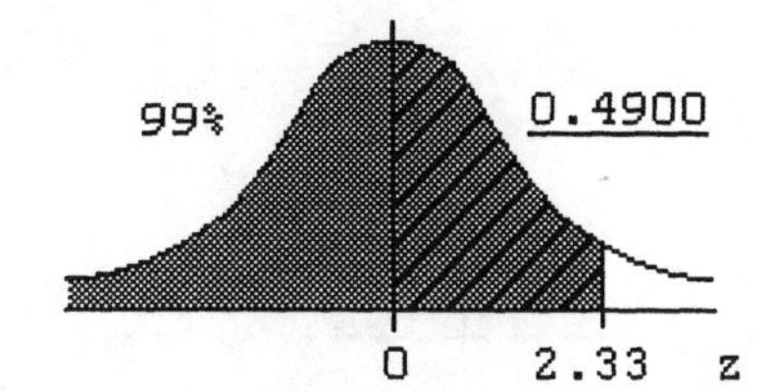

$z = (x - \mu)/\sigma$

$2.33 = (6 - \mu)/0.2$

$0.466 = 6 - \mu$

$\mu = 5.534$

b. Everyone's generated values will vary.

c. Answers will vary, but most of them will be 0 or 1 of the 40 overflow.

d. Yes, the setting seems to work.

6.109 a. $P(3 \text{ wrong in } 5) = P[x = 3|B(n = 5, p = 0.05)] = \underline{0.001}$
(Table 2, Appendix B, ES9-p711)

b. P(no more than 3 wrong in 5)
$= P[x = 0, 1, 2, 3|B(n = 5, p = 0.05)]$
$= 0.774 + 0.204 + 0.021 + 0.001 = \underline{1.000}$
(Table 2)

c. P(no more than 3 wrong in 15)
$= P[x = 0, 1, 2, 3|B(n = 15, p = 0.05)]$
$= 0.463 + 0.366 + 0.135 + 0.031 = \underline{0.995}$
(Table 2)

d. P(no more than 3 wrong in 150) = $P[x \le 3 | B(150, 0.05)]$

$\mu = np = (150)(0.05) = \underline{7.5}$

$\sigma = \sqrt{npq} = \sqrt{(150)(0.05)(0.95)} = \underline{2.67}$

$$P[x \le 3] = P[x < 3.5]$$
$$= P[z < (3.5 - 7.5)/2.67]$$
$$= P[z < -1.50]$$
$$= 0.5000 - 0.4332 = \underline{0.0668}$$

e. $P[x > 100] = P[x > 100.5]$
$$= P[z > (100.5 - 116)/6.98]$$
$$= P[z > -2.22]$$
$$= 0.5000 + 0.4868 = \underline{0.9868}$$

6.111

Nation	$\mu = np$	$\sigma = \sqrt{npq}$	$P(x \ge 70)$*
China	66	7.989	0.3300
Germany	10	3.154	0.0000+
India	130	11.024	0.9999+
Japan	8	2.823	0.0000+
Mexico	64	7.871	0.2420
Russia	34	5.781	0.0000+
S. Africa	98	9.654	0.9984
United States	14	3.729	0.0000+

*China: $z = (69.5 - 66)/7.989 = 0.44$;
$P(z > 0.44) = 0.5000 - 0.1700) = \underline{0.3300}$

Germany: $z = (69.5 - 10)/3.154 = 18.86$;
$P(z > 18.86) = 0.5000 - 0.4999^{+} = \underline{0.0000+}$

India: $z = (69.5 - 130)/11.024 = -5.49$;
$P(z > -5.49) = 0.5000 + 0.4999^{+} = \underline{0.9999+}$

Japan: $z = (69.5 - 8)/2.823 = 21.78$;
$P(z > 21.78) = 0.5000 - 0.4999^{+} = \underline{0.0000+}$

Mexico: $z = (69.5 - 64)/7.871 = 0.70$;
$P(z > 0.70) = 0.5000 - 0.2580 = \underline{0.2420}$

Russia: $z = (69.5 - 34)/5.781 = 6.14$;
$P(z > 6.14) = 0.5000 - 0.4999^{+} = \underline{0.0000+}$

S. Africa: $z = (69.5 - 98)/9.654 = -2.95$;
$P(z > -2.95) = 0.5000 + 0.4984) = \underline{0.9984}$

United States: $z = (69.5 - 14)/3.729 = 14.88$;
$P(z > 14.88) = 0.5000 - 0.4999^{+} = \underline{0.0000+}$

6.113 a. $\mu = np = (50)(0.75) = 37.5$;

$\sigma = \sqrt{npq} = \sqrt{(50)(0.75)(0.25)} = 3.062$

at $x = 35.5$: $z = (35.5 - 37.5)/3.062 = -0.65$

Therefore, $P(x > 35) = P(z > -0.65)$

$= 0.5000 + 0.2422 = \underline{0.7422}$

b. $\mu = np = (50)(0.47) = 23.5$;

$\sigma = \sqrt{npq} = \sqrt{(50)(0.47)(0.53)} = 3.529$

at $x = 24.5$: $z = (24.5 - 23.5)/3.529 = 0.28$

Therefore, $P(x < 25) = P(z < 0.28)$

$= 0.5000 + 0.1103 = \underline{0.6103}$

c. at $x = 29.5$: $z = (29.5 - 37.5)/3.062 = -2.61$

at $x = 40.5$: $z = (40.5 - 37.5)/3.062 = 0.98$

Therefore, $P(30 \le x \le 40) = P(-2.61 \le z \le 0.98)$

$= 0.4955 + 0.3365 = \underline{0.8320}$

d. at $x = 19.5$: $z = (19.5 - 23.5)/3.529 = -1.13$

at $x = 30.5$: $z = (30.5 - 23.5)/3.529 = 1.98$

Therefore, $P(20 \le x \le 30) = P(-1.13 \le z \le 1.98)$

$= 0.3708 + 0.4761 = \underline{0.8469}$

CHAPTER 7 ∇ SAMPLE VARIABILITY

Chapter Preview

Chapters 1 and 2 introduced the concept of a sample and its various measures. Measures of central tendency, measures of dispersion, and the shape of the distribution of the data give a single "snapshot" of the population from which the sample was taken. If repeated samples are taken and statistics noted, a clearer picture of the population from which the sample came will develop. These combined statistics will enable us to better predict the population's parameters. Chapter 7 works with this illustration of repeated sampling in the form of a sampling distribution. A sampling distribution is basically a probability distribution for a sample statistic. Therefore, there can be sampling distributions for the sample mean, for the sample standard deviation or for the sample range, to name a few. The significant results that will surface for the probability distribution of the mean specifically, will be contained in the Central Limit Theorem. This theorem justifies the use of the normal distribution in solving a wide range of problems.

Data on the age of the population, from the 2000 census, taken in the United States is presented in this chapter's case study.

CHAPTER 7 CASE STUDY

7.1 a.

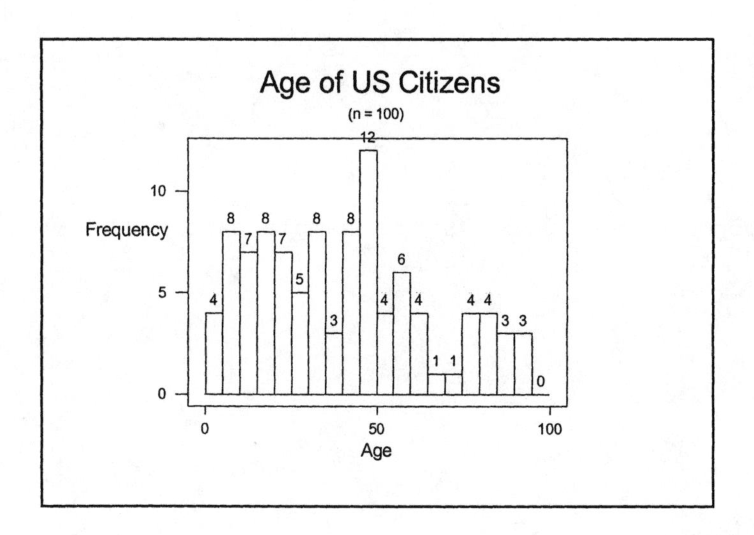

$\overline{x}$ = 39.31, median = 39.50, s = 25.16
Mounded from 0 to 60 with a tail extending from 60 to 100 making the distribution skewed to the right.

b. The sample looks very much like the population.
c. Skewed to the right.
d. Not exactly but fairly close.

SECTION 7.1 ANSWER NOW EXERCISES

7.2 a. No
b. Variability

7.3 Each digit has the same chance of being selected; therefore, each sample combination has the same probability as any other combination.
P(0) = 0.04, since there is only one way out of the 25 samples to get an $\overline{x}$ of 0; i.e. {0,0} and 1/25 = 0.04.
P(2) = 0.12, since there are 3 ways out of the 25 samples to get an $\overline{x}$ of 2; i.e., {0,4},{2,2},{4,0} and 3/25 = 0.12.

7.4

x	$P(x)$	$xP(x)$	$x^2P(x)$
1	0.2	0.2	0.2
2	0.2	0.4	0.8
3	0.2	0.6	1.8
4	0.2	0.8	3.2
5	0.2	1.0	5.0
Σ	1.0ck	3.0	11.0

$\mu = \Sigma xP(x) = \underline{3.0}$

$\sigma = \sqrt{\Sigma x^2P(x) - (\Sigma xP(x))^2}$

$= \sqrt{11.0 - (3.0)^2}$

$= \sqrt{2} = \underline{1.41}$

7.5

Class	Freq.
1.8-2.2	3
2.2-2.6	5
2.6-3.0	6
3.0-3.4	6
3.4-3.8	5
3.8-4.2	4
4.2-4.6	1
Σ	30

7.6

$\overline{x}$	f	$\overline{x}f$	$\overline{x}^2f$
2.0	3	6.0	12.00
2.2	3	6.6	14.52
2.4	2	4.8	11.52
2.6	1	2.6	6.76
2.8	5	14.0	39.20
3.0	4	12.0	36.00
3.2	2	6.4	20.48
3.4	3	10.2	34.68
3.6	2	7.2	25.92
3.8	4	15.2	57.76
4.4	1	4.4	19.36
Σ	30	89.4	278.20

$\overline{\overline{x}}$ = mean of the sample means

$\overline{\overline{x}} = \Sigma\overline{x}f/\Sigma f = 89.4/30 = \underline{2.98}$

$s_{\overline{x}}$ = standard deviation of the sample means

$$s_{\overline{x}} = \sqrt{\frac{\Sigma\,\overline{x}^2f - (\Sigma\,\overline{x}f)^2 / \Sigma\,f}{\Sigma\,f - 1}} = \sqrt{\frac{278.2 - (89.4)^2 / 30}{29}}$$

$= 0.63756 = \underline{0.638}$

7.7 a.

Classes	Freq.
4.5-6.5	1
6.5-8.5	1
10.5-12.5	2
12.5-14.5	1
14.5-16.5	3
Σ	12

b. Samples are not all the same size; the airlines have different size fleets.

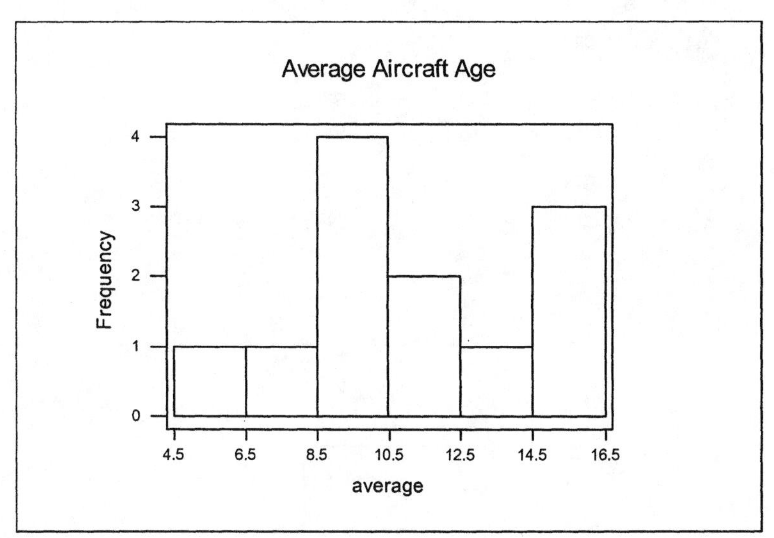

SECTION 7.1 EXERCISES

> Use a tree diagram to find all possible samples, for exercise 7.9a. Each sample will have a probability of 1/n, where n is the number of samples. Remember $\sum P(\text{statistic}) = 1$.

7.9 a.

11	31	51	71	91
13	33	53	73	93
15	35	55	75	95
17	37	57	77	97
19	39	59	79	99

b.

$\bar{x}$	1	2	3	4	5	6	7	8	9
$P(\bar{x})$	0.04	0.08	0.12	0.16	0.20	0.16	0.12	0.08	0.04

c.

R	0	2	4	6	8
P(R)	0.20	0.32	0.24	0.16	0.08

7.11 Every student will have different results, however they should be at least similar to these.

samples obtained:

899	091	031	982	159	612	720	534	758	178
337	520	185	893	601	968	560	959	368	943

a.

8.7	3.3	1.3	6.3	5.0	3.0	3.0	4.0	6.7	5.3
4.3	2.3	4.7	6.7	2.3	7.7	3.7	7.7	5.7	5.3

b.

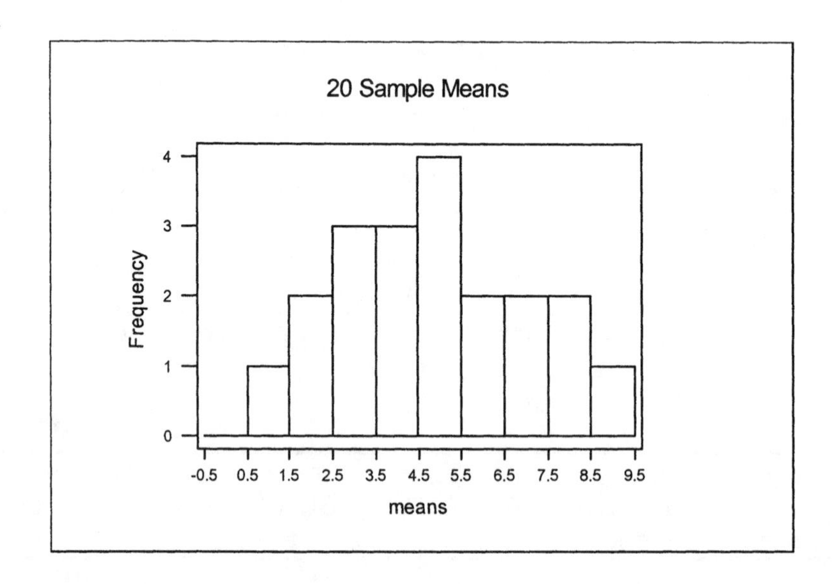

c. The shape is somewhat symmetric, centered around the 4.5 to 5.5 class.

d.

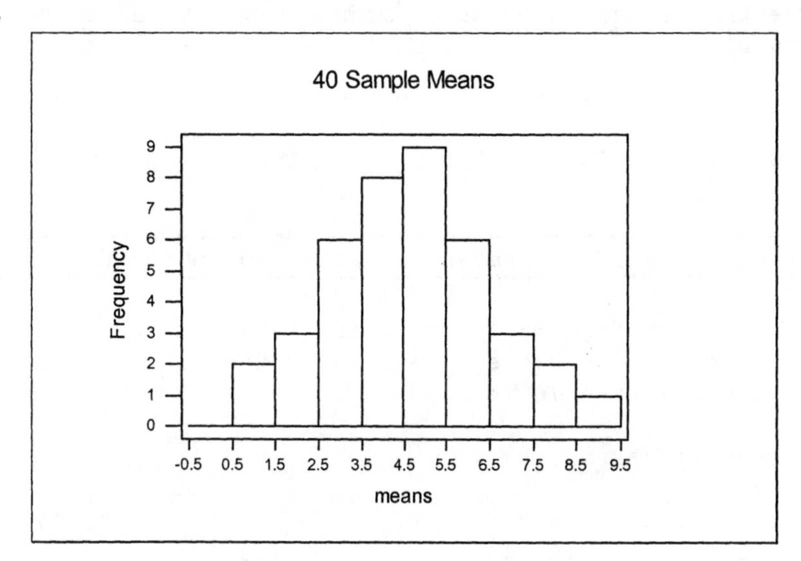

The shape is appearing more like a normal distribution and centered near 4.5.

7.13 Answers will vary somewhat, depending on what values the student selects from the table.

Set	1	2	3	4	5	Mean
1	1	3	0	9	1	2.8
2	0	7	8	9	2	5.2
3	0	5	4	0	8	3.4
4	9	4	2	1	0	3.2
5	7	2	2	9	7	5.4
6	3	0	6	0	9	3.6
7	2	4	8	2	9	5.0
8	5	8	9	5	9	7.2
9	3	0	5	2	7	3.4
10	3	5	3	9	0	4.0
11	7	6	1	0	8	4.4
12	6	4	9	9	0	5.6
13	5	8	6	3	1	4.6
14	2	9	4	7	5	5.4
15	0	4	5	6	7	4.4
16	1	7	0	7	3	3.6
17	3	4	9	0	6	4.4
18	5	2	3	7	1	3.6
19	8	9	0	1	4	4.4
20	6	6	3	5	7	5.4
			Grand Mean			4.45

The theoretical population mean = μ = 4.5, as found in Exercise 5.33.

Absolute difference = (4.50 - 4.45) = 0.05

% Error = (0.05 ÷ 4.5)(100) = 1.11%

Computer commands for repeated sampling can be found in ES9 p.318.

7.15 a. Every student will have different results; however, the means of the 200 samples, each of size 10, should resemble those listed in (b).

b. sample means

91.860	109.866	98.004	86.827	102.329	100.573	98.011
114.985	107.411	96.460	86.894	105.660	96.999	93.765
106.034	94.516	105.026	102.436	103.894	90.158	86.329
102.932	98.434	97.764	95.966	104.754	97.277	96.445
86.349	101.668	101.512	93.906	94.665	95.768	108.146
89.229	109.310	100.526	98.969	101.932	108.459	100.079
104.962	101.606	103.315	91.811	93.948	94.520	106.383
107.004	97.663	101.009	99.208	95.563	92.886	107.509
107.101	105.658	103.223	96.658	95.939	108.203	108.573
90.982	101.484	101.294	103.938	101.708	100.146	97.664
96.985	98.901	97.347	95.219	102.195	95.477	92.300
100.841	112.838	102.455	99.470	96.042	107.587	95.336
95.144	92.109	109.939	96.739	91.548	102.796	94.154
109.485	102.767	101.439	88.470	101.822	104.096	87.914
91.202	95.397	94.413	99.975	94.377	89.091	94.101
104.353	90.017	96.030	104.042	94.446	99.678	89.653
96.662	92.773	98.930	102.418	107.959	98.822	101.498
105.041	96.652	105.297	102.878	96.347	104.832	94.467
109.017	105.082	89.613	110.447	115.052	102.291	90.511
92.783	93.481	102.061	100.769	102.865	104.078	87.550
97.086	100.175	89.797	122.981	99.870	104.534	99.702
99.968	98.243	98.681	105.884	96.934	98.235	97.535
103.662	107.472	82.100	97.276	94.818	101.765	99.148
85.795	107.241	104.025	88.025	104.061	97.676	88.778
102.363	110.035	103.318	110.855	97.457	89.955	110.174
95.746	94.286	106.459	101.952	99.420	94.427	103.308
96.455	102.120	95.651	103.804	111.829	102.809	105.131
101.082	89.315	101.396	101.146	101.080	96.246	94.004
103.730	100.669	101.901	97.239			

c.

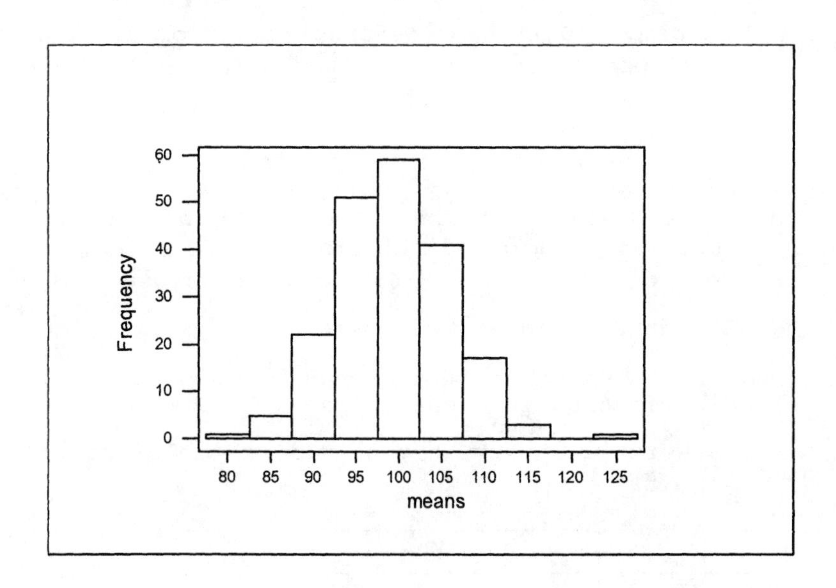

d. The shape is approximately normal, being mounded, approximately symmetrical, and centered near 100.

SECTION 7.2 ANSWER NOW EXERCISES

7.16 $25/\sqrt{16} = \underline{6.25}$ $25/\sqrt{36} = \underline{4.167}$ $25/\sqrt{100} = \underline{2.50}$

7.17 a. 1.0 or one

b. $\sigma_{\bar{x}} = \sigma / \sqrt{n}$; as n increases the value of this fraction, the standard deviation of sample mean, gets smaller.

7.18 b. Answers will vary, but should resemble the following results from one simulation:
Mean of xbars = $\bar{\bar{x}}$ = 65.27; very close to μ = 65.15

c. Standard deviation of xbars = $s_{\bar{x}}$ = 1.383; less than σ
$s_{\bar{x}}$ is approximately $2.754/\sqrt{4}$ = 1.377.

d. The shape of the histogram is approximately normal.

e. Took many (1001) samples of size 4 from an approximately normal population and
   1. mean of the xbars μ
   2. $s_{\bar{x}}$ $\sigma / \sqrt{n}$
   3. approximately normal distribution

7.19 b. Answers will vary, but should resemble the following results from one simulation:
Mean of xbars $= \bar{\bar{x}} = 6.105$; very close to $\mu = 6.029$

Standard deviation of xbars $= s_{\bar{x}} = 5.318$; less than σ

$s_{\bar{x}}$ is approximately $10.79/\sqrt{4} = 5.395$
distribution is skewed right

c. Answers will vary, but should resemble the following results:

	$\bar{\bar{x}}$	μ	$s_{\bar{x}}$	$\sigma / \sqrt{n}$	shape
25	5.974	6.029	2.087	$10.79/\sqrt{25} = 2.158$	Skewed right
100	6.018	6.029	1.137	$10.79/\sqrt{100} = 1.079$	Sl. lopsided
1000	6.03	6.029	0.335	$10.79/\sqrt{1000} = 0.341$	normal

d. See part 'c'.

SECTION 7.2 EXERCISES

The most important sampling distribution is the sampling distribution of the sample means. It provides the information that makes up the sampling distribution of sample means and the central limit theorem.

SAMPLING DISTRIBUTION OF SAMPLE MEANS & CENTRAL LIMIT THEOREM

If all possible random samples, each of size n, are taken from any population with a mean μ and standard deviation σ, the sampling distribution of sample means ($\bar{x}$'s) will result in the following:

1. The mean of the sample means (x bars) will equal the mean of the population, $\mu_{\bar{x}} = \mu$.
2. The standard deviation of the sample means (x bars) will be equal to the population standard deviation divided by the square root of the sample size, $\sigma_{\bar{x}} = \sigma / \sqrt{n}$. . . .

3. A normal distribution when the parent population is normally distributed or becomes approximately normal distributed a the sample size increases when the parent population is not normally distributed.

In essence, $\overline{x}$ is normally distributed when n is large enough, no matter what shape the population is. The further the population shape is from normal, the larger the sample size needs to be.

$\sigma_{\overline{x}}$ = the standard deviation of the $\overline{x}$'s is now referred to as the standard error of the mean.

7.21 a. Approximately normal
b. 2.4 televisions
c. $1.2/\sqrt{80} = 0.134$

7.23 a. The mean value for the sampling distribution is the same as the population mean, 43.3 cents.
b. The standard error of the mean is $\sigma/\sqrt{n} = 7.5/\sqrt{150} =$ 0.61 cents per pound.
c. The sampling distribution will be bell-shaped, approximately normal.

7.25 a. Every student will have different results; however, the means of the 100 samples, each of size 6, should resemble those listed in (b).
b. sample means

17.0148	21.0960	18.8767	18.4458	23.7957	17.4572	19.9137
22.9338	21.0164	18.9116	15.8072	23.1245	20.1439	20.8047
18.7836	19.5104	16.7224	18.1819	19.7173	19.4121	19.7335
18.9738	20.0119	19.7525	18.8456	19.3583	21.3249	20.7735
20.7584	18.9180	19.1935	22.2808	21.8498	19.2428	23.8004
19.1168	20.8396	22.2579	20.6043	17.9834	18.9368	20.2062
22.0984	19.4569	21.1661	19.9014	20.5755	17.0555	23.0511
21.3578	20.1454	19.3213	20.1349	18.3778	17.4479	21.5321
22.9323	18.8017	21.7476	18.1265	20.4839	20.3481	18.8614
21.2243	20.4734	17.4328	18.1990	23.0886	17.1433	18.4057
20.8029	20.7676	18.8139	18.2665	20.3459	20.8185	18.1914
19.1677	24.5807	22.6314	20.0144	17.6009	18.1102	19.1101
17.4663	17.1651	20.8377	22.2157	21.4562	19.4697	23.6880
18.8945	19.4205	19.7740	17.9856	20.2497	19.3615	21.2845
21.2142	17.3701					

c.

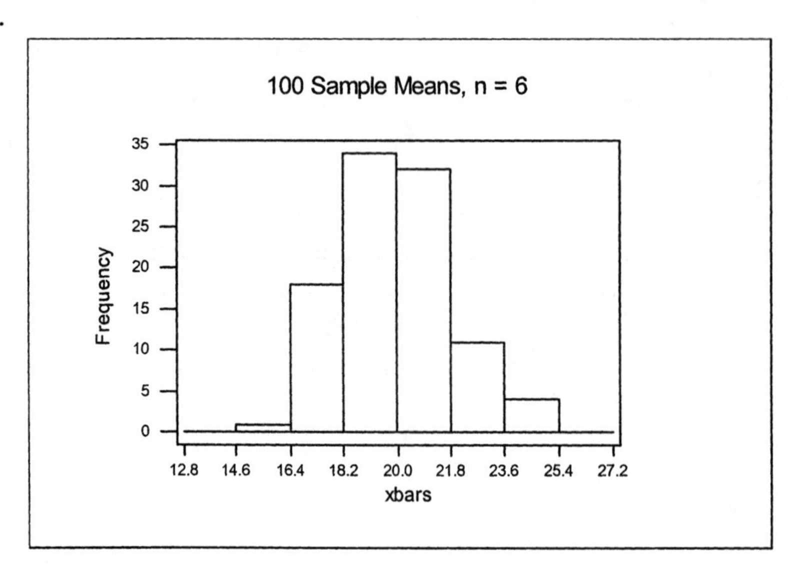

Mean of xbars $= \bar{\bar{x}} = 19.924$
Standard deviation of xbars $= s_{\bar{x}} = 1.8023$

d. $\bar{\bar{x}}$ is approximately 20.
$s_{\bar{x}}$ is approximately $4.5/\sqrt{6} = 1.84$.
The shape of the histogram is approximately normal.

SECTION 7.3 ANSWER NOW EXERCISES

7.27 Both 90 and 110 are two standard errors away from the mean and the $P(0 < z < 2) = 0.4772$ or $P(-2 < z < 0) = 0.4772$.

7.28 $P(\bar{x} < 39.75) = P[z < (39.75 - 39.0)/(2/\sqrt{25})]$
$= P[z < 1.88]$
$= 0.5000 + 0.4699 = \underline{0.9699}$

7.29 If $z = -0.67$, then
$-0.67 = (\bar{x} - 39.0)/(2/\sqrt{25})$
$-0.268 = \bar{x} - 39.0$
$\bar{x} = 38.732 = \underline{38.73 \text{ inches}}$

SECTION 7.3 EXERCISES

> It is helpful to draw a normal curve, locating μ and shading in the desired portion for each problem. A new z formula must now be used to determine probabilities about $\overline{x}$.
>
> $$z = \frac{\overline{x} - \mu}{\sigma / \sqrt{n}}$$

7.31 a. Heights are approximately normally distributed with a $\mu = 69$ and $\sigma = 4$.

b. $P(x > 70) = P[z > (70 - 69)/4]$
$= P[z > 0.25]$
$= 0.5000 - 0.0987 = \underline{0.4013}$

c. The distribution of $\overline{x}$'s will be approximately normally distributed.

d. $\mu_{\overline{x}} = \underline{69}$; $\sigma_{\overline{x}} = 4/\sqrt{16} = \underline{1.0}$

e. $P(\overline{x} > 70) = P[z > (70 - 69)/1.0]$
$= P[z > +1.00]$
$= 0.5000 - 0.3413 = \underline{0.1587}$

f. $P(\overline{x} < 67) = P[z < (67 - 69)/1.0]$
$= P[z < -2.00]$
$= 0.5000 - 0.4772 = \underline{0.0228}$

> Watch the wording of the various probability problems.
>
> If the probability for an individual item or person (x) is desired, use $z = (x - \mu)/\sigma$.
>
> If the probability for a sample mean ($\overline{x}$) is desired, use $z = (\overline{x} - \mu)/(\sigma / \sqrt{n})$.

7.33 a. $P(38 < x < 40) = P[(38 - 39)/2 < z < (40 - 39)/2]$
$= P[-0.50 < z < +0.50]$
$= 0.1915 + 0.1915 = \underline{0.3830}$

b. $P(38 < \overline{x} < 40) =$

$= P[(38 - 39)/(2/\sqrt{30}) < z < (40 - 39)/(2/\sqrt{30})]$
$= P[-2.74 < z < +2.74]$
$= 0.4969 + 0.4969 = \underline{0.9938}$

c. $P(x > 40) = P[z > (40 - 39)/2]$
$= P[z > 0.50]$
$= 0.5000 - 0.1915 = \underline{0.3085}$

d. $P(\overline{x} > 40) = P[z > (40 - 39)/(2/\sqrt{30})]$
$= P[z > 2.74]$
$= 0.5000 - 0.4969 = \underline{0.0031}$

7.35 $\mu = 11.3$ mph, $\sigma = 3.5$ mph

a. $P(x > 13.5) = P[z > (13.5 - 11.3)/3.5] = P(z > 0.63)$
$= 0.5000 - 0.2357 = \underline{0.2643}$

b. $P(\overline{x} > 13.5) = P[z > (13.5 - 11.3)/(3.5/\sqrt{9}]$
$= P[z > 1.89]$
$= 0.5000 - 0.4706 = \underline{0.0294}$

c. No, especially for (a). The wind speed distribution will be skewed to the right. Samples of size 9 are not large enough for the CLT to suggest the sampling distribution is approximately normal.

d. The actual probabilities are most likely not nearly as high as those found in (a) and (b).

7.37 a. $P(4500 < \overline{x} < 4750)$

$= P((4500 - 4651)/(310/\sqrt{50}) < z < (4750 - 4651)/(310/\sqrt{50}))$
$= P(-3.44 < z < 2.26) = 0.4997 + 0.4881 = 0.9878$

b. $P(\overline{x} > 4800) = P(z > (4800 - 4651)/(310/\sqrt{50}))$
$= P(z > 3.40) = 0.5000 - 0.4997 = 0.0003$

c. Yes, it is reasonable to assume normality for the sampling distribution since the sample size is 50.

7.39 a. Using the computer commands in ES9 p.330:

$P(4 < \overline{x} < 6) = 0.841345 - 0.158655 = \underline{0.68269}$

Using Table 3: $P(4 < \overline{x} < 6) =$

$= P[(4 - 5)/(2/\sqrt{4}) < z < (6 - 5)/(2/\sqrt{4})]$

$= P[-1.00 < z < +1.00]$

$= 0.3413 + 0.3413 = \underline{0.6826}$

Everybody will get different answers, however, the results should be similar to the results below.

b. sample means

4.45967	4.04628	4.56959	4.74174	5.00541	4.99137	5.27322
4.18889	6.16806	5.10562	4.21561	5.61553	4.73023	3.23896
5.14986	5.83925	3.87701	4.15342	6.95931	6.69219	4.92261
4.00410	5.33392	5.69277	4.04213	6.00010	5.13663	7.04077
4.95368	6.07187	4.61556	5.96161	5.95970	5.93136	4.20692
6.13040	3.54086	4.73198	4.18744	5.05378	3.01671	3.47408
4.93583	4.18053	6.02229	3.94990	4.81555	5.11444	3.43246
4.81337	5.51779	3.50610	3.70504	5.49885	4.25768	4.01218
6.07272	4.52822	5.57177	4.35903	3.49724	4.20362	4.52965
5.47631	4.86655	3.91419	4.71412	5.39523	4.63979	5.28885
3.65115	4.72093	3.48571	3.73255	5.33954	3.99134	4.62782
6.29056	4.51388	5.25607	6.30200	6.23775	5.14931	3.57207
6.28936	4.45702	4.57191	5.31639	6.45744	3.70814	5.61942
4.47878	3.06674	5.69816	4.75256	5.66117	4.08780	3.88444
4.93384	5.66094					

c.

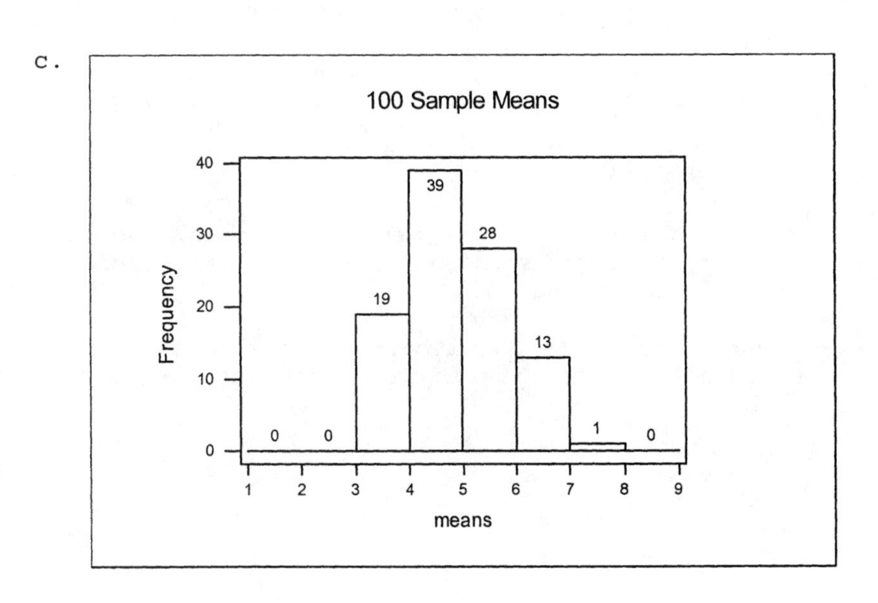

Inspecting the histogram, we find 67 of the sample means are between 4 and 6; 67/100 = 67%

d. 67% is very close to the expected 0.6826.

RETURN TO CHAPTER CASE STUDY

7.41 a.

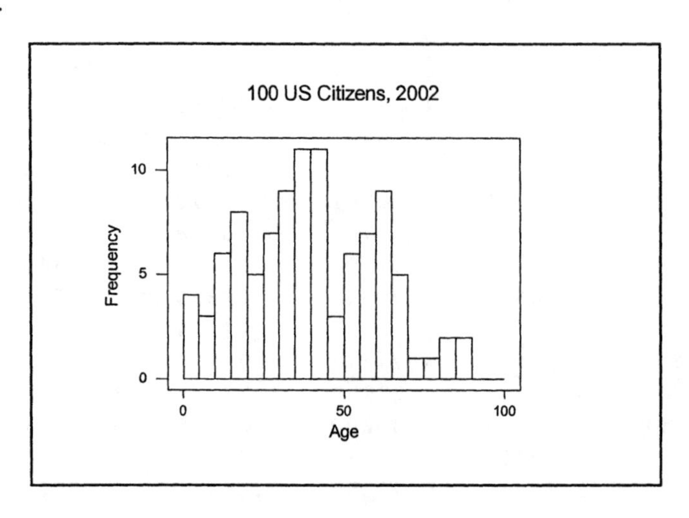

$\bar{x}$ = 39.28, $\tilde{x}$ = 39.00, s = 20.85

Descriptive Statistics: Age

Variable	N	Mean	Median	StDev
Age (Return)	100	39.28	39.00	20.85

Variable	Minimum	Maximum	Q1	Q3
Age (Return)	3.00	87.00	23.00	57.50

b. Reasonably well. There are differences, the histogram looks similar, the numerical statistics are similar in value.

c. Mounded from 0 to 70 with a tail extending from 70 to 90 making the distribution somewhat skewed to the right.

d.

Variable	N	Mean	Median	StDev
Sample (CCS)	100	39.31	39.50	25.16
Age (Return)	100	39.28	39.00	20.85

Variable	Minimum	Maximum	Q1	Q3
Sample (CCS)	*2.00*	*94.00*	*17.50*	*55.00*
Age (Return)	3.00	87.00	23.00	57.50

The numerical statistics are very similar in value.

e. The graphs are very similar to each other, as well as similar to the population's histogram.

CHAPTER EXERCISES

7.45 a. $P(2.63 - e < x < 2.63 + e) = 0.95$

$P(-1.96 < z < +1.96) = 0.95$, using Table 3 (Appendix B, ES9-p714)

$z = (x - \mu)/\sigma$

$+1.96 = [(2.63 + e) - 2.63]/0.25$ therefore $e = \underline{0.49}$

b. $P(2.63 - E < \overline{x} < 2.63 + E) = 0.95$

$P(-1.96 < z < +1.96) = 0.95$, using Table 3 (Appendix B, ES9-p714)

$z = (\overline{x} - \mu_{\overline{x}})/\sigma_{\overline{x}}$

$+1.96 = [(2.63 + E) - 2.63]/(0.25/\sqrt{100})$

therefore $E = \underline{0.049}$

7.47 a. $P(x > 15.0 \text{ million}) = P(z > (15.0 - 11.7)/2.8)$
$= P(z > 1.18)$
$= 0.5000 - 0.3810 = \underline{0.1190}$

b. $P(\overline{x} < 10 \text{ million}) = P[z < (10.0 - 11.7)/(2.8/\sqrt{20})]$
$= P[z < -2.72]$
$= 0.5000 - 0.4967 = \underline{0.0033}$

7.49 a. $P(245 < x < 255)$

$= P[(245 - 235)/\sqrt{400} < z < (255 - 235)/\sqrt{400}]$

$= P[+0.50 < z < +1.00]$

$= 0.3413 - 0.1915 = \underline{0.1498}$

b. $P(\overline{x} > 250) = P[z > (250 - 235)/(20/\sqrt{10})]$

$= P[z > +2.37]$

$= 0.5000 - 0.4911 = \underline{0.0089}$

7.51 $P(\overline{x} < 680) = P[z < (680 - 700)/(120/\sqrt{144})]$

$= P[z < -2.00]$

$= 0.5000 - 0.4772 = \underline{0.0228}$

7.53 $P(\sum x > 38{,}000) = P[z > (\sum x - n\mu)/(\sigma\sqrt{n})]$

$= P[z > (38{,}000 - (50)(750))/(25\sqrt{50})]$

$= P[z > 500/176.777]$

$= P[z > +2.83]$

$= 0.5000 - 0.4977 = \underline{0.0023}$

7.55 a. Let $\sum x$ represent the total weight:

$P(\sum x > 4000) = P(\sum x/n > 4000/25)$

$= P(\overline{x} > 160)$

$= P[z > (160 - 300)/(50/\sqrt{25})]$

$= P[z > -14.0]$

$=$ approximately $\underline{1.000}$

b. $P(\sum x < 8000) = P(\overline{x} < 320)$

$= P[z < (320 - 300)/(50/\sqrt{25})]$

$= P[z < 2.00]$

$= 0.5000 + 0.4772 = \underline{0.9772}$

7.57 a. Every student will have different results, however the totals and means of the 50 samples, each of size 10, should resemble those listed below.

sums

1357.86	1344.27	1357.23	1323.87	1335.53	1357.25	1339.91
1365.17	1308.99	1357.99	1346.36	1367.63	1315.55	1354.79
1378.82	1392.20	1320.65	1333.62	1405.80	1373.16	1400.35
1399.88	1414.90	1316.60	1337.11	1309.65	1339.62	1295.95
1350.44	1333.20	1338.42	1389.96	1304.70	1312.90	1312.14
1343.50	1315.46	1390.69	1387.76	1403.76	1339.83	1314.40
1416.16	1355.12	1299.10	1352.39	1367.93	1354.54	1305.35
1325.34						

xbars

135.786	134.427	135.723	132.387	133.553	135.725	133.991
136.517	130.899	135.799	134.637	136.763	131.555	135.479
137.882	139.220	132.065	133.362	140.580	137.316	140.035
139.988	141.490	131.660	133.711	130.965	133.962	129.595
135.044	133.320	133.842	138.996	130.470	131.290	131.214
134.350	131.546	139.069	138.776	140.376	133.983	131.440
141.616	135.512	129.910	135.239	136.793	135.454	130.535
132.534						

b.

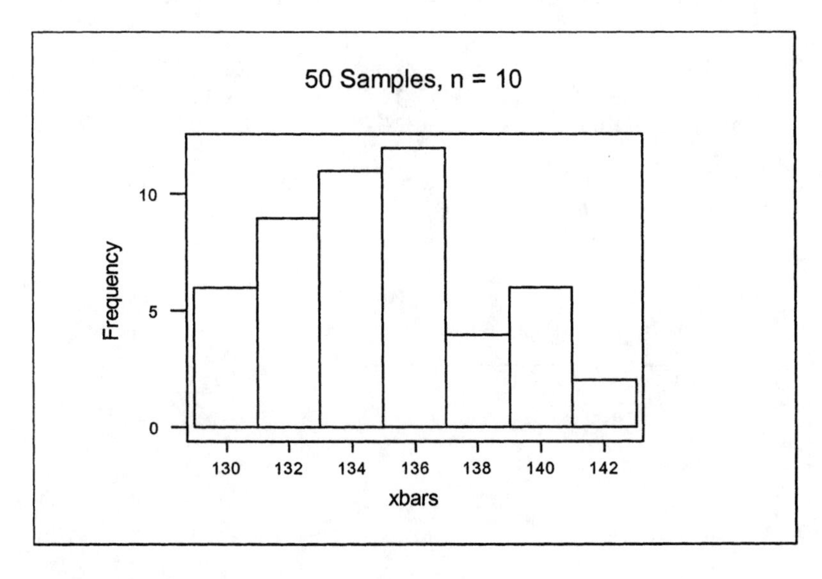

Mean of xbars = 134.93
Standard deviation of xbars = 3.2776

c.

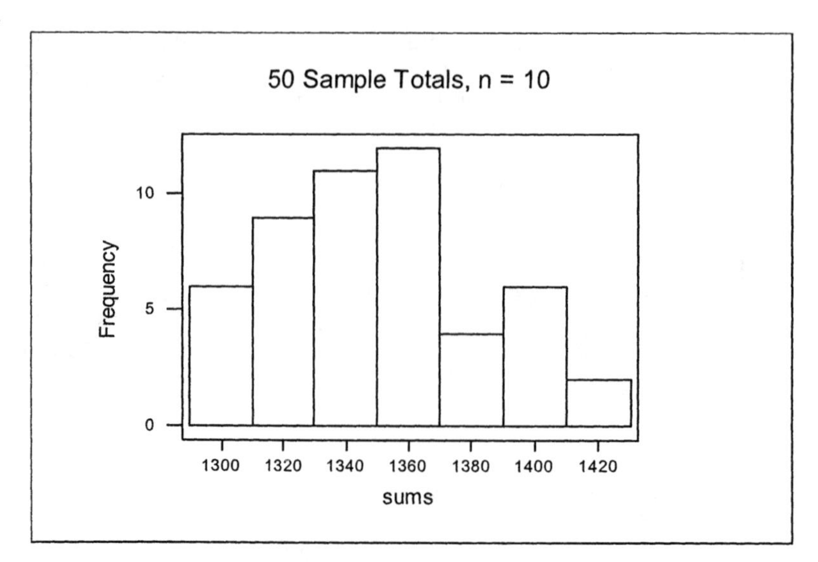

Mean of sums = 1349.3
Standard deviation of sums = 32.776

d. The histograms in (b) and (c) are identical in shape. Using the DESCribe command, it is evident that the xbar results are just the sum results divided by 10, the sample size.

Variable	N	Mean	Median	TrMean	StDev	SEMean
sums	50	1349.3	1345.3	1348.3	32.8	4.6
xbars	50	134.93	134.53	134.83	3.28	0.46

Variable	Min	Max	Q1	Q3
sums	1296.0	1416.2	1319.6	1369.2
xbars	129.60	141.62	131.96	136.92

7.59 a. $\mu = np = 200(0.3) = \underline{60}$

$\sigma = \sqrt{npq} = \sqrt{200(0.3)(0.7)} = \sqrt{42} = \underline{6.48}$

b. (all other x values have a probability equal to zero)

x	probability	x	probability	x	probability
28	0.0000001	51	0.0238909	74	0.0061875
29	0.0000002	52	0.0293386	75	0.0044550
30	0.0000004	53	0.0351114	76	0.0031403
31	0.0000009	54	0.0409633	77	0.0021673
32	0.0000021	55	0.0466024	78	0.0014647
33	0.0000045	56	0.0517144	79	0.0009694
34	0.0000096	57	0.0559916	80	0.0006284
35	0.0000194	58	0.0591635	81	0.0003990
36	0.0000382	59	0.0610258	82	0.0002481
37	0.0000725	60	0.0614617	83	0.0001512
38	0.0001334	61	0.0604542	84	0.0000903
39	0.0002374	62	0.0580861	85	0.0000528
40	0.0004096	63	0.0545298	86	0.0000303
41	0.0006850	64	0.0500263	87	0.0000170
42	0.0011114	65	0.0448587	88	0.0000093
43	0.0017501	66	0.0393242	89	0.0000050
44	0.0026763	67	0.0337065	90	0.0000027
45	0.0039762	68	0.0282539	91	0.0000014
46	0.0057421	69	0.0231647	92	0.0000007
47	0.0080633	70	0.0185791	93	0.0000003
48	0.0110151	71	0.0145791	94	0.0000002
49	0.0146440	72	0.0111947	95	0.0000001
50	0.0189535	73	0.0084125		

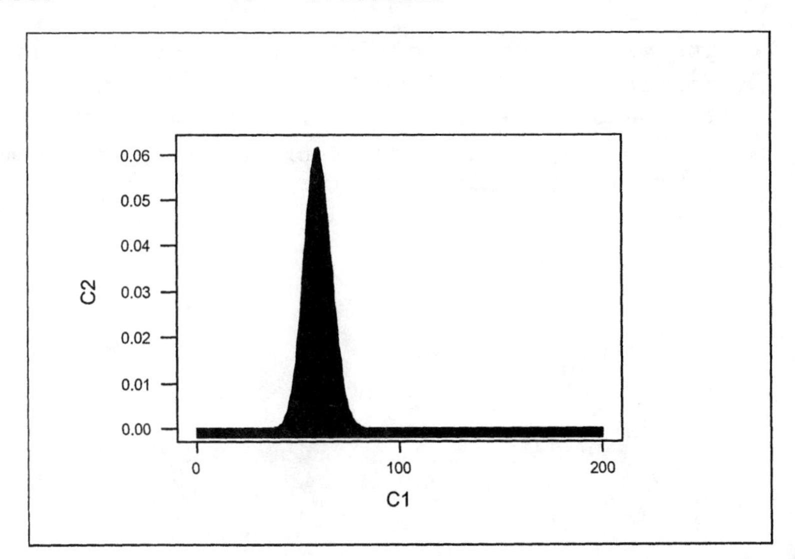

c. Every student will have different results, however the histograms should resemble that in (d)

d.

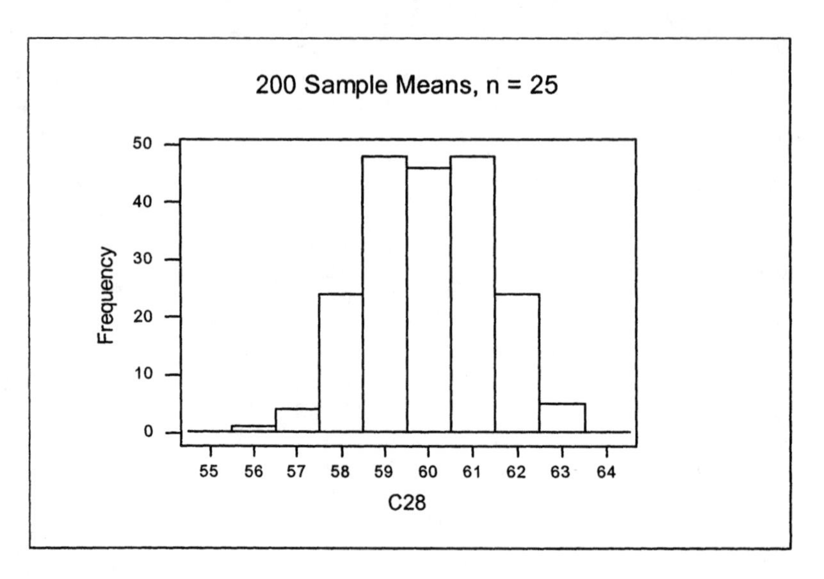

Mean of C28 = $\bar{\bar{x}}$ = 59.979

Standard deviation of C28 = $s_{\bar{x}}$ = 1.368

e. The mean of the sample means, $\bar{\bar{x}}$ = 59.979, is approximately equal to μ = 60. The standard deviation of the sample means, $s_{\bar{x}}$ = 1.368, is approximately equal to $6.48/\sqrt{25} = 1.296$. The distribution of the sample means is approximately normally distributed.

CHAPTER 8 ∇ INTRODUCTION TO STATISTICAL INFERENCES

Chapter Preview

Chapter 8 introduces inferential statistics. Generalizations about population parameters are made based on sample data in inferential statistics. These generalizations can be made in the form of hypothesis tests or confidence interval estimations. Each is calculated with a degree of uncertainty. The integral elements and procedure for obtaining a confidence interval and for completing a hypothesis test will be presented in this chapter. They will be performed with respect to the population mean, μ. The population standard deviation, σ, will be considered as a known quantity.

Statistical information on the average height of women reported by The National Center for Health Statistis is presented in the Case Study for this chapter.

CHAPTER 8 CASE STUDY

8.1 a. female health professionals
b. $\bar{x} = 64.7$, $s = 3.5$
Distribution is mounded about center, approximately symmetrical

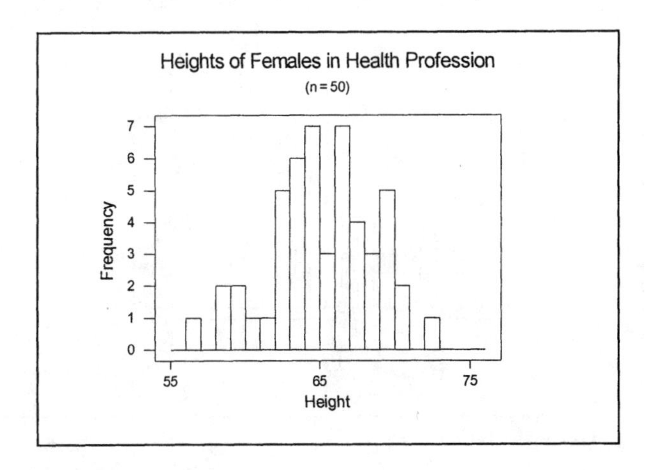

c. The distribution of the sample data is expected to look like the distribution of the population. The mean and standard deviation of the sample are expected to approximate the mean and standard deviation of the population.
The sampling distribution to which the mean of this sample belongs: has a mean value that is approximated by the mean of the sample, has a standard error approximated by the standard deviation of the sample divided by square root of 50, and has an approximately normal distribution.

d. Using $z = (\overline{x} - \mu)/(\sigma/\sqrt{n})$ with $\mu = 63.7$, $\sigma = 2.75$, $n = 50$ and $z = \pm 1.65$ (middle 90%):

$-1.65 = (\overline{x} - 63.7)/(2.75/\sqrt{50})$
$\overline{x} - 63.7 = (-1.65)(0.3889)$
$\overline{x} = 63.7 - 0.64$
$\overline{x} = 63.06$
and
$+1.65 = (\overline{x} - 63.7)/(2.75/\sqrt{50})$
$\overline{x} - 63.7 = (1.65)(0.3889)$
$\overline{x} = 63.7 + 0.64$
$\overline{x} = 64.34$

<u>63.06 to 64.34</u>

e.

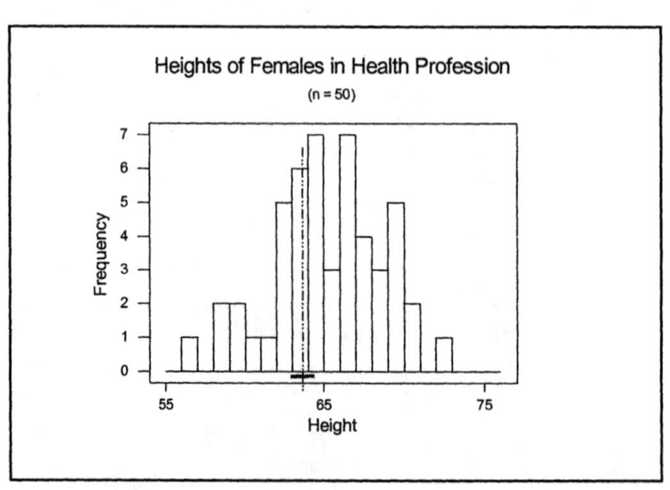

No, the sample mean, 64.72, falls outside of the interval 63.06 to 64.34. This means that this sample mean is not one of the 90% that are within the interval constructed in (d).

f. $P(\overline{x} \geq 64.72) = P(z > (64.72 - 63.7)/(2.75/\sqrt{50}))$
$= P(z > 2.62) = 0.5000 - 0.4956 = 0.0044$
This means that only 0.0044, or 44 in 10,000, samples ($n = 50$) drawn randomly from a population with mean 63.7 and standard deviation 2.75 will have a mean as large as 64.72.

g. It would seem highly unlikely that this sample of 50 data is from the population described.

SECTION 8.1 ANSWER NOW EXERCISES

8.2 Difficulty and collector fatigue in obtaining and also in evaluating a very large sample; cost of sampling; destruction of product in cases like the rivets illustration.

8.3 $\sigma_{\overline{x}} = \sigma / \sqrt{n} = 18 / \sqrt{36} = 18 / 6 = 3$

8.4
$$\begin{aligned} P(\mu - 2\sigma_{\overline{x}} < \overline{x} < \mu + 2\sigma_{\overline{x}}) &= P(-2\sigma_{\overline{x}} < \overline{x} - \mu < +2\sigma_{\overline{x}}) \\ &= P(-2 < (\overline{x} - \mu)/\sigma_{\overline{x}} < +2) \\ &= P(-2 < z < 2) \\ &= 0.4772 + 0.4772 = 0.9544 = \underline{95.44\%} \end{aligned}$$

8.5 a. Between 10:14 AM and 10:34 AM, the next eruption should occur.

b. Yes; the snapshot was recorded on 10/27/02 at 10:18:02AM
which is within the interval from 10:14 to 10:34 AM.

c. 90% of the eruptions occur within predicted interval

SECTION 8.1 EXERCISES

ESTIMATION

Point Estimate for a Population Parameter - the value of the corresponding sample statistic

ex.) $\overline{x}$ is the point estimate of μ
s is the point estimate of σ
s^2 is the point estimate of σ^2

Confidence intervals are used to estimate a population parameter on an interval with a degree of certainty. We could begin by taking a sample and finding $\overline{x}$ to just estimate μ. The sample statistic, $\overline{x}$, is a point estimate of the population parameter μ. How good an estimate it is depends not only on the sample size and variability of the data, but also whether or not the sample statistic is unbiased.

Unbiased Statistic - a sample statistic whose sampling distribution has a mean value equal to the corresponding population parameter

One would assume that $\overline{x}$ is not exactly equal to μ, but hopefully relatively close. It is by this reasoning that we work with interval estimates of population parameters.

Level of Confidence = $1 - \alpha$ = the probability that the interval constructed, based on the sample, contains the true population parameter.

8.7
a. 24 = sample size = n; 4'11" = sample mean = $\overline{x}$
b. 16 = population standard deviation = σ
c. 190 = sample variance = s^2
d. 69 = population mean = μ

8.9
a. II has the lower variability; both have a mean value equal to the parameter.
b. II has a mean value equal to the parameter, I does not.
c. Neither is a good choice; I is negatively biased with less variability, while II is only slightly positively biased with a larger variability.

8.11 $\bar{x}$ is unbiased since $\mu_{\bar{x}} = \mu$; the variability $\sigma_{\bar{x}} = \sigma / \sqrt{n}$ decreases as n increases

> The level of confidence depends on the number of standard errors from the sample mean.
>
> For $\bar{x} \pm z\sigma_{\bar{x}}$: $\bar{x}$ = mean (given)
>
> $\sigma_{\bar{x}}$ = standard error of the mean (given)
>
> z = number of standard errors
>
> Find the corresponding probability for the z-value using Table 3, Appendix B. Multiply the probability by 2 to cover both parts of the interval. This total probability is equal to the level of confidence.

8.13 19(\$17,320) = \$ 329080
6(\$20,200) = 121200
\$ 450280 for 25 projects;
therefore 174(\$450280/25) = \$3,133,948.80

SECTION 8.2 ANSWER NOW EXERCISES

8.15 Performance of an entire line tend to be normally distributed. The normal distribution can then be used to calculate probabilities.

8.16 The point estimate is the center of the confidence interval. The maximum error is based on the level of confidence, sample size and standard deviation. The confidence interval is bounded by the two numbers found by adding and subtracting the point estimate and the maximum error.

8.17 a. 75.92

b. $E = z(\alpha/2)\cdot\sigma/\sqrt{n} = (2.33)(0.5/\sqrt{10}) = 0.3684 = 0.368$

c. $\bar{x} \pm E = 75.92 \pm 0.368$
75.552 to 76.288, the 0.98 confidence interval for μ

8.18 The numbers are calculated for samples of snow for some of the months; it would be impossible to gather this information from every inch of snow.

8.19 a. 15.9; ≈ 68%

b. 31.4; ≈ 95%

c. 41.2; ≈ 99%

d. higher level makes for a wider width; to be more certain the parameter is contained in the interval

8.20 Each student will have different results; however, 4.5 should be in the interval about 90% of the time.

8.21 $n = [z(\alpha/2)\cdot\sigma/E]^2 = [(2.33)(3)/1]^2 = 48.8601 = \underline{49}$

SECTION 8.2 EXERCISES

ESTIMATION OF THE POPULATION MEAN - μ

Point estimate of μ: $\bar{x}$

Interval estimate of μ = confidence interval

$1 - \alpha$ = level of confidence, the probability or degree of certainty desired (ex.: 95%, 99% ...)

A $(1-\alpha)$ confidence interval estimate for μ is:

$$\bar{x} - z(\alpha/2)\cdot\sigma/\sqrt{n} \quad \text{to} \quad \bar{x} + z(\alpha/2)\cdot\sigma/\sqrt{n} \quad **$$

$\bar{x} - z(\alpha/2)\cdot\sigma/\sqrt{n}$ = lower confidence limit

$\bar{x} + z(\alpha/2)\cdot\sigma/\sqrt{n}$ = upper confidence limit

$E = z(\alpha/2)\cdot\sigma/\sqrt{n}$ = maximum error of the estimate

**To find $z(\alpha/2)$:

(suppose for example that a 95% confidence interval is desired)

$95\% = 1 - \alpha$, that is,

$.95 = 1 - \alpha$

solving for alpha, α, gives

$\alpha = .05$

dividing both sides by 2 gives

$\alpha/2 = .025$

Now determine the probability associated with z(.025) using Table 4B (Appendix B, ES9-p715), the Critical Values of Standard Normal Distribution for Two-Tailed Situations. (This table conveniently gives the most popular critical values for z.)
The Student Suite CD contains the video clip: "Estimation".

8.23 Either the sampled population is normally distributed or the random sample is sufficiently large for the Central Limit Theorem to hold.

THE CONFIDENCE INTERVAL: A FIVE-STEP PROCEDURE

Step 1: The Set-Up:
Describe the population parameter of concern.
Step 2: The Confidence Interval Criteria:
a. Check the assumptions.
b. Identify the probability distribution and the formula to be used.
c. Determine the level of confidence, $1 - \alpha$.
Step 3: The sample evidence:
Collect the sample information.
Step 4: The Confidence Interval:
a. Determine the confidence coefficient.
b. Find the maximum error of estimate.
c. Find the lower and upper confidence limits.
Step5: The Results:
State the confidence interval.

8.25 a. Step 1: The mean, μ
Step 2: a. normality indicated
b. z, $\sigma = 6$ c. $1-\alpha = 0.95$
Step 3: $n = 16$, $\bar{x} = 28.7$
Step 4: a. $\alpha/2 = 0.05/2 = 0.025$; $z(0.025) = 1.96$
b. $E = z(\alpha/2)\cdot\sigma/\sqrt{n} = (1.96)(6/\sqrt{16})$
$= (1.96)(1.5) = 2.94$
c. $\bar{x} \pm E = 28.7 \pm 2.94$
Step 5: 25.76 to 31.64, the 0.95 confidence interval for μ

b. Yes; the sampled population is normally distributed.

8.27 a. Step 1: The mean, μ

Step 2: a. normality assumed because of CLT with n = 86.

b. z, $\sigma = 16.4$ c. $1-\alpha = 0.90$

Step 3: $n = 86$, $\bar{x} = 128.5$

Step 4: a. $\alpha/2 = 0.10/2 = 0.05$; $z(0.05) = 1.65$

b. $E = z(\alpha/2)\cdot\sigma/\sqrt{n} = (1.65)(16.4/\sqrt{86})$

$= (1.65)(1.76845) = 2.92$

c. $\bar{x} \pm E = 128.5 \pm 2.92$

Step 5: 125.58 to 131.42, the 0.90 confidence interval for μ

b. Yes; the sample size is sufficiently large to satisfy the CLT.

8.29 a. Speed readings for the Channel Tunnel train.

b. Step 1: The mean speed of the Channel Tunnel train

Step 2: a. normality indicated

b. z, $\sigma = 19$ c. $1-\alpha = 0.90$

Step 3: $n = 20$, $\bar{x} = 184$

Step 4: a. $\alpha/2 = 0.10/2 = 0.05$; $z(0.05) = 1.65$

b. $E = z(\alpha/2)\cdot\sigma/\sqrt{n} = (1.65)(19/\sqrt{20})$

$= (1.65)(4.25) = 7.01$

c. $\bar{x} \pm E = 184 \pm 7.01$

Step 5: 176.99 to 191.01, the 0.90 confidence interval for μ

c. Step 1-3: as shown in (b), except 2c. $1-\alpha = 0.95$

Step 4: a. $\alpha/2 = 0.05/2 = 0.025$; $z(0.025) = 1.96$

b. $E = z(\alpha/2)\cdot\sigma/\sqrt{n} = (1.96)(19/\sqrt{20})$

$= (1.96)(4.25) = 8.33$

c. $\bar{x} \pm E = 184 \pm 8.33$

Step 5: 175.67 to 192.33, the 0.95 confidence interval for μ

Computer and/or calculator commands to calculate a confidence interval for μ, provided σ is known, can be found in ES9-p.357 The output will also contain the sample mean and standard deviation.

REMEMBER: variance = σ^2, therefore $\sigma = \sqrt{\sigma^2}$

8.31 a. 25.3

b. Step 1: The mean age of night school students
Step 2: a. normality assumed, CLT with $n = 60$.
b. z, $\sigma^2 = 16$ or $\sigma = 4$ c. $1-\alpha = 0.95$
Step 3: $n = 60$, $\bar{x} = 25.3$
Step 4: a. $\alpha/2 = 0.05/2 = 0.025$; $z(0.025) = 1.96$
b. $E = z(\alpha/2)\cdot\sigma/\sqrt{n} = (1.96)(4/\sqrt{60})$
$= (1.96)(0.516) = 1.01$
c. $\bar{x} \pm E = 25.3 \pm 1.01$
Step 5: 24.29 to 26.31, the 0.95 confidence interval for μ

c. Step 1-3: as shown in (b), except 2c. $1-\alpha = 0.99$
Step 4: a. $\alpha/2 = 0.01/2 = 0.005$; $z(0.005) = 2.58$
b. $E = z(\alpha/2)\cdot\sigma/\sqrt{n} = (2.58)(4/\sqrt{60})$
$= (2.58)(0.516) = 1.33$
c. $\bar{x} \pm E = 25.3 \pm 1.33$
Step 5: 23.97 to 26.63, the 0.99 confidence interval for μ

8.33 Step 1: The mean geometry score for all eighth-grade students in the US
Step 2: a. normality assumed, standard error given
b. z, $\sigma/\sqrt{n} = 4.4$ c. $1-\alpha = 0.95$
Step 3: $\bar{x} = 473$
Step 4: a. $\alpha/2 = 0.05/2 = 0.025$; $z(0.025) = 1.96$
b. $E = z(\alpha/2)\cdot\sigma/\sqrt{n} = (1.96)(4.4) = 8.6$
c. $\bar{x} \pm E = 473 \pm 8.6$
Step 5: 464.4 to 481.6, the 0.95 confidence interval for μ

8.35 a. No answer. Directs student to code data.
b.

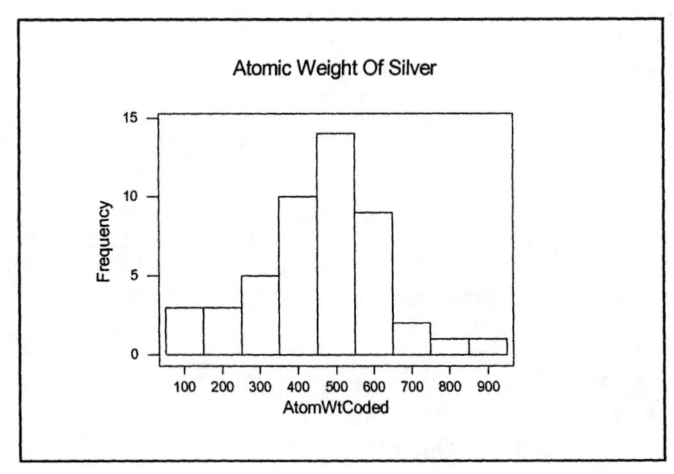

The units are "ten millionths"of the original unit of measure.

c. Mean = 450.6 and standard deviation = 173.4

d. Mean = 107.86814506 and standard deviation = 0.00001734

e. The histogram definitely shows an approximately normal distribution. Also, the Normal Probability Plot shows a straight-line pattern.

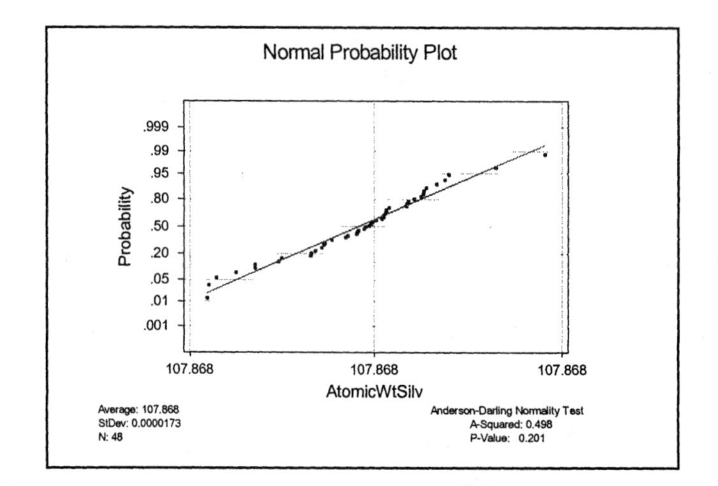

f. The SDSM and CLT both apply. The SDSM defines the mean and standard error for the sampling distribution and says that if the population is normal then the sampling distribution is normal. The histogram for the sample strongly suggests that the population has a normal distribution. The CLT tells us that the sampling distribution approaches a normal distribution for larger samples.

g. Sigma is not known.

h. A value is needed for the standard deviation of the population, and if it is not available then the next best thing is to estimate it with a reliable point estimate. The sample standard deviation is the value that will be used.

i. Step 1: The mean value of all such observations
Step 2: a. normality assumed based on histogram and normality plot
b. z, s = 173.4 c. $1-\alpha = 0.95$
Step 3 $n = 48$, $\overline{x} = 450.6$

Step 4: a. $\alpha/2 = 0.05/2 = 0.025; \quad z(0.025) = 1.96$

b. $E = z(\alpha/2)\cdot\sigma/\sqrt{n} = (1.96)(173.4/\sqrt{48})$

$= (1.96)(25.03) = 49.055$

c. $\bar{x} \pm E = 450.6 \pm 49.055$

Step 5: 401.5 to 499.7, the 0.95 confidence interval for μ

or

107.86814015 to 107.86814997

To find the sample size n required for a 1-α confidence interval, use the formula: $n = [z(\alpha/2)\cdot\sigma/E]^2$, where

- z = standard normal distribution
- α = calculated from the 1-α confidence interval desired
- σ = population standard deviation
- E = maximum error of the estimate

The maximum error of the estimate, E, is the amount of error that is tolerable or allowed. Quite often, finding the word "within" in an exercise will locate the acceptable value for E.

8.37 $n = [z(\alpha/2)\cdot\sigma/E]^2 = [(2.58)(1.0)/0.5]^2 = 26.6 = \underline{27}$

8.39 $n = [z(\alpha/2)\cdot\sigma/E]^2 = [(1.96)(0.4)/0.15]^2 = 27.318 = \underline{28}$

SECTION 8.3 ANSWER NOW EXERCISES

8.41 Lobsters are or are not active at night

8.42 H_o: The system is reliable
H_a: The system is not reliable

8.43. H_a: Teaching techniques have a significant effect on student's exam scores.

8.44 Type A correct decision:

Truth of situation:	the party will be a dud.
Conclusion:	the party will be a dud.
Action:	did not go [avoided dud party]

Type B correct decision:

Truth of situation:	the party will be a great time.
Conclusion:	the party will be a great time.
Action:	did go [party was great time]

Type I error:

Truth of situation:	the party will be a dud.
Conclusion:	the party will be a great time.
Action:	did go [party was a dud]

Type II error:

Truth of situation:	the party will be a great time.
Conclusion:	the party will be a dud.
Action:	did not go [missed great party]

Remember; the truth of the situation is not known before the decision is made, the conclusion reached and the resulting actions take place. Only after the party is over can the evaluation be made.

8.45 You missed a great time.

8.46 α is the probability of rejecting a TRUE null hypothesis; $1-\beta$ is the probability of rejecting a FALSE null hypothesis; they are two distinctly different acts that both result in rejecting the null hypothesis.

8.47 The smaller the probability of an event is, the less often it occurs.

8.48 a. Two groups of students, one is coached and the other is not.
b. The coached group receives higher SAT scores.

SECTION 8.3 EXERCISES

DEFINITIONS FOR HYPOTHESIS TESTS

Hypothesis - a statement that something is true.

Null Hypothesis, H_o - a statement that specifies a value for a population parameter

ex.: H_o: The mean weight is 40 pounds

Alternative Hypothesis, H_a - opposite of H_o, a statement that specifies an "opposite" value for a population parameter

ex.: H_a: The mean weight is not 40 pounds

Type I Error - the error resulting from rejecting a true null

α (alpha) - the probability of a type I error, that is the probability of rejecting H_o when it is true.

Type II Error - the error resulting from not rejecting a false null hypothesis

β (beta) - the probability of a type II error, that is the probability of not rejecting H_o when it is false.

Keep α and β as small as possible, depending on the severity of the respective error.

The Student Suite CD has two video clips: "Hypothesis Testing" and "Formulating a Hypothesis".

8.49 a. H_o: Special delivery mail does not take too much time
H_a: Special delivery mail takes too much time

b. H_o: The new design is not more comfortable
H_a: The new design is more comfortable

c. H_o: Cigarette smoke has no effect on the quality of a person's life
H_a: Cigarette smoke has an effect on the quality of a person's life

d. H_o: The hair conditioner is not effective on "split ends"
H_a: The hair conditioner is effective on "split ends"

8.51 a. H_o: The victim is alive
H_a: The victim is not alive

b. Type A correct decision: The victim is alive and is treated as though alive.

Type I error: The victim is alive, but is treated as though dead.

Type II error: The victim is dead, but treated as if alive.

Type B correct decision: The victim is dead and treated as dead.

c. The type I error is very serious. The victim may very well be dead shortly without the attention that is not being received.

The type II error is not as serious. The victim is receiving attention that is of no value. This would be serious only if there were other victims that needed this attention.

8.53 a. A type I error occurs when it is determined that the majority of Americans do not favor laws against assault weapons when, in fact, the majority do favor such laws. A type II error occurs when it is determined that the majority of Americans do favor laws against assault weapons when, in fact, they do not favor such laws.

b. A type I error occurs when it is determined that the fast food is low salt when, in fact, it is not low salt. A type II error occurs when it is determined that the fast food is not low salt when, in fact, it is low salt.

c. A type I error occurs when it is determined that the building must be demolished when, in fact, it should not be demolished. A type II error occurs when it is determined that the building must not be demolished when, in fact, it should be demolished.

d. A type I error occurs when it is determined that there is waste in government spending when, in fact, there is not waste.
A type II error occurs when it is determined that there is no waste in government spending when, in fact, there is waste.

TERMINOLOGY FOR DECISIONS IN HYPOTHESIS TESTS

Reject H_o: use when the evidence disagrees with the null hypothesis.

Fail to reject H_o: use when the evidence does not disagree with the null hypothesis.

Note: The purpose of the hypothesis test is to allow the evidence a chance to discredit the null hypothesis.
Remember: If one believes the null hypothesis to be true, generally there is no test.

8.55 a. Type I   b. Type II   c. Type I   d. Type II

8.57 a. The type I error is very serious and, therefore, we are willing to allow it to occur with a probability of 0.001; that is, only 1 chance in 1000.

b. The type I error is somewhat serious and, therefore, we are willing to allow it to occur with a probability of 0.05; that is, 1 chance in 20.

c. The type I error is not at all serious and, therefore, we are willing to allow it to occur with a probability of 0.10; that is, 1 chance in 10.

d. The type II error is not at all serious and, therefore, we are willing to allow it to occur with a probability of 0.10; that is, 1 chance in 10.

8.59 a. α b. β

8.61 a. When the test procedure begins, the experimenter is thoroughly convinced the alternative hypothesis can be shown to be true; thus when the decision *reject H_o* is attained, the experimenter will want to say something like "see I told you so." Thus the statement of the conclusion is a fairly strong statement like; "the evidence shows beyond a shadow of a doubt (is significant) that the alternative hypothesis is correct."

b. When the test procedure begins, the experimenter is thoroughly convinced the alternative hypothesis can be shown to be true; thus when the decision *fail to reject H_o* is attained, the experimenter is disappointed and will want to say something like "okay so this evidence was not significant, I'll try again tomorrow." Thus the statement of the conclusion is a fairly mild statement like, "the evidence was not sufficient to show the alternative hypothesis to be correct."

The **power** of a test is equal to 1 - β. It is the measure of the ability of a hypothesis test to reject a false null hypothesis.

8.63 a. α = P(rejecting H_o when the H_o is true)
$= P(x \geq 86 | \mu=80) = P(z > (86 - 80)/5) = P(z > 1.20)$
$= 0.5000 - 0.3849 = \underline{0.1151}$

b. β = P(accepting H_o when the H_o is false)
$= P(x < 86 | \mu=90) = P(z < (86 - 90)/5) = P(z < -0.80)$
$= 0.5000 - 0.2881 = \underline{0.2119}$

SECTION 8.4 ANSWER NOW EXERCISES

8.65 H_o: The mean shearing strength is at least 925 lbs.
H_a: The mean shearing strength is less than 925 lbs.

8.66 H_o: $\mu = 54.4$
H_a: $\mu \neq 54.4$

8.67 a. H_o: $\mu = 1.25$ ($\leq$)
H_a: $\mu > 1.25$

b. H_o: $\mu = 335$ ($\geq$)
H_a: $\mu < 335$

c. H_o: $\mu = 230,000$
H_a: $\mu \neq 230,000$

8.68 Type A correct decision: The mean shearing strength is at least 925 lbs and it is decided that it is.
Type I error: The mean shearing strength is at least 925 lbs and it is decided that it is less than 925 lbs.
Type II error: The mean shearing strength is less than 925 lbs and it is decided that it is greater than or equal to 925 lbs.
Type B correct decision: The mean shearing strength is less than 925 lbs and it is decided that it is less than 925 lbs.

Type II error; you buy and use weak rivets.

8.69 $z = (\overline{x} - \mu)/(\sigma/\sqrt{n})$

$z^* = (54.3 - 56)/(7/\sqrt{36}) = \underline{-1.46}$

8.70 a. How likely are the sample results to have happened if the null hypothesis is true?
b. Probability

8.71 a. p-value $= P(z < -2.3) = P(z > 2.3)$
$= 0.5000 - 0.4893 = \underline{0.0107}$

b. p-value $= P(z > 1.8) = 0.5000 - 0.4641 = \underline{0.0359}$

8.72 a. Fail to reject H_o b. Reject H_o

8.73 a. $z^* = (\overline{x} - \mu)/(\sigma/\sqrt{n}) = (24.5 - 22.5)/(6/\sqrt{36}) = 2.0$

p-value $= P(z > 2.0) = 0.5000 - 0.4772 = \underline{0.0228}$

b. $z* = (\bar{x} - \mu)/(\sigma/\sqrt{n}) = (192.5 - 200)/(40/\sqrt{50}) = -1.33$

p-value $= P(z < -1.33) = P(z > 1.33)$
$= 0.5000 - 0.4082 = \underline{0.0918}$

c. $z* = (\bar{x} - \mu)/(\sigma/\sqrt{n}) = (11.52 - 12.4)/(2.2/\sqrt{16}) = -1.6$

p-value $= 2 \cdot P(z < -1.6) = 2 \cdot P(z > 1.6)$
$= 2(0.5000 - 0.4452) = 2(0.0548) = \underline{0.1096}$

8.74 a. Directives - no answer.
b. ≈ 0.0000
c. no means down by 1451 or lower; The probability of taking a sample of 24 and having a mean less than 1451 when the true mean is equal to 1500 is 0.0000.
d. Reject Ho.

8.75 p-value $= 2 \cdot P(z > 1.1) = 2(0.5000 - 0.3643) = 2(0.1357) = \underline{0.2714}$

8.76 a. 4 - 3.6 = .04, due to '≠', 4 + .04 = 4.4
b. ≈ 0.0277
c. green areas in the tails; The probability of taking a sample of 100 and having a mean less than 3.6 or greater than 4.4 when the true mean is equal to 4 is 0.0277.
d. Reject Ho.

8.77 The p-value measures the likeliness of the sample results based on a true null hypothesis.

8.78 N = n and is the number of data values
MEAN calculated using formula $\Sigma x/n$
STDEV calculated using formula $\sqrt{\Sigma(x - \bar{x})^2/(n - 1)}$
SEMEAN calculated using formula $\sigma/\sqrt{n}$
z calculated using formula $(\bar{x} - \mu)/(\sigma/\sqrt{n})$
p-value calculated using formula $P(z < -1.50)$

8.79 Results will vary; however, expect your results to be similar to those shown in Table 8.7a.

SECTION 8.4 EXERCISES

THE PROBABILITY-VALUE HYPOTHESIS TEST: A FIVE-STEP PROCEDURE

Step 1: The Set-Up:
- a. Describe the population parameter of concern.
- b. State the null hypothesis (H_o) and the alternative hypothesis (H_a).

Step 2: The Hypothesis Test Criteria:
- a. Check the assumptions.
- b. Identify the probability distribution and the test statistic formula to be used.
- c. Determine the level of significance, α.

Step 3: The Sample Evidence:
- a. Collect the sample information.
- b. Calculate the value of the test statistic.

Step 4: The Probability Distribution:
- a. Calculate the p-value for the test statistic.
- b. Determine whether or not the p-value is smaller than α.

Step 5: The Results:
- a. State the decision about H_o.
- b. State the conclusion about H_a.

The Null and Alternative Hypotheses, H_o and H_a

H_o: $\mu = 100$ versus H_a: $\mu \neq 100$ (= and ≠ form the opposite of H_a is a two-sided alternative)

possible wording for this combination:
- a) mean is different from 100 (≠)
- b) mean is not 100 (≠)
- c) mean is 100 (=)

OR

H_o: $\mu = 100$ (≤) versus H_a: $\mu > 100$ (≤ and > form the opposite of each other)

H_a is a one-sided alternative
possible wording for this combination:
- a) mean is greater than 100 (>)
- b) mean is at most 100 (≤)
- c) mean is no more than 100 (≤) . . .

OR

H_o: μ = 100 (≥) versus H_a: μ < 100 (≥ and < form the opposite of H_a is a one-sided alternative

possible wording for this combination:
a) mean is less than 100 (<)
b) mean is at least 100 (≥)
c) mean is no less than 100 (≥)

Always show equality (=) in the null hypothesis, since the null hypothesis must specify a single specific value for μ (like μ = 100).
The null hypothesis could be rejected in favor of the alternative hypothesis for three different reasons; 1) $\mu \neq 100$ or 2) $\mu > 100$ or 3) $\mu < 100$. Together, the two opposing statements, H_o and H_a, must contain or account for all numerical values around and including μ. This allows for the addition of ≥ or ≤ to the null hypothesis. Therefore, if the alternative hypothesis is < or >, ≥ or ≤, respectively, may be added to the null hypothesis. If ≥ or ≤ is being tested, the appropriate symbol should be written in parentheses after the amount stated for μ.
Sometimes, depending on the wording, it is easier to write the alternative hypothesis first. The alternative hypothesis can only contain >, < or ≠.

Hint: Sometimes it is helpful to either: 1) remove the word "no" or "not" when it is included in the claim, or 2) insert "no" or "not" when it is not in the claim, to form the opposite of the claim.

8.81 a. H_o: μ = 210 lbs (≤) vs. H_a: μ > 210

b. H_o: μ = 570 lbs/unit vs. H_a: $\mu \neq 570$

c. H_o: μ = \$9.00 (≤) vs. H_a: μ > \$9.00

ERRORS

Type I error - occurs when H_o is rejected and it is a true statement.

Type II error - occurs when H_o is accepted and it is a false statement.

Since we work under the assumption that H_o is a true statement, all decisions are made <u>based on</u> or <u>pertaining to</u> H_o.

If we are unable to reject H_o, the terminology "fail to reject H_o" is used; whereas if we are able to reject H_o, "reject H_o" is used. After this decision statement, include an additional statement, explaining how the test results support or did not support the experimenter's convictions, to form the conclusion.

8.83 a. A type I error would be committed if a decision of reject H_o was reached and interpreted as 'mean hourly charge is less than $60 per hour' when in fact the mean hourly charge is at least $60 per hour.

b. A type II error would be committed if a decision of fail to reject H_o was reached and interpreted as 'mean hourly charge is at least $60 per hour' when in fact the mean hourly charge is less than $60 per hour.

Use the sample information (sample mean and size) and the population parameter in the null hypothesis (μ) to calculate the test statistic z^*.

$$z^* = (\bar{x} - \mu)/(\sigma/\sqrt{n})$$

8.85 a. $z = (\bar{x} - \mu)/(\sigma/\sqrt{n})$

$z^* = (10.6 - 10)/(3/\sqrt{40}) = \underline{1.26}$

b. $z = (\bar{x} - \mu)/(\sigma/\sqrt{n})$

$z^* = (126.2 - 120)/(23/\sqrt{25}) = \underline{1.35}$

c. $z = (\bar{x} - \mu)/(\sigma/\sqrt{n})$

$z^* = (18.93 - 18.2)/(3.7/\sqrt{140}) = \underline{2.33}$

d. $z = (\bar{x} - \mu)/(\sigma/\sqrt{n})$

$z^* = (79.6 - 81)/(13.3/\sqrt{50}) = \underline{-0.74}$

DECISIONS AND CONCLUSIONS

Since the null hypothesis, H_o, is usually thought to be the statement whose truth is being challenged by the experimenter, all decisions are about the null hypothesis. The alternative hypothesis, H_a, however is usually thought to express the experimenter's viewpoint. Thus, the interpretation of the decision or conclusion is expressed from the experimenter and alternative hypothesis point of view.

Decision:

1) If the p-value is less than or equal to the specified level of significance (α), the null hypothesis will be rejected (if **P** $\leq \alpha$, reject H_o).
2) If the p-value is greater than the specified level of significance (α), fail to reject the null hypothesis (if **P** $> \alpha$, fail to reject H_o).

Conclusion:

1) If the decision is "reject H_o," the conclusion should read "There is sufficient evidence at the α level of significance to show that ... (the meaning of the alternative hypothesis)."
2) If the decision is "fail to reject H_o," the conclusion should read "There is not sufficient evidence at the α level of significance to show that ... (the meaning of the alternative hypothesis)."

The Student Suite CD has a video clip: "The p-value".

8.87 a. *Reject H_o* or *Fail to reject H_o*

b. When the calculated p-value is smaller than or equal to α , the decision will be *reject H_o*.

When the calculated p-value is larger than α, the decision will be *fail to reject H_o*.

8.89 a. Fail to reject H_o b. Reject H_o

The p-value approach uses the calculated test statistic to find the area under the curve that contains the calculated test statistic and any values "beyond" it, in the direction of the alternative hypothesis. This probability (area under the curve), based on the position of the calculated test statistic, is compared to the level of significance (α) for the test and a decision is made.

Rules for calculating the p-value

1) If H_a contains <, then the p-value = $P(z < z^*)$.
2) If H_a contains >, then the p-value = $P(z > z^*)$.
3) If H_a contains $\neq$, then the p-value = $2P(z > |z^*|)$.

The p-value can then be calculated by using the z* value with Table 3, or it can be found directly using Table 5 (Appendix B, in ES9-p716), or it can be found using a computer and/or calculator.

8.91 a. p-value = $P(z > 1.48) = 0.5000 - 0.4306 = \underline{0.0694}$

b. p-value = $P(z < -0.85) = 0.5000 - 0.3023 = \underline{0.1977}$

c. p-value = $2 \cdot P(z > 1.17) = 2(0.5000 - 0.3790) = \underline{0.2420}$

d. p-value = $P(z < -2.11) = 0.5000 - 0.4826 = \underline{0.0174}$

e. p-value = $2 \cdot P(z > 0.93) = 2(0.5000 - 0.3238) = \underline{0.3524}$

8.93 a. **P** = $P(z > z^*) = 0.0582$

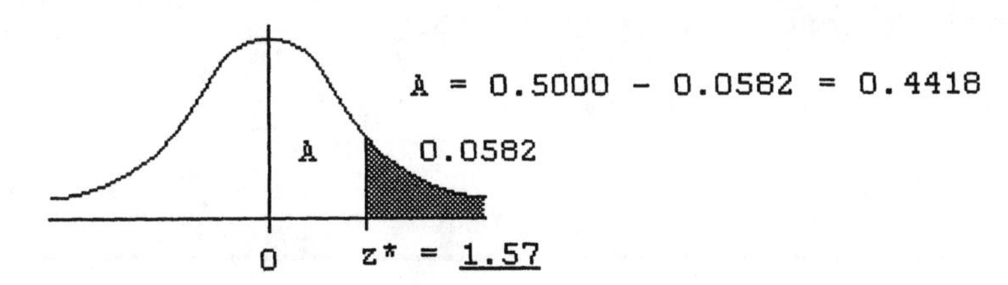

b. **P** = $P(z < z^*) = 0.0166$

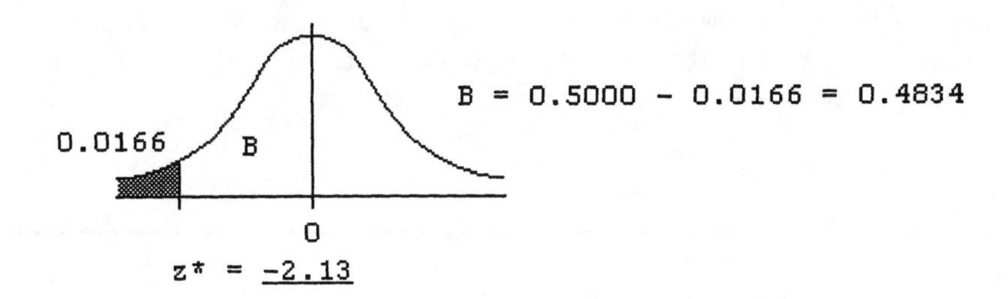

c. $\mathbf{P} = P(z < -z^*) + P(z > +z^*) = 2 \cdot P(z > +z^*) = 0.0042$

$P(z > +z^*) = 0.0021$

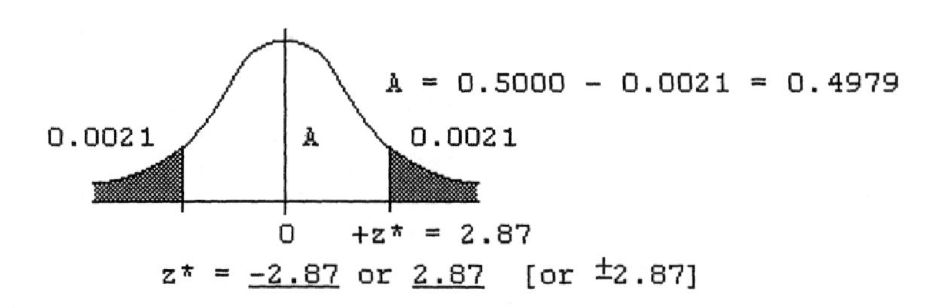

$z^* = \underline{-2.87}$ or $\underline{2.87}$ [or ± 2.87]

Computer and calculator commands to complete a hypothesis test for a mean μ with a standard deviation σ can be found in ES9-p.383

Compare the calculated p-value to the given level of significance, α. Using the rules for comparison as stated in ST-p244, a decision can be made about the null hypothesis.

8.95 a. H_o: $\mu = 525$ vs. H_a: $\mu < 525$

b. Fail to reject H_o; the population mean is not significantly less than 525.

c. $\sigma_{\bar{x}} = \sigma/\sqrt{n} = 60.0/\sqrt{38} = 9.7333 = \underline{9.733}$

WORD PROBLEMS

1. Look for the key words that indicate the need for a hypothesis test for μ using a z-test. Statements that mention; "testing a <u>mean</u> amount" or "make a decision about the mean value" and the fact that the population standard deviation (σ) or variance (σ^2) are given are examples of these key words. . . .

---

2. Write down the values needed: $\mu, \sigma, \overline{x}, n, \alpha$. If the mean is mentioned in a sentence with no reference to a sample, then it is most likely μ (population mean). If the mean is mentioned in a sentence involving the sample or thereafter, then it is usually $\overline{x}$ (sample mean). Often the sample size (n) is mentioned in the same sentence. If the standard deviation or variance is given in a sentence without any reference to a sample, then it also is usually the population σ or σ^2 respectively.

3. Proceed with the hypothesis steps as outlined in:
ES9-pp370-377 or ST-pp241-245.

8.97 a. The mean test score for all elementary education majors.

b. H_o: $\mu = 35.70$ (≥)
H_a: $\mu < 35.70$

c. $z* = (32.63 - 35.70)/(6.73/\sqrt{165}) = -5.86$
p-value = $P(z < -5.86) = P(z > 5.86) \approx +0.0000$

d. Reject H_o, the mean test score is less than 35.70 at the 0.001 level of significance.

8.99 a. The mean accuracy of quartz watches measured in seconds in error per month.

b. H_o: $\mu = 20$ (≤)
H_a: $\mu > 20$

c. normality is assumed, CLT with n = 36
use z with σ = 9.1; an α = 0.05 is given

d. $n = 36$, $\overline{x} = 22.7$

e. $z = (\overline{x} - \mu)/(\sigma/\sqrt{n})$
$z* = (22.7 - 20)/(9.1/\sqrt{36}) = 1.78$
P = P(z > 1.78);
Using Table 3, Appendix B, ES9-p714:
P = 0.5000 - 0.4625 = 0.0375
Using Table 5, Appendix B, ES9-p716:
0.0359 < **P** < 0.0401

f. **P** < α; Reject H_o
At the 0.05 level of significance, there is sufficient evidence to support the contention that the wrist watches priced under $25 exhibit greater error (less accuracy) than watches in general.
$z* = (24.47 - 25.12)/(5.3/\sqrt{100}) = -1.2264$;
p-value = 0.1100

SECTION 8.5 ANSWER NOW EXERCISES

8.101 H_o: The mean shearing strength is at least 925 lbs.
H_a: The mean shearing strength is less than 925 lbs.

8.102 H_o: $\mu = 9$ ($\le$)
H_a: $\mu > 9$

8.103 a. H_o: $\mu = 1.25$ ($\ge$)
H_a: $\mu < 1.25$

b. H_o: $\mu = 335$
H_a: $\mu \ne 335$

c. H_o: $\mu = 230{,}000$ ($\le$)
H_a: $\mu > 230{,}000$

8.104 Type A correct decision: The mean shearing strength is at least 925 lbs and it is decided that it is at least 925 lbs.
Type I error: The mean shearing strength is at least 925 lbs and it is decided that it is less than 925 lbs.
Type II error: The mean shearing strength is less than 925 lbs and it is decided that it is greater than or equal to 925 lbs.
Type B correct decision: The mean shearing strength is less than 925 lbs and it is decided that it is less than 925 lbs.

Type II error; you buy and use weak rivets.

8.105 $z = (\bar{x} - \mu)/(\sigma/\sqrt{n})$
$z^* = (354.3 - 356)/(17/\sqrt{120}) = \underline{-1.10}$

8.106 a. Reject H_o b. Fail to reject H_o

8.107 $z \le -2.33$

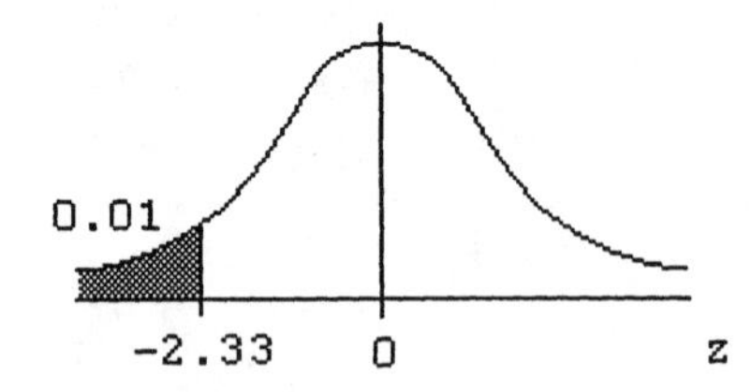

8.108 $z \geq 2.05$

1 kg ≈ 2.2046 lbs

8.109 54.4 kg = 54.4(2.2046lbs) ≈ 119.9 or 120 lbs.

8.110 Results will vary, however expect your results to be similar to those shown in Table 8.10.

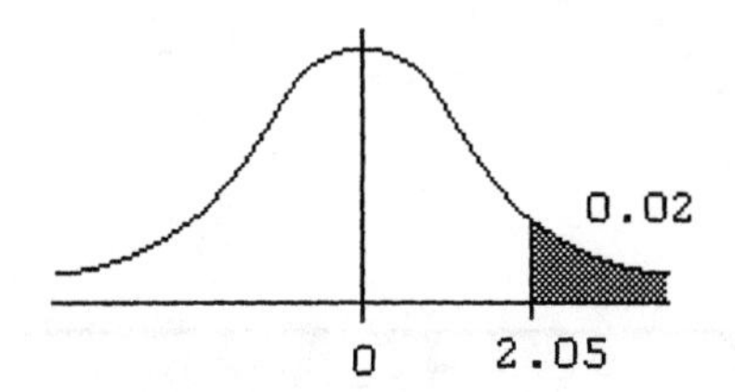

SECTION 8.5 EXERCISES

THE CLASSICAL HYPOTHESIS TEST: A FIVE-STEP PROCEDURE

Step 1: The Set-Up:
- a. Describe the population parameter of concern.
- b. State the null hypothesis (H_o) and the alternative hypothesis (H_a).

Step 2: The Hypothesis Test Criteria:
- a. Check the assumptions.
- b. Identify the probability distribution and the test statistic formula to be used.
- c. Determine the level of significance, α.

Step 3: The Sample Evidence:
- a. Collect the sample information.
- b. Calculate the value of the test statistic.

Step 4: The Probability Distribution:
- a. Determine the critical region and critical value(s).
- b. Determine whether or not the calculated test statistic is in the critical region.

Step 5: The Results:
- a. State the decision about H_o.
- b. State the conclusion about H_a.

Review "The Null and Alternative Hypotheses, H_o and H_a" in ST-pp241&242, if necessary.

8.111 a. H_o: μ = 16 yrs ($\geq$) vs. H_a: μ < 16

b. H_o: μ = 6 ft 6 in ($\leq$) vs. H_a: μ > 6 ft 6 in

c. H_o: μ = 285 ft ($\geq$) vs. H_a: μ < 285

d. H_o: μ = 0.375 inches ($\leq$) vs. H_a: μ > 0.375

e. H_o: μ = 200 units vs. H_a: $\mu \neq 200$

Review "Errors" and "H_o Decisions", if necessary, in: ES9-pp366&367, or ST-pp242&243.

8.113 H_o: μ = 85 ($\leq$) vs. H_a: μ > 85

a. It is decided that the mean minimum plumber's call is greater than \$85 when in fact it is not more than \$85.

b. It is decided that the mean minimum plumber's call is at most \$85 when in fact it is greater than \$85.

Critical region - that part under the curve where H_o will be rejected (size based on α)
Noncritical region - the remaining part under the curve where H_o will not be rejected
Critical value(s) - the $z(\alpha)$ or boundary point values of z, separating the critical and noncritical regions

See ES9-pp.392&393 for a visual display of these regions and value(s).

8.115 a. The critical region is the set of all values of the test statistic that will cause us to reject H_o.

b. The critical value(s) is the value(s) of the test statistic that forms the boundary between the critical region and the non-critical region. The critical value is in the critical region.

Determining the test criteria

1. Draw a picture of the standard normal (z) curve.
   (0 is at the center)

2. Locate the critical region (based on α and H_a)
   a) if H_a contains $<$, all of the α is placed in the left tail
   b) if H_a contains $>$, all of the α is placed in the right tail
   c) if H_a contains $\neq$, place $\alpha/2$ in each tail.

3. Shade in the critical region (the area where you will reject H_o).

4. Find the appropriate critical value(s) using the $z(\alpha)$ concept and the Standard Normal Distribution (Table 3, Appendix B, ES9-p714, or Table 4(a) for one-tail and Table 4(b) for two-tails, Appendix B, ES9-p715).
   If H_a contains $>$, the critical value is $z(\alpha)$.
   If H_a contains $<$, the critical value is $z(1-\alpha)$ <u>or</u> $-z(\alpha)$.
   If H_a contains $\neq$, the critical values are $\pm z(\alpha/2)$ <u>or</u> $z(\alpha/2)$ with $z(1-\alpha/2)$.

   Remember this boundary value divides the area under the curve into critical and noncritical regions and is part of the critical region.

8.117 a.

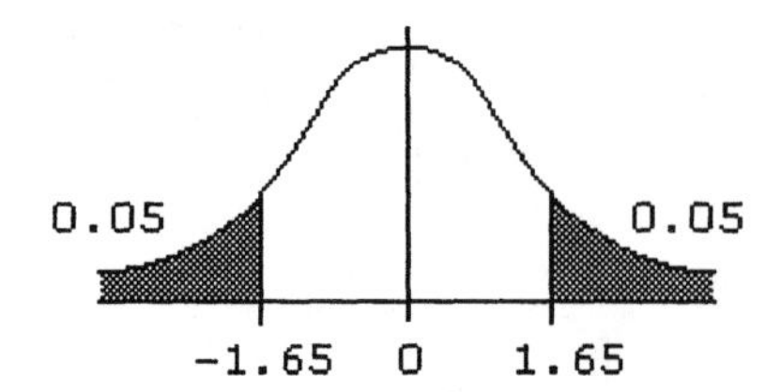

b.

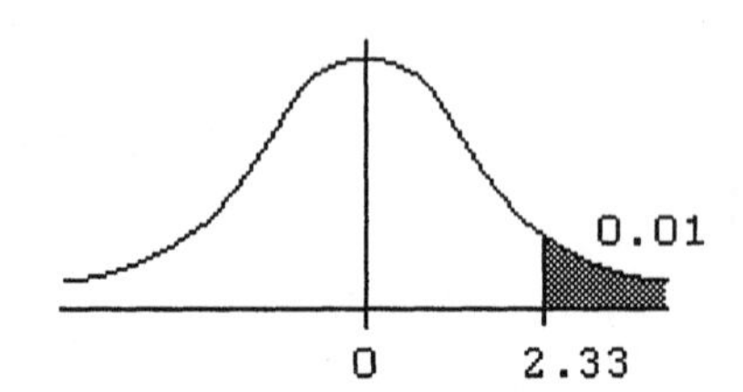

c.

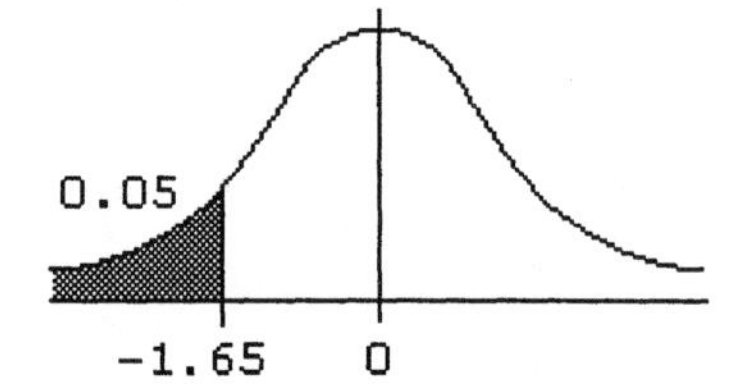

d.

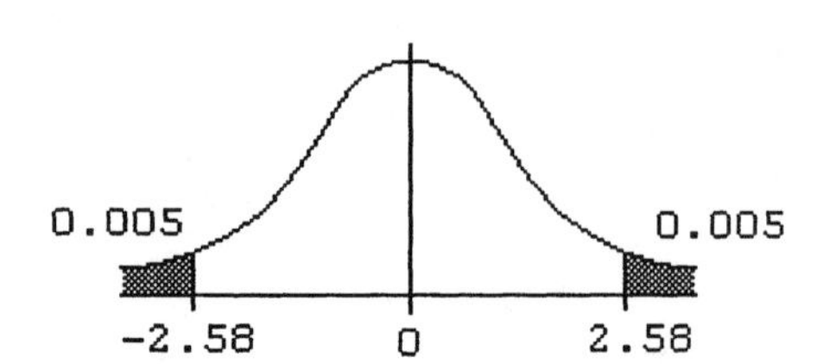

For exercise 8.119 use $z^* = (\overline{x} - \mu)/(\sigma/\sqrt{n})$, substituting the given values, then solving for the required unknown.

8.119 $z^* = (\overline{x} - \mu)/(\sigma/\sqrt{n})$

$-1.18 = (\overline{x} - 250)/(22.6/\sqrt{85})$

$-1.18 = (\overline{x} - 250)/2.451314$

$-2.89255 = \overline{x} - 250$

$\overline{x} = 247.107449 = \underline{247.1}$

$\overline{x} = \Sigma x/n$

$247.107449 = \Sigma x/85$

$\Sigma x = \underline{21{,}004.133}$

NOTE: Standard Error $= \sigma_{\overline{x}} = \sigma / \sqrt{n}$ and $z = (\overline{x} - \mu)/(\sigma/\sqrt{n})$

8.121 a. z = n(standard errors from mean):

$z = (\overline{x} - \mu)/(\sigma/\sqrt{n})$

$z = (4.8 - 4.5)/(1.0/\sqrt{100}) = \underline{3.0}$

$\overline{x}$ = 4.8 is 3.0 standard errors **above** the mean $\mu = 4.5$

b. If $\alpha = 0.01$, the critical region is $z \geq 2.33$. Since z* is equal to 3.00, it is in the critical region. Therefore, yes, reject H_o.

RESULTS, DECISIONS AND CONCLUSIONS

Since the null hypothesis, H_o, is usually thought to be the statement whose truth is being challenged by the experimenter, all decisions are about the null hypothesis. The alternative hypothesis, H_a, however is usually thought to express the experimenter's viewpoint. Thus, the conclusion (interpretation of the decision) is expressed from the experimenter and alternative hypothesis point of view.

The two possible outcomes are:

1. z^* falls in the critical region or
2. z^* falls in the noncritical region.

Decision and Conclusion:

If z^* falls in the critical region, we **reject H_o**. The conclusion is very strong and proclaims the alternative to be the case, that is, there is sufficient evidence to overturn H_o in favor of H_a. It should read something like "There is sufficient evidence at the α level of significance to show that ...(the meaning of the H_a)."

If z^* falls in the noncritical (acceptance) region, we **fail to reject H_o**. The conclusion is much weaker, that is, it suggests that the data does not provide sufficient evidence to overturn H_o. This does not necessarily mean that we have to accept H_o at this point, but only that this sample did not provide sufficient evidence to reject H_o. It should read something like "There is not sufficient evidence at the α level of significance to show that ...(the meaning of the H_a)."

8.123 a. *Reject H_o* or *Fail to reject H_o*

b. When the calculated test statistic falls in the critical region, the decision will be *reject H_o*.

When the calculated test statistic falls in the non-critical region, the decision will be *fail to reject H_o*.

Computer and calculator commands to complete a hypothesis test using the classical approach can be found in ES9-p.383 It is the same command used for the probability approach. Compare the calculated z value (test statistic) with the corresponding critical value(s). The locations of z*, relative to the critical value of z, will determine the decision you must make about the null hypothesis.

8.125 a. H_o: $\mu = 15.0$ vs. H_a: $\mu \neq 15.0$

b. Critical values: $\pm z(0.005) = \pm 2.58$
Decision: reject H_o
Conclusion: There is sufficient evidence to conclude that the mean is different than 15.0.

c. $\sigma_{\bar{x}} = \sigma/\sqrt{n} = 0.5/\sqrt{30} = 0.091287 = \underline{0.0913}$

See ST-p246 for information on "Word Problems", if necessary.

Hint for writing the hypotheses for exercise 8.127

Look at the fourth sentence of the exercise, "Is there sufficient evidence to conclude ... scored higher than the state average?" "higher than" indicates greater than (>). Therefore, the alternative hypothesis is greater than (>). Equality (=) is used in the null hypothesis (as usual), but it stands for less than or equal to (≤).

8.127 Step 1: a. The mean score for the Emergency Medical Services Certification Examiniation

b. H_o: $\mu = 79.68$ (≤)
H_a: $\mu > 79.68$

Step 2: a. normality is assumed, CLT with n = 50

b. z, $\sigma = 9.06$; c. $\alpha = 0.05$

Step 3: a. $n = 50$, $\bar{x} = 81.05$

b. $z = (\bar{x} - \mu)/(\sigma/\sqrt{n})$

$z^* = (81.05 - 79.68)/(9.06/\sqrt{50}) = 1.07$

Step 4: a. z(0.05) = 1.65

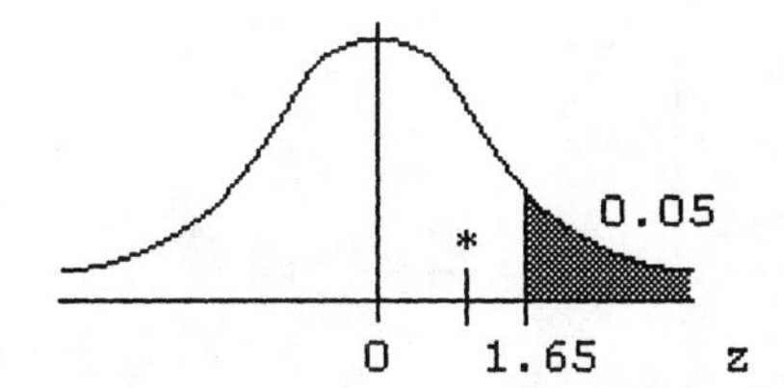

b. z* falls in the noncritical region, see Step 4a

Step 5. a. Fail to reject Ho

b. There is not sufficient evidence conclude that the sample average is higher than the state average, at the 0.05 level of significance.

<u>Hint for writing the hypotheses for exercise 8.129</u>

Look at the first sentence in the exercise, "The manager at Air Express feels ... are <u>less than</u> in the past." Since <u>less than</u> (<) does not include the equal to, it must be used in the alternative hypothesis. The negation is "...NOT less than...", which is > or =. Therefore, equality (=) is used in the null hypothesis (as usual), but it stands for greater than or equal to (≥).

8.129 Step 1: a. The mean weight of packages shipped by Air Express

b. H_o: $\mu = 36.5$ (≥)

H_a: $\mu < 36.5$

Step 2: a. normality assumed, CLT with n = 64

b. z, $\sigma = 14.2$ c. $\alpha = 0.01$

Step 3: a. n = 64, $\bar{x} = 32.11$

b. $z = (\bar{x} - \mu)/(\sigma/\sqrt{n})$

$z* = (32.11 - 36.5)/(14.2/\sqrt{64}) = -2.47$

Step 4: a. $-z(0.01) = -2.33$

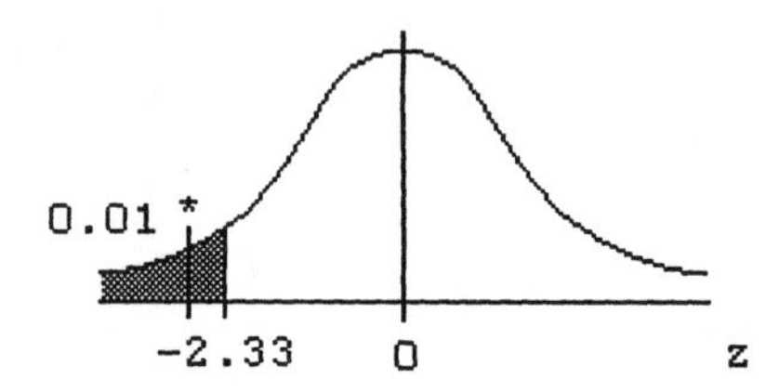

b. z* falls in the critical region, see Step 4a

Step 5: a. Reject H_o

b. At the 0.01 level of significance, the population mean is significantly less than the claimed mean of 36.5.

RETURN TO CHAPTER CASE STUDY

8.131 a. No, but it can be expected to be reasonably close in value.

b. See graph in part (e). $\bar{x} = 64.72$

c. The distribution of the sample data is expected to look like the distribution of the population. The mean and standard deviation of the sample are expected to approximate the mean and standard deviation of the population.
The sampling distribution to which the mean of this sample belongs: has a mean value that is approximated by the mean of the sample, has a standard error approximated by the standard deviation of the sample divided by square root of 50, and has an approximately normal distribution.

d. Using $z = (\bar{x} - \mu)/(\sigma/\sqrt{n})$ with $\mu = 63.7$, $\sigma = 2.75$, $n = 50$ and $z = \pm 1.96$ (middle 95%):

$-1.96 = (\bar{x} - 63.7)/(2.75/\sqrt{50})$
$\bar{x} - 63.7 = (-1.96)(0.3889)$
$\bar{x} = 63.7 - 0.76$
$\bar{x} = 62.94$
and

$+1.96 = (\bar{x} - 63.7)/(2.75/\sqrt{50})$
$\bar{x} - 63.7 = (1.96)(0.3889)$
$\bar{x} = 63.7 + 0.76$
$\bar{x} = 64.46$

62.94 to 64.46

e.

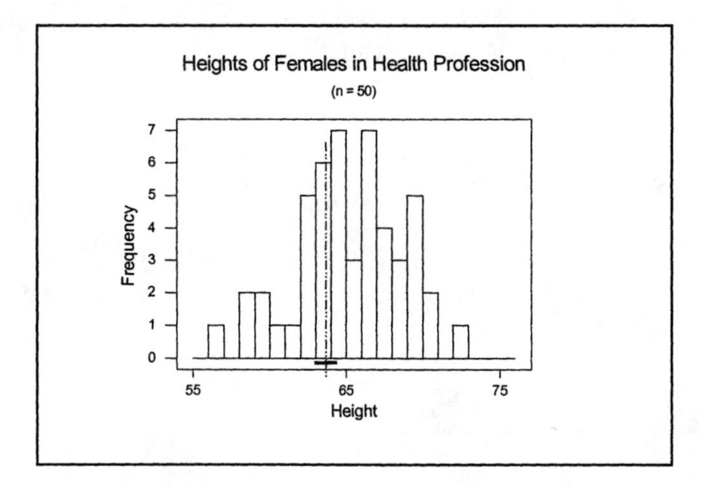

No, the sample mean, 64.72, falls outside of the interval 62.94 to 64.46. This means that this sample mean is not one of the 95% that are within the interval constructed in (d).

8.133 Answers will vary - everybody will be collecting their own data.

CHAPTER EXERCISES

8.135 a. It determines the value of z, the number of standard errors, that is used as the confidence coefficient.

b. The decrease in level of confidence will narrows the resulting interval estimate.

8.137 a. H_o: $\mu = 100$ b. H_a: $\mu \neq 100$

c. α = <u>0.01</u> d. μ = <u>100</u>

e. $\overline{x}$ = <u>96</u> f. σ = <u>12</u>

g. $\sigma_{\overline{x}} = 12/\sqrt{50} = 1.697 = \underline{1.70}$

h. $z* = (\overline{x} - \mu)/(\sigma/\sqrt{n}) = (96-100)/1.7 = \underline{-2.35}$

i. p-value = $2 \cdot P(z < -2.35) = 2 \cdot P(z > 2.35) = 2(0.5000 - 0.4906) = 2(0.0094) = \underline{0.0188}$

j. Fail to reject H_o

k. p-value = 0.0188

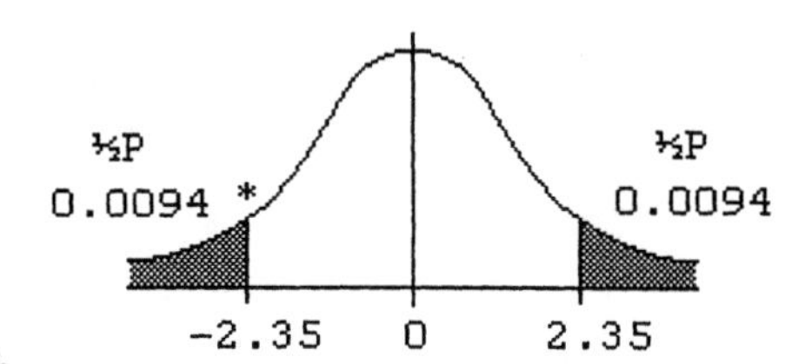

8.139 a. Step 1: The mean, μ

Step 2: a. normality assumed, CLT with n = 100

b. z, $\sigma = 5.0$ c. $1-\alpha = 0.95$

Step 3: $n = 100$, $\bar{x} = 40.6$

Step 4: a. $\alpha/2 = 0.05/2 = 0.025$; $z(0.025) = 1.96$

b. $E = z(\alpha/2) \cdot \sigma/\sqrt{n} = (1.96)(5/\sqrt{100})$

$= (1.96)(0.50) = 0.98$

c. $\bar{x} \pm E = 40.6 \pm 0.98$

Step 5: $\underline{39.6}$ $\underline{to}$ $\underline{41.6}$, the 0.95 confidence interval for μ

b. Step 1: a. The mean, μ

b. H_o: $\mu = 40$

H_a: $\mu \neq 40$

Step 2: a. normality assumed, CLT with n = 100

b. z, $\sigma = 5.0$ c. $\alpha = 0.05$

Step 3: a. $n = 100$, $\bar{x} = 40.6$

b. $z = (\bar{x} - \mu)/(\sigma/\sqrt{n})$

$z* = (40.6 - 40)/(5/\sqrt{100}) = 1.20$

Step 4: a. **P** = $2P(z > 1.20)$;

Using Table 3, Appendix B, ES9-p714:

P = $2(0.5000 - 0.3849) = 2(0.1151) = 0.2302$

Using Table 5, Appendix B, ES9-p716:

P = $2(0.1151) = 0.2302$

b. **P** > α

Step 5: a. Fail to reject H_o

b. At the 0.05 level of significance, there is not sufficient evidence to support the contention that the mean is not equal to 40.

c. Step 1: a. The mean, μ

b. H_o: $\mu = 40$

H_a: $\mu \neq 40$

Step 2: a. normality assumed, CLT with n = 100

b. z, $\sigma = 5.0$ c. $\alpha = 0.05$

Step 3: a. n = 100, $\overline{x} = 40.6$

b. $z = (\overline{x} - \mu)/(\sigma/\sqrt{n})$

$z^* = (40.6 - 40)/(5/\sqrt{100}) = 1.20$

Step 4: a. $\pm z(0.025) = \pm 1.96$

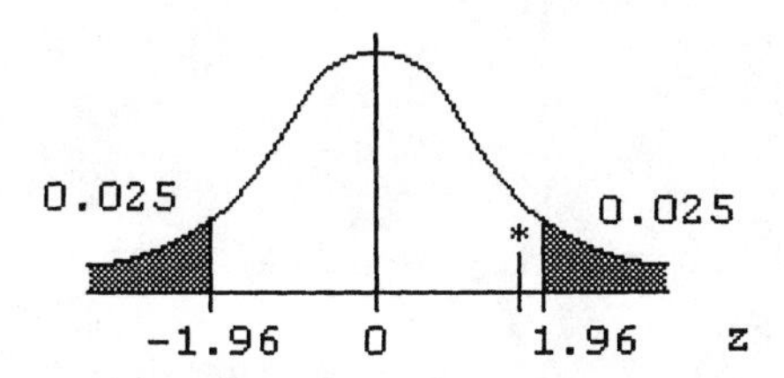

b. z* falls in the noncritical region, see * in Step 4a.

Step 5: a. Fail to reject H_o

b. At the 0.05 level of significance, there is not sufficient evidence to support the contention that the mean is not equal to 40.

d.

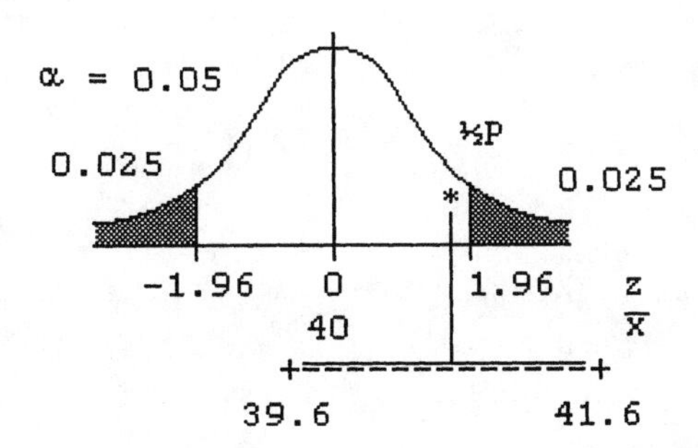

$z^* = 1.20$ is in the noncritical region or P = 0.2302 is greater than α, and $\mu = 40$ is within the interval estimate of 39.6 to 41.6.

8.141 a. Step 1: The mean, μ

Step 2: a. normality assumed, CLT with $n = 100$

b. z, $\sigma = 5.0$ c. $1-\alpha = 0.95$

Step 3: $n = 100$, $\bar{x} = 40.9$

Step 4: a. $\alpha/2 = 0.05/2 = 0.025$; $z(0.025) = 1.96$

b. $E = z(\alpha/2)\cdot\sigma/\sqrt{n} = (1.96)(5/\sqrt{100})$

$= (1.96)(0.50) = 0.98$

c. $\bar{x} \pm E = 40.9 \pm 0.98$

Step 5: 39.9 to 41.9, the 0.95 confidence interval for μ

b. Step 1: a. The mean, μ

b. H_o: $\mu = 40$ ($\leq$)

H_a: $\mu > 40$

Step 2: a. normality assumed, CLT with $n = 100$

b. z, $\sigma = 5.0$ c. $\alpha = 0.05$

Step 3: a. $n = 100$, $\bar{x} = 40.9$

b. $z = (\bar{x} - \mu)/(\sigma/\sqrt{n})$

$z* = (40.9 - 40)/(5/\sqrt{100}) = 1.80$

Step 4: a. **P** $= P(z > 1.80)$;

Using Table 3, Appendix B, ES8-p714:

P $= (0.5000 - 0.4641) = 0.0359$

Using Table 5, Appendix B, ES8-p716:

P $= 0.0359$

b. **P** $< \alpha$

Step 5: a. Reject H_o

b. At the 0.05 level of significance, there is sufficient evidence to support the contention that the mean is greater than 40.

c. Step 1: a. The mean, μ

b. H_o: $\mu = 40$ ($\leq$)

H_a: $\mu > 40$

Step 2: a. normality assumed, CLT with $n = 100$

b. z, $\sigma = 5.0$ c. $\alpha = 0.05$

Step 3: a. $n = 100$, $\bar{x} = 40.9$

b. $z = (\bar{x} - \mu)/(\sigma/\sqrt{n})$

$z* = (40.9 - 40)/(5/\sqrt{100}) = 1.80$

Step 4: a. z(0.05) = 1.65

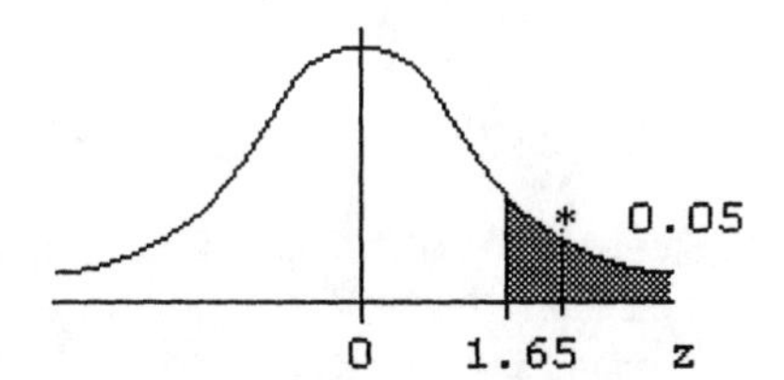

b. z* falls in the critical region, see Step 4a

Step 5: a. Reject H_o

b. At the 0.05 level of significance, there is sufficient evidence to support the contention that the mean is greater than 40.

d.

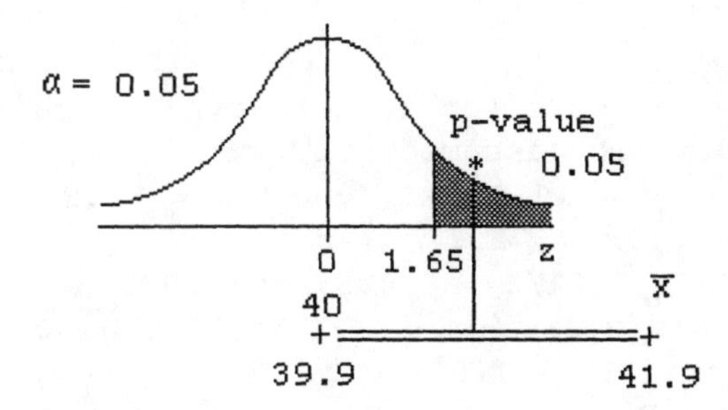

z^* = 1.80 is in the critical region and P = 0.0359 is less than α, and μ = 40 is within the interval estimate of 39.9 to 41.9. Since the interval is two-sided and the hypothesis tests are one-sided, it is hard to compare the estimate and the hypothesis tests.

> Exercise 8.143 shows the effect of the level of confidence (1 - α) on the width of a confidence interval.

8.143 a. Step 1: The mean weights of full boxes of a certain kind of cereal

Step 2: a. normality indicated

b. z, σ = 0.27 c. 1-α = 0.95

Step 3: n = 18, $\overline{x}$ = 9.87

Step 4: a. $\alpha/2 = 0.05/2 = 0.025$; $z(0.025) = 1.96$
b. $E = z(\alpha/2)\cdot\sigma/\sqrt{n} = (1.96)(0.27/\sqrt{18})$
$= (1.96)(0.0636) = 0.12$
c. $\bar{x} \pm E = 9.87 \pm 0.12$
Step 5: <u>9.75 to 9.99</u>, the 0.95 confidence interval for μ

b. Step 1: The mean weights of full boxes of a certain kind of cereal
Step 2: a. normality indicated
b. z, $\sigma = 0.27$ c. $1-\alpha = 0.99$
Step 3: $n = 18$, $\bar{x} = 9.87$
Step 4: a. $\alpha/2 = 0.01/2 = 0.005$; $z(0.005) = 2.58$
b. $E = z(\alpha/2)\cdot\sigma/\sqrt{n} = (2.58)(0.27/\sqrt{18})$
$= (2.58)(0.0636) = 0.16$
c. $\bar{x} \pm E = 9.87 \pm 0.16$
Step 5: <u>9.71 to 10.03</u>, the 0.99 confidence interval for μ

c. The increased confidence level widened the interval.

8.145 a. Step 1: The mean score for a clerk-typist position
Step 2: a. normality assumed, CLT with n = 100
b. z, $\sigma = 10.5$ c. $1-\alpha = 0.99$
Step 3: $n = 100$, $\bar{x} = 72.6$
Step 4: a. $\alpha/2 = 0.01/2 = 0.005$; $z(0.005) = 2.58$
b. $E = z(\alpha/2)\cdot\sigma/\sqrt{n} = (2.58)(10.5/\sqrt{100})$
$= (2.58)(1.05) = 2.71$
c. $\bar{x} \pm E = 72.6 \pm 2.71$
Step 5: <u>69.89 to 75.31</u>, the 0.99 confidence interval for μ

b. <u>Yes.</u> 75.0 falls within the interval.

8.147 Step 1: The mean efficacy expectation score for preoperative patients
Step 2: a. normality assumed, CLT with n = 200
b. z, $\sigma = 0.94$ c. $1-\alpha = 0.95$
Step 3: $n = 200$, $\bar{x} = 4.00$
Step 4: a. $\alpha/2 = 0.05/2 = 0.025$; $z(0.025) = 1.96$
b. $E = z(\alpha/2)\cdot\sigma/\sqrt{n} = (1.96)(0.94/\sqrt{200})$
$= (1.96)(0.066) = 0.13$
c. $\bar{x} \pm E = 4.00 \pm 0.13$
Step 5: <u>3.87 to 4.13</u>, the 0.95 confidence interval for μ

8.149 $n = [z(\alpha/2)\cdot\sigma/E]^2 = [(2.58)(\sigma)/(\sigma/3)]^2 = 59.9 = \underline{60}$

8.151 $z = (\overline{x} - \mu)/(\sigma/\sqrt{n})$

$z^* = (12.5 - 10.0)/(7.5/\sqrt{75}) = \underline{2.89}$

$\mathbf{P} = P(z > 2.89) = 0.5000 - 0.4981 = \underline{0.0019}$

8.153 a. H_o: $\mu = 0.50$
H_a: $\mu \neq 0.50$

b. $z^* = (0.51 - 0.50)/(0.04/\sqrt{25}) = 1.25$

$\mathbf{P} = 2\cdot P(z > 1.25) = 2(0.5000 - 0.3944) = \underline{0.2112}$

c. $z = \pm\ 2.33$

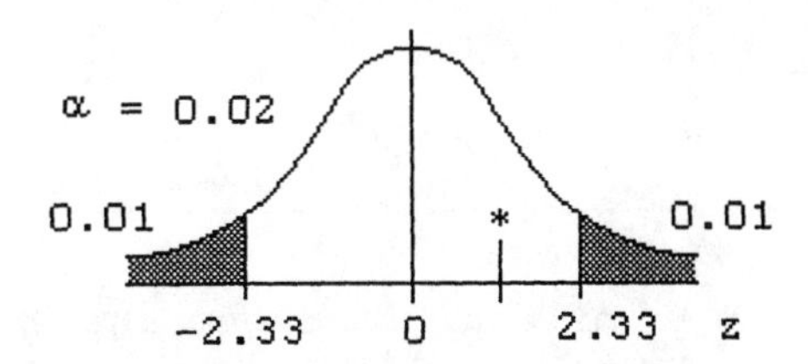

> **NOTE:** When both methods of hypothesis testing are asked for; steps 1, 2, 3 and 5 are the same, and step 4 is shown twice, p-value approach followed by the classical. Dashed lines are used to separate the answers.

8.155 Step 1: a. The mean weight of one load of pollen and nectar being carried by a worker bee to the hive after collecting it

b. H_o: $\mu = 0.0113$ ($\leq$)
H_a: $\mu > 0.0113$

Step 2: a. normality indicated, CLT with n = 200

b. z, $\sigma = 0.0063$ c. $\alpha = 0.01$

Step 3: a. $n = 200$, $\overline{x} = 0.0124$

b. $z = (\overline{x} - \mu)/(\sigma/\sqrt{n})$

$z^* = (0.0124 - 0.0113)/(0.0063/\sqrt{200}) = 2.47$

Step 4: -- using p-value approach ----------------------

a. $\mathbf{P} = P(z > 2.47)$;

Using Table 3, Appendix B, ES9-p714:
$\mathbf{P} = 0.5000 - 0.4932 = 0.0068$
Using Table 5, Appendix B, ES9-p716:
$0.0062 < P < 0.0071$

b. $\mathbf{P} < \alpha$

--- using classical approach ---------------------

a.

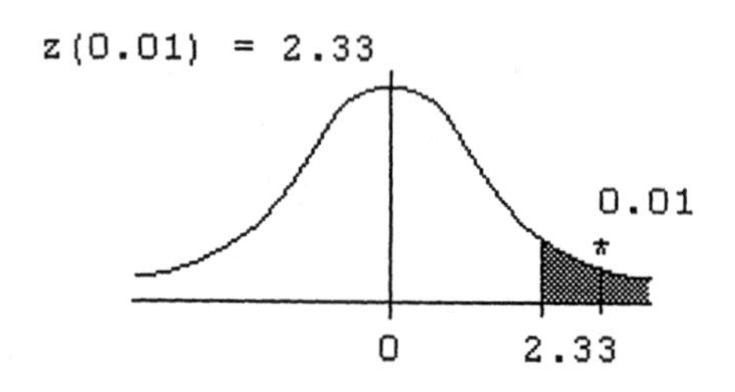

Step 5: a. Reject H_o

b. At the 0.01 level of significance, the sample does provide sufficient evidence to conclude that the mean load of pollen and nectar carried by Fuzzy's strain of Italian worker bees is greater than the rest of the honey bee population.

8.157 Step 1: a. The mean customer checkout time at a large supermarket

b. H_o: $\mu = 9$ ($\leq$)
H_a: $\mu > 9$

Step 2: a. normality indicated

b. z, $\sigma = 2.5$ c. $\alpha = 0.02$

Step 3: a. $n = 24$, $\overline{x} = 10.6$

b. $z = (\overline{x} - \mu)/(\sigma/\sqrt{n})$
$z^* = (10.6 - 9.0)/(2.5/\sqrt{24}) = 3.14$

Step 4: -- using p-value approach ----------------------

a. $\mathbf{P} = P(z > 3.14)$;

Using Table 3, Appendix B, ES9-p714:
$\mathbf{P} = 0.5000 - 0.4992 = 0.0008$
Using Table 5, Appendix B, ES9-p716:
$0.0008 < P < 0.0010$

b. $\mathbf{P} < \alpha$

-- using classical approach ----------------------

a.

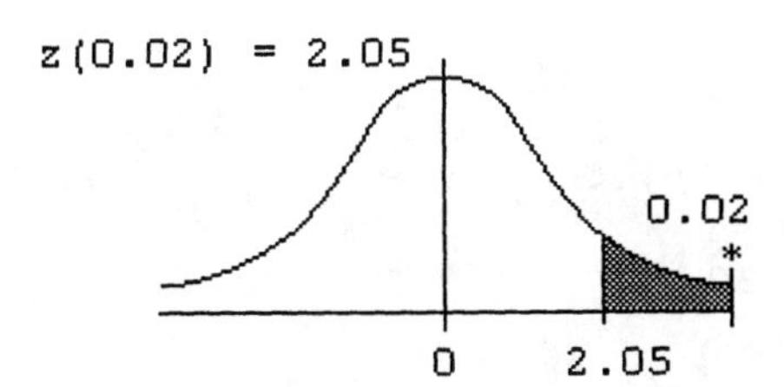

b. z* falls in the critical region, see Step 4a

Step 5: a. Reject H_o

b. At the 0.02 level of significance, the sample does provide sufficient evidence to conclude the mean waiting time is more than the claimed 9 minutes.

8.159 a. Step 1: a. The mean annual consumption of natural gas by residential customers

b. H_o: $\mu = 129.2$ ($\geq$)

H_a: $\mu < 129.2$

Step 2: a. normality assumed, CLT with n = 300

b. z, $\sigma = 18$ c. $\alpha = 0.01$

Step 3: a. n = 300, $\bar{x} = 127.1$

b. $z = (\bar{x} - \mu)/(\sigma/\sqrt{n})$

$z^* = (127.1 - 129.2)/(18/\sqrt{300}) = -2.02$

Step 4. a. **P** = P(z < -2.02) = P(z > 2.02);

Using Table 3, Appendix B, ES9-p714:

P = 0.5000 - 0.4783 = 0.0217

Using Table 5, Appendix B, ES9-p716:

0.0202 < P < 0.0228

b. **P** > α

Step 5: a. Fail to reject H_o

b. At the 0.01 level of significance, the sample does not provide sufficient evidence to conclude that the mean annual consumption has declined.

b. The p-value indicates the likelihood (approximately 0.02) of being wrong when you state that a significant reduction has not occurred.

8.161 a. (2) $H_a: r > A$ Failure to reject H_o will result in the drug being marketed. Because of the high current mortality rate, burden of proof is on the old ineffective drug.

b. (1) $H_a: r < A$ Failure to reject H_o will result in the new drug not being marketed. Because of the low current mortality rate, burden of proof is on the new drug.

8.163 a. 125.10 to 132.95, the 0.95 confidence interval for μ

b. $\sigma_{\bar{x}} = \sigma / \sqrt{n} = 10.0 / \sqrt{25} = 2.00$

$\bar{x} \pm z(\alpha/2) \cdot \sigma/\sqrt{n} = 129.02 \pm (1.96)(2.00)$
129.02 ± 3.92

<u>125.10 to 132.94</u>, the 0.95 confidence interval for μ

8.165 a. $H_a: \mu \neq 18$; Fail to reject H_o; The population mean is not significantly different from 18.

b. $\sigma_{\bar{x}} = \sigma / \sqrt{n} = 4.00 / \sqrt{28} = 0.756$

$z = (\bar{x} - \mu)/(\sigma/\sqrt{n})$

$z* = (17.217 - 18)/(0.756) = -1.04$

p-value $= 2 \cdot P(z < -1.04) = 2 \cdot P(z > 1.04)$
$= 2(0.5000 - 0.3508) = 2(0.1492) = 0.2984 = \underline{0.30}$

8.167 Every student will have different results, but they should be similar to the following.

a. <u>Using Minitab</u>:
The commands needed to obtain 50 rows/samples of 28 data per row/sample:
Calc > Random Data > Normal >
<u>50</u> rows in columns <u>C1-C28</u>; Mean <u>19</u>; Standard deviation <u>4</u>
The commands needed to obtain the 50 sample means:
Calc > Row Statistic ... >
Mean of <u>**C1-C28**</u>; Store results in <u>**C29**</u>
The commands to calculate z*:
Calc > Calculator ... >
Store result (z*) in <u>**C30**</u>;
Using expression <u>**(C29-18)/(4/SQRT(28))**</u>
To sort the z* values into ranked order:
Manip > Sort ... >
Sort column <u>**C30**</u>; Store in <u>**C31**</u>; Sort by <u>**C30**</u>

Using Excel:

The commands needed to obtain 50 rows/samples of 28 data per row/sample:

Tools > Data Analysis > Random Number Generation > OK

Number of Variables **28**; Number of Random Numbers **50**; Distribution Normal; Mean **19**; Standard deviation **4**; Output Range **A1**

The commands needed to obtain the 50 sample means:

Activate cell **AC1**

Insert function f_x > Statistical > AVERAGE > OK

Number1 **A1:AB1** > OK

Drag right corner of average value box down to give other averages

The commands to calculate z*:

Activate cell **AD1**

Edit Formula (=) > **(AC1-18)/(4/SQRT(28))**

b. Examine column C31 or AD1, count the z*'s that are less than -1.04 and greater than 1.04. In one run of the commands, 30/50 or 60% of the values were more extreme than the given z-values; On average, 30% should be more extreme (see 8.153b); an empirical value for the probability that z* is more extreme than the -1.04 in a two-tailed test when the true mean is actually 19.

c. Critical values = ±1.65; Examine column C31 or AD1, count the z*'s that are between -1.65 and 1.65. For one run of the commands, 33/50 = 66% fell in the noncritical region; an empirical β, probability of type II error.

CHAPTER 9 ∇ INFERENCES INVOLVING ONE POPULATION

Chapter Preview

Chapter 9 continues the work of inferential statistics started in Chapter 8. The concepts of hypothesis tests and confidence intervals will still be presented but on samples where the population standard deviation (σ) is unknown. Also inferences regarding the population binomial probability (p) and the population variance/standard deviation (σ^2/σ) will be introduced.

An article published by the U.S. Department of Health and Human Services about exercise is the topic of the Case Study in this chapter.

CHAPTER CASE STUDY

9.1 $n = 40$, $\Sigma x = 2140$, $\Sigma x^2 = 140{,}200$

a. $\bar{x} = 53.5$ minutes, $s = 25.68$

b. approximately 53.5 minutes

c. Not really! 53.5 minutes is more than a half hour less than the recommended 90 minutes (least amount).

SECTION 9.1 ANSWER NOW EXERCISES

9.2 Pick any 3 numbers; the fourth must be the negative of the sum of the first three. For example; 4, 3, 1, whose sum is 8; the fourth is required to be -8 for the sum to be zero. Three numbers were chosen freely, however there was no choice for the last number.

9.3 The bottom row of Table 6 is identical to the $z(\alpha)$ values in Table 4A. If σ is unknown, Table 6 is used, but note that for df > 100, the critical values for Student-t are approximately the same as those of the standard normal. Remember, the t-distribution approaches the standard normal distribution as the number of degrees of freedom increase.

9.4 a. t(12, 0.01) = 2.68 b. t(22, 0.025) = 2.07

9.5 a. t(18, 0.90) = -t(18, 0.10) = -1.33
b. t(9, 0.99) = -t(9, 0.01) = -2.82

9.6 $\alpha/2 = 0.05/2 = 0.025$; $\pm t(12, 0.025) = \pm 2.18$

9.7 Use the cumulative probability distribution function, Student's t distribution with 18 DF

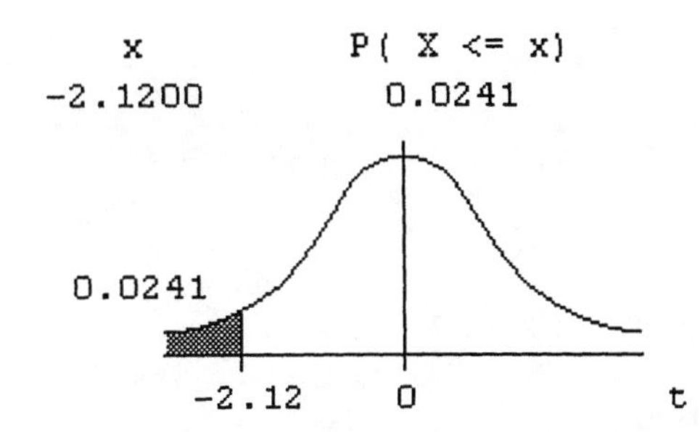

9.8 Use the cumulative probability distribution function, then find the complement; Student's t distribution with 15 DF

x	P(X <= x)
1.1200	0.8598

1.0 - 0.8598 = 0.1402

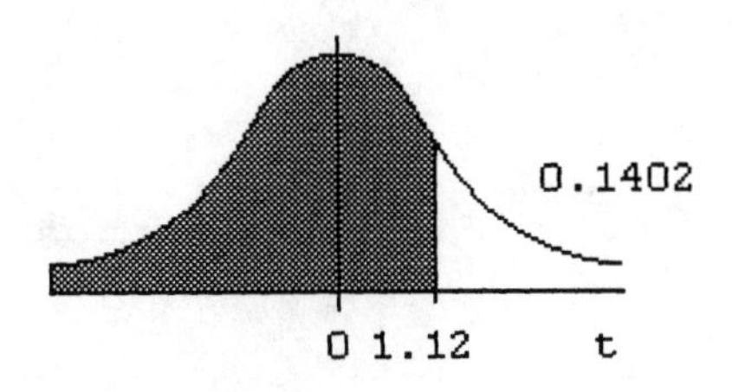

9.9 Step 1: The mean, μ

Step 2: a. normality assumed

b. t   c. $1-\alpha = 0.95$

Step 3: $n = 24$, $\overline{x} = 16.7$, $s = 2.6$

Step 4: a. $\alpha/2 = 0.05/2 = 0.025$; df = 23; (23, 0.025) = 2.07

b. $E = t(df,\alpha/2)\cdot(s/\sqrt{n}) = (2.07)(2.6/\sqrt{24})$

$= 1.0986 = 1.10$

c. $\overline{x} \pm E = 16.7 \pm 1.10$

Step 5: <u>15.60 to 17.8</u>, the 0.95 confidence interval for μ

9.10

Variable	N	Mean	StDev	SE Mean	98.0 % CI
C1	12	7.750	2.137	0.617	6.073, 9.427)

9.11 $\mu = 32$, $n = 16$, $\overline{x} = 32.93$, $s = 3.1$

$t = (\overline{x} - \mu)/(s/\sqrt{n})$

$t^* = (32.93 - 32)/(3.1/\sqrt{16}) = \underline{1.20}$

9.12 $\mathbf{P} = P(t > 1.20 | df = 15)$;

Using Table 6, Appendix B, ES9-p717:

$0.10 < \mathbf{P} < 0.25$

Using Table 7, Appendix B, ES9-p718:

$\mathbf{P} = 0.124$

$\alpha = 0.05$; $\mathbf{P} > \alpha$; <u>Fail to reject H_o</u>

9.13 $t(15, 0.05) = 1.75$; $t^* = 1.20$

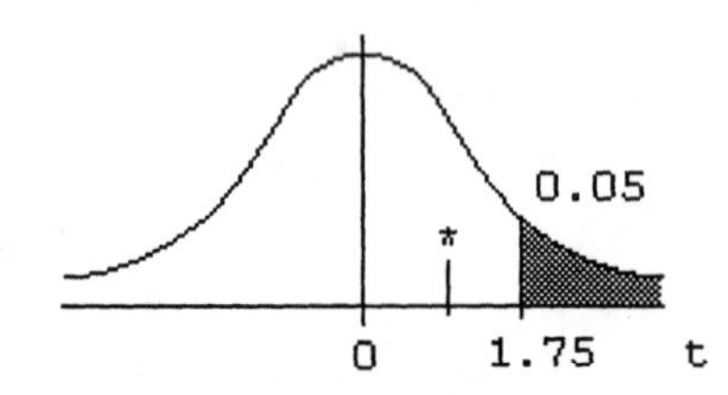

t^* falls in the noncritical region; <u>Fail to reject H_o</u>

9.14 $\mu = 73$, $n = 12$, $\overline{x} = 71.46$, $s = 4.1$

$t = (\overline{x} - \mu)/(s/\sqrt{n})$

$t^* = (71.46 - 73)/(4.1/\sqrt{12}) = \underline{-1.30}$

9.15 $\mathbf{P} = P(t < -1.30 | df = 11) + P(t > 1.30 | df = 11)$
$= 2P(t > 1.30 | df = 11)$;
Using Table 6, Appendix B, ES9-p717:
$0.20 < \mathbf{P} < 0.50$
Using Table 7, Appendix B, ES9-p718:
$\mathbf{P} = 2(0.110) = 0.220$
$\alpha = 0.05$; $\mathbf{P} > \alpha$; <u>Fail to reject H_o</u>

9.16 $\pm t(11, 0.025) = \pm 2.20$; $t^* = -1.30$

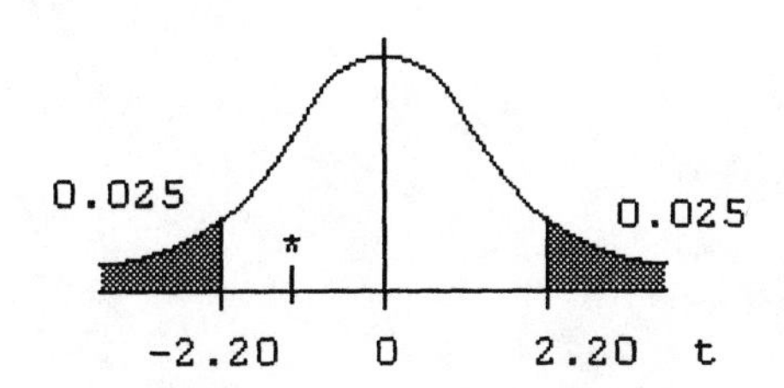

t^* falls in the noncritical region; <u>Fail to reject H_o</u>

9.17 Use the cumulative probability distribution function to find ½**P**, Student's t distribution with 11 DF

x	P(X <= x)
-1.3000	0.1101

$\mathbf{P} = 2P(t^* < -1.30 | df = 11) = 2(0.1101) = \underline{0.2202}$

9.18 Test of mu = 52.00 vs mu < 52.00

Variable	N	Mean	StDev	SE Mean	T	P
C1	12	49.75	5.48	1.58	-1.42	0.091

$\alpha = 0.01$; $P > \alpha$; Fail to reject H_o
OR
$-t(11, 0.01) = -2.72$;
$t^* = -1.42$ falls in the noncritical region; Fail to reject H_o

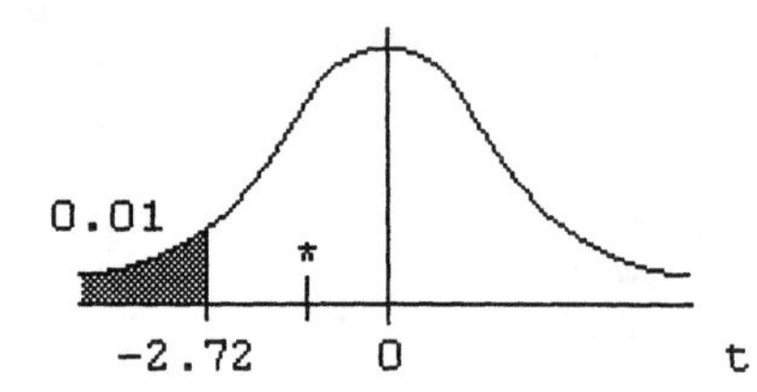

9.19 a. $\mathbf{P} = P(t > 1.92 | df = 44)$
using Table 6: $0.025 < \mathbf{P} < 0.05$
using Table 7: $0.026 < \mathbf{P} < 0.033$
$\alpha = 0.05$, $\mathbf{P} \le \alpha$, therefore it is significant

b. $\mathbf{P} = P(t > 3.41 | df = 44)$
using Table 6: $\mathbf{P} < 0.005$ [not enough information given]
using Table 7: $\mathbf{P} = 0.001$
$\alpha = 0.001$, $\mathbf{P} \le \alpha$, therefore it is significant

c. $\mathbf{P} = P(t > 1.81 | df = 44)$ - two-tailed
using Table 6: $0.050 < \mathbf{P} < 0.100$
using Table 7: $0.064 < \mathbf{P} < 0.080$

To be significant, the smallest possible p-value is desired. For a two-tailed test and Table 6, $0.05 < p < 0.10$, i.e., $p < 0.10$ would be reported. For a one-tailed test, $0.025 < p < 0.05$, i.e., $p < 0.05$ would be reported.

SECTION 9.1 EXERCISES

t-Distribution
(used when σ is unknown)

Key facts about the t-distribution:

1. The total area under the t-distribution is 1.
2. It is symmetric about 0.
3. Its shape is a more "spread out" version of the normal shape.
4. A different curve exists for each sample size.
5. The shape of the distribution approaches the normal distribution shape as n increases [For df > 100, t is approximately normal.].
6. Critical values are determined based on α and degrees of freedom(df) - Table 6 (Appendix B, ES9-p717).
7. Degrees of freedom is abbreviated as *df*, where df = n - 1 for this application.

Explore the t-distribution for different degrees of freedom using Interactivity 9A.

Notation: $t(df,\alpha)$ = t(degrees of freedom, area to the right)
↑ Table 6 ↑ row id # ↑ column id #

ex.: $t(13,.025)$ means df = 13 (row) and α = .025 (column), using

Table 6, $t(13,.025) = 2.16$ (df = n-1)

For $t(df,\alpha)$, consider the α given as the amount in one tail and use the top row label - "Amount of α in One-Tail". For two-tailed tests, an additional row label is given - "Amount of α in Two-Tails". Note that it is twice the amounts in the one-tail row, therefore α does not have to be divided by two.
For $\alpha > 0.5000$, use the $1-\alpha$ amount and negate the t-value.

ex.: $t(14,0.90)$; $\alpha = 0.90$, $1-\alpha = 0.10$,
$t(14,0.90) = -t(14,0.10) = -1.35$
(Table 6 is in Appendix B, ES9-p717.)

9.21 a. -1.72 b. -2.06 c. -2.47 d. 2.01

For a two-sided test:

1. divide α by 2 and use the top row of column labels of Table 6 (Appendix B, ES9-p717) identified as "Amount of α ($\alpha/2$, in this case) in One-Tail"

or

2. use the second row of column labels of Table 6 identified as "Amount of α in Two-Tails."

Drawing a diagram of a t-curve and labeling the regions with the given information will be helpful in answering exercises 9.23 and 9.25.

9.23 df = 7

9.25 a. -2.49 b. 1.71 c. -0.685

9.27 a. Symmetric about mean: mean is 0

b. Standard deviation of t-distribution is greater than 1; t-distribution is different for each different sample size while there is only one z-distribution, t has df.

Estimating μ - the population mean

(σ unknown)

1. point estimate: $\bar{x}$

2. confidence interval: $\bar{x} \pm t(df, \alpha/2) \cdot (s/\sqrt{n})$, where df = n-1

Review steps for constructing a confidence interval for μ: ES9-p350, ST-p229. The t-distribution is used when σ is unknown, and sampling is from an approximately normal distribution or the sample size is large.

9.29 Step 1: The mean age of onset of offending for those convicted of antitrust offenses

Step 2: a. Age distribution presumed normal for this offense

b. t c. $1-\alpha = 0.90$

Step 3: $n = 35$, $\bar{x} = 54$, $s = 7.5$

Step 4: a. $\alpha/2 = 0.10/2 = 0.05$; df = 34; $t(34, 0.05) = 1.70$

b. $E = t(df,\alpha/2)\cdot(s/\sqrt{n}) = (1.70)(7.5/\sqrt{35})$
$= 2.15514 = 2.16$

c. $\bar{x} \pm E = 54 \pm 2.16$

Step 5: <u>51.84 to 56.16</u>, the 0.90 confidence interval for μ

9.31 a. $\bar{x} = \Sigma x/n = 1962.26/41 =$ <u>\$47.86</u>

b. $s = \sqrt{[\Sigma(x-\bar{x})^2 / (n-1)} = \sqrt{8202.496/40} =$ <u>\$14.32</u>

Computer and/or calculator commands to calculate a confidence interval for the population mean, μ, when the population standard deviation, σ, is unknown can be found in ES9-p423.

9.33 Step 1: The mean percentage change in number of adoptions from 1993 to 1997

Step 2: a. The sampled population appears to be normally distributed.

b. t c. $1-\alpha = 0.95$

Step 3: $n = 20$, $\bar{x} = 3.95$, $s = 14.02$

Step 4: a. $\alpha/2 = 0.05/2 = 0.025$; df = 19; $t(19, 0.025) = 2.09$

b. $E = t(df,\alpha/2)\cdot(s/\sqrt{n}) = (2.09)(14.02/\sqrt{20}) = 6.55$

c. $\bar{x} \pm E = 3.95 \pm 6.55$

Step 5: <u>-2.60 to 10.50</u>, the 0.95 confidence interval for μ

9.35 Verify - answers given in exercise.

Hypotheses are written the same way as before. Sample size and standard deviation have no effect on the stating of hypotheses.

9.37 a. H_o: $\mu = 11$ ($\geq$) vs. H_a: $\mu < 11$
b. H_o: $\mu = 54$ ($\leq$) vs. H_a: $\mu > 54$
c. H_o: $\mu = 75$ vs. H_a: $\mu \neq 75$

Calculating the **P**-value using the t-distribution

Table 6 or Table 7 (Appendix B, ES9-pp717&718) or a computer and/or calculator, can be used to estimate the p-value

1. Using Table 6 to place bounds on the value of **P**
   a) locate df row
   b) locate the absolute value of the calculated t-value between two critical values in the df row
   c) the p-value is in the interval between the two H_a.

2. Using Table 7 to estimate or place bounds on the value of **P**
   a) locate the absolute value of the calculated t-value and the df directly for the corresponding probability value

   OR

   b) locate the absolute value of the calculated t-value and its df between appropriate bounds. From the box formed, use the upper left and lower right values for the interval. (see ES9-p424)

3. Using a computer and/or calculator
   a) the p-value is calculated directly and given in the output when completing the hypothesis test (see ES9-p424)

   OR

   b) the p-value is calculated using the cumulative probability commands:
   Subtract the probability value from 1 or multiply it by 2, depending on the exercise. The cumulative probability given is P(t ≤ t-value). (see ES9-pp416&417)

Draw a picture as before, of an "approximately" normal distribution curve. Shade in the critical regions based on the alternative hypothesis (H_a). Using α and df, find the critical value(s) using Table 6 (Appendix B, ES9-p717).

9.39 a.

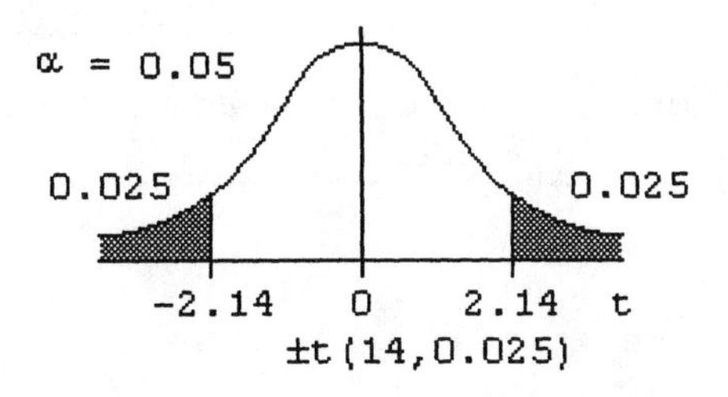

b.

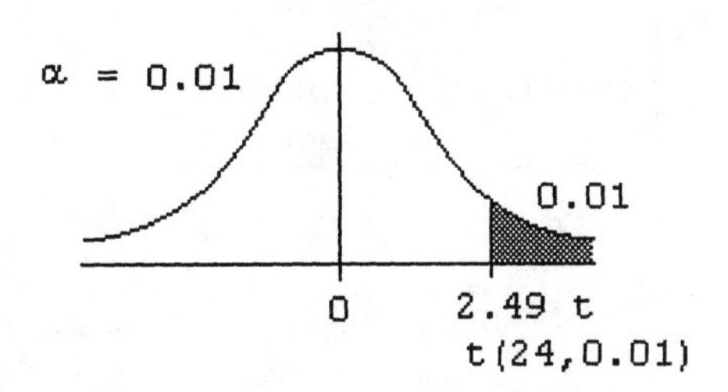

c.

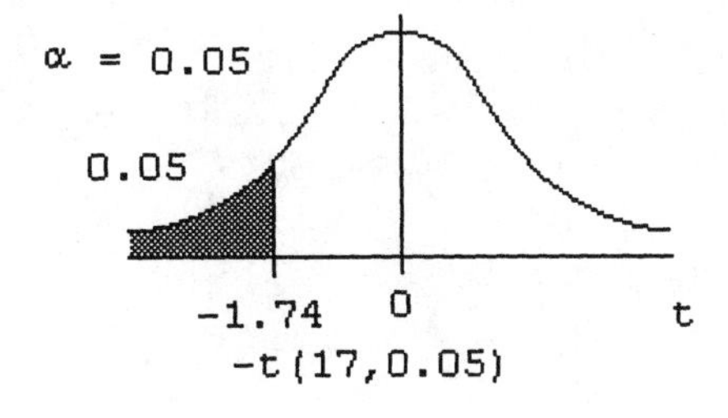

d.

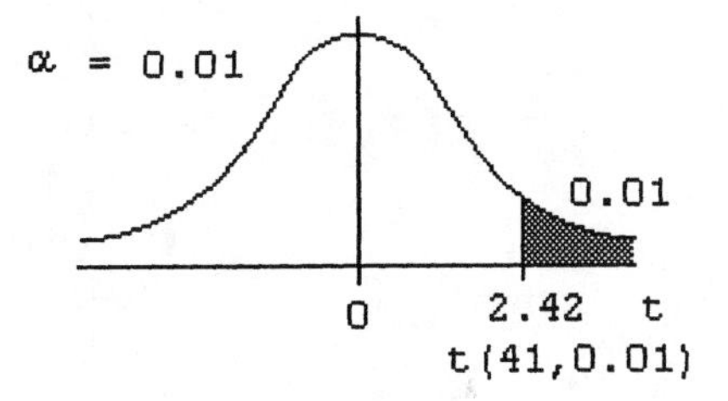

Hint for writing the hypotheses for exercise 9.41

Look at the first sentence of the exercise, "A student group maintains that the average student must travel for at least 25 minutes ...". "At least" indicates greater than or equal to (≥). Since the greater than or equal to includes the equality, it belongs in the null hypothesis. Continue to write the null hypothesis with the equal sign (=), but include the greater-than or equal-to sign (≥) in parentheses after it. The negation is "..less than", hence the alternative hypothesis must have a less-than sign (<).

Hypothesis tests, (p-value approach), will be completed in the same format as before. You may want to review: ES9-p370, ST-p241. The only differences are in:

1. the calculated test statistic, which is t, where
   $t = (\bar{x} - \mu)/(s/\sqrt{n})$

2. using Table 6 or Table 7 (Appendix B, ES9-pp717&718) to estimate the p-value

Remember: σ (population standard deviation) is unknown, therefore s (sample standard deviation) is used.

9.41 a. P-value approach:

Step 1: a. The mean travel time to college
  b. H_o: $\mu = 25$ (at least) ($\geq$)
     H_a: $\mu < 25$ (less than)

Step 2: a. Travel times are mounded; CLT is satisfied with n = 31
  b. t   c. $\alpha = 0.01$

Step 3: a. $n = 31$, $\bar{x} = 19.4$, $s = 9.6$
  b. $t = (\bar{x} - \mu)/(s/\sqrt{n})$
     $t* = (19.4 - 25.0)/(9.6/\sqrt{31}) = -3.25$

Step 4: a. $\mathbf{P} = P(t < -3.25 | df = 30) = P(t > 3.25 | df = 30)$;
     Using Table 6, Appendix B, ES9-p717:
     $\mathbf{P} < 0.005$
     Using Table 7, Appendix B, ES9-p718:
     $0.001 < \mathbf{P} < 0.002$
  b. $\mathbf{P} < \alpha$

Step 5: a. Reject H_o
  b. At the 0.01 level of significance, the sample does provide sufficient evidence to justify the contention that mean travel time is less than 25 minutes.

Hypothesis tests (classical approach) will be completed using the same format as before. You may want to review: ES9-pp.386-387, ST-p249. The only differences are in:

1. finding the critical value(s) of t
   remember you need α (column) and degrees of

2. the calculated test statistic, which is t, where
   $t = (\bar{x} - \mu)/(s/\sqrt{n})$.

b. Classical approach:

Step 1: a. The mean travel time to college

b. H_o: $\mu = 25$ (at least) ($\geq$)

H_a: $\mu < 25$ (less than

Step 2: a. Travel times are mounded; assume normality, CLT with n = 31

b. t   c. $\alpha = 0.01$

Step 3: a. $n = 31$, $\bar{x} = 19.4$, $s = 9.6$

b. $t = (\bar{x} - \mu)/(s/\sqrt{n})$

$t* = (19.4 - 25.0)/(9.6/\sqrt{31}) = -3.25$

Step 4: a. $-t(30, 0.01) = -2.46$

b. t* falls in the critical region, see Step 4a

Step 5: a. Reject H_o

b. At the 0.01 level of significance, the sample does provide sufficient evidence to justify the contention that mean travel time is less than 25 minutes.

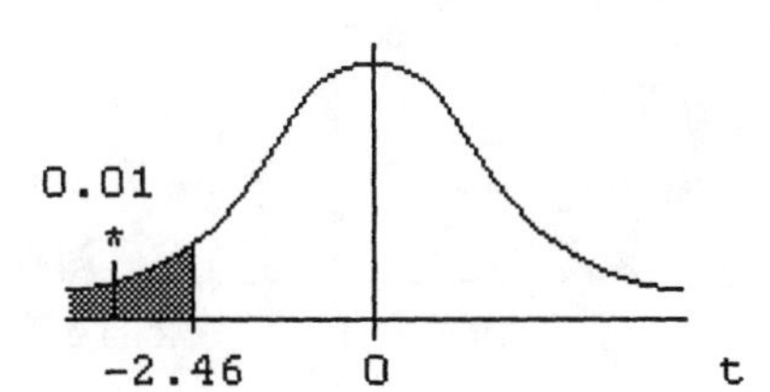

<u>Hint for writing the hypotheses for exercise 9.43</u>

Look at the last sentence of the exercise, "Assuming that the sample .. test the hypothesis of '<u>different from</u>' using a level of significance of 0.05". The words "different from" indicate a different than or <u>not equal to</u> ($\neq$). The negation becomes "equal to" and the null hypothesis would be written with an equality sign (=).

9.43 a. The "population" data ranged from 6% to 71.6%, therefore the midrange is 38.8%. When the midrange is close in value to the mean, the distribution is approximately symmetrical; therefore, the assumption of normality is reasonable.

b & c.

Step 1: a. The mean percentage intake of kilocalories from fat

b. H_o: $\mu = 38.4\%$

H_a: $\mu \neq 38.4\%$

Step 2: a. normality indicated

b. t   c. $\alpha = 0.05$

Step 3: a. $n = 15$, $\bar{x} = 40.5\%$, $s = 7.5\%$

b. $t = (\bar{x} - \mu)/(s/\sqrt{n})$

$t* = (40.5 - 38.4)/(7.5/\sqrt{15}) = 1.08$

Step 4: -- using p-value approach --------------------

a. $\mathbf{P} = 2P(t > 1.08 | df = 14)$;

Using Table 6, Appendix B, ES9-p717:

$0.20 < \mathbf{P} < 0.50$

Using Table 7, Appendix B, ES9-p718:

$0.144 < \frac{1}{2}\mathbf{P} < 0.169$; $0.288 < \mathbf{P} < 0.338$

b. $\mathbf{P} > \alpha$

-- using classical approach -----------------

a. $\pm t(14, 0.025) = \pm 2.14$

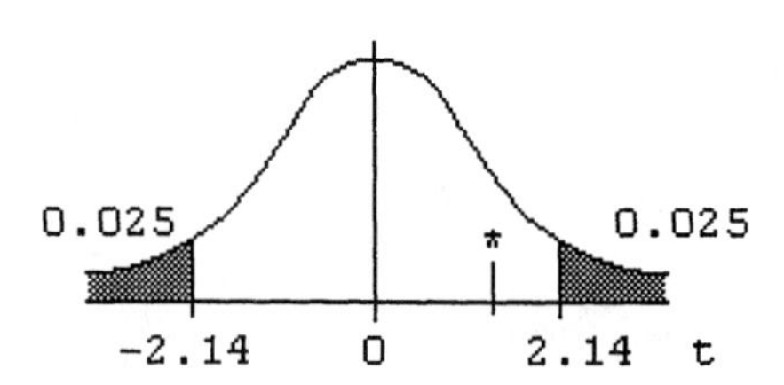

b. t* falls in the noncritical region, see Step 4a

Step 5: a. Fail to reject H_o

b. The sample does not provide sufficient evidence to justify the contention that the mean percentage is different than 38.4%, at the 0.05 level of significance.

Hint for writing the hypotheses for exercise 9.45

Look at the first and second sentences in the exercise, "It is claimed ... average of 35 on a given test" and "Is the claim reasonable ...?". The words "average of 35" indicate equal to (=) 35. Equal to belongs in the null hypothesis. The negation becomes "NOT equal to," hence the alternative hypothesis must have a not-equal-to sign (≠). Since neither sentence gave any indication of greater or less than, the two-sided, not equal to (≠), approach is assumed.

9.45 Sample statistics: $n = 6$, $\Sigma x = 222$, $\Sigma x^2 = 8330$, $\bar{x} = 37.0$, $s = 4.817$

Step 1: a. The mean test score at a certain university

b. H_o: $\mu = 35$ (reasonable)

H_a: $\mu \neq 35$ (not reasonable)

Step 2: a. normality indicated

b. t c. $\alpha = 0.05$

Step 3: a. $n = 6$, $\bar{x} = 37.0$, $s = 4.817$

b. $t = (\bar{x} - \mu)/(s/\sqrt{n})$

$t* = (37.0 - 35.0)/(4.817/\sqrt{6}) = 1.02$

Step 4: -- using p-value approach --------------------

a. **P** $= 2P(t > 1.02 | df = 5)$;

Using Table 6, Appendix B, ES9-p717:

$0.20 <$ **P** < 0.50

Using Table 7, Appendix B, ES9-p718:

$0.161 <$ ½**P** $< 0.182]$ $0.322 <$ **P** < 0.364

b. **P** $> \alpha$

-- using classical approach -----------------

a. $\pm t(5, 0.025) = \pm 2.57$

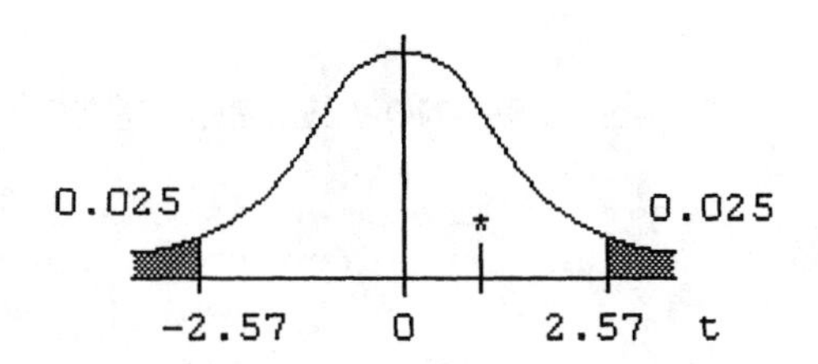

b. t* falls in the noncritical region, see Step 4a

Step 5: a. Fail to reject H_o

b. The sample does not provide sufficient evidence to reject the claim that the mean score is 35, at the 0.05 level of significance.

> Computer and/or calculator commands to perform a hypothesis test for a population mean if the population standard deviation (σ) is unknown can be found in ES9-p418.
>
> The alternative command works the same as in the z-distribution (σ known) command. The output will also look the same except a t-value will be calculated in place of the z-value.

9.47 Verify - answers given in exercise.

9.49 Summary of results:
$n = 35$, $\bar{x} = 18.8142$, $s = 0.0296$, se mean = 0.0050
H_o: $\mu = 18.810$ vs. H_a: $\mu \neq 18.810$
$t* = \underline{0.84}$, $P = \underline{0.41}$; $t \leq -2.75$, $t \geq 2.75$

Fail to reject H_o; the sample does not provide sufficient evidence to reject the claim that mean length is 18.810, at the 0.01 level.

> Computer commands to perform the simulation from the various distributions can be found on pages 266-267 of this manual. Slight variations need to be made for t* versus the z* that was used in Chapter 8.

SECTION 9.2 ANSWER NOW EXERCISES

9.51 a. Yes, it seems likely that the mean of the observed proportions would be the true proportion for the population.

b. Unbiased because the mean of the p′ distribution is p, the parameter being estimated.

9.52 $\sqrt{npq}/n = \sqrt{npq}/\sqrt{n^2} = \sqrt{npq/n^2} = \sqrt{pq/n}$

9.53 a. $\sqrt{p'q'/n} = \sqrt{(0.23)(0.77)/400} = \underline{0.02104}$

b. Step 1: The proportion of convertibles driven by students

Step 2: a. The sample was randomly selected and each subject's response was independent of those of the others surveyed.

b. $n = 400$; $n > 20$, $np = (400)(92/400) = 92$, $nq = (400)(308/400) = 308$, np and nq both > 5

c. $1 - \alpha = 0.95$

Step 3: $n = 400$, $x = 92$, $p' = x/n = 92/400 = 0.23$

Step 4: a. $z(\alpha/2) = z(0.025) = 1.96$

b. $E = z(\alpha/2) \cdot \sqrt{p'q'/n} = 1.96 \cdot \sqrt{(0.23)(0.77)/400}$
$= (1.96)(0.02104) = 0.041$

c. $p' \pm E = 0.23 \pm 0.041$

Step 5: $\underline{0.189 \text{ to } 0.271}$ is the 0.95 interval for p = P(drives convertible)

9.54 a. $E = z(\alpha/2) \cdot \sqrt{p'q'/n} = 1.96 \cdot \sqrt{(0.5)(0.5)/1000}$
$= (1.96)(0.015811) = 0.031$

b. The value of the margin of error is set by using a rounded value of the maximum error.

c. No bias results from the interviewing process;

d. In section, the "success" outcome is clearly identifiable, whereas in polling the outcomes are more subjective and possible error may easily be introduced;

e. Measuring the error caused by non-response, interviewer error, and so on, is not possible.

9.55 Step 1: $1 - \alpha = 0.95$; $z(\alpha/2) = z(0.025) = 1.96$

Step 2: $E = 0.02$

Step 3: no estimate given, $p^* = 0.5$ and $q^* = 0.5$

Step 4: $n = \{[z(\alpha/2)]^2 \cdot p^* \cdot q^*\}/E^2$

$= (1.96^2)(0.5)(0.5)/(0.02^2) = \underline{2401}$

9.56 Step 1: $1 - \alpha = 0.90$; $z(\alpha/2) = z(0.05) = 1.65$

Step 2: $E = 0.02$

Step 3: $p^* = 0.25$ and $q^* = 0.75$

Step 4: $n = \{[z(\alpha/2)]^2 \cdot p^* \cdot q^*\}/E^2$
$= (1.65^2)(0.25)(0.75)/(0.02^2) = 1276.17 = \underline{1277}$

9.57 $p = 0.70$, $n = 300$, $x = 224$, $p' = x/n = 224/300 = 0.747$

$z = (p' - p)/\sqrt{pq/n}$

$z* = (0.747 - 0.70)/\sqrt{(0.7)(0.3)/300} = 1.78$

9.58 $\mathbf{P} = P(z > 1.78)$;

Using Table 3, Appendix B, ES9-p714:

$\mathbf{P} = 0.5000 - 0.4625 = 0.0375$

Using Table 5, Appendix B, ES9-p716:

$0.0359 < \mathbf{P} < 0.0401$

$\alpha = 0.05$; $\mathbf{P} < \alpha$; Reject H_o

9.59 $z(0.05) = 1.65$

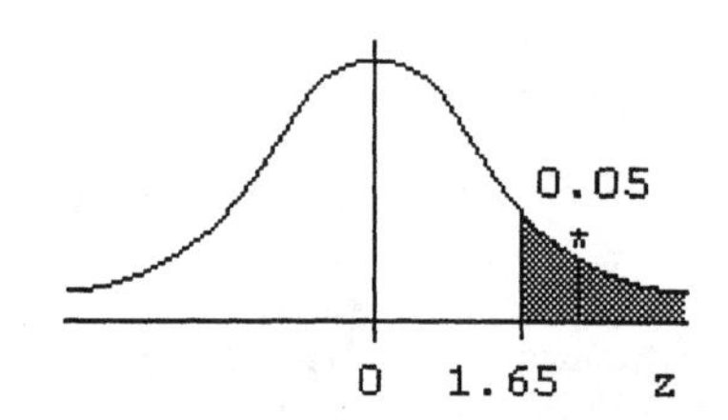

z* falls in the critical region; Reject H_o

9.60 The 90% confidence interval for Illustration 9.7 for p was $0.052 < p < 0.118$. 15%, the claimed p, is not contained within this 90% confidence interval.

9.61 a. $P(x \geq 140|B(250, 0.50)) = 1 - P(x \leq 139)$

$= 1.0000 - 0.9668 = \underline{0.0332}$

b. $P(x \leq 110|B(250, 0.50)) = \underline{0.0332}$

c. P(more extreme) $= 0.0332 + 0.0332 = \underline{0.0664}$

d. Coenen's claim: "struck evenly or balanced" or H_o: $P(H) = 0.5$,

Blight's suspicion: "not balanced" or H_a: $P(H) \neq 0.5$;

e. "If the coin were unbiased" is the statement of the null hypothesis. The probability of "getting a result as extreme" is the definition of the p-value.

f. 0.560 ± 0.062 or 0.498 to 0.622

g. the 6% is a maximum error of estimate since the results are from a binomial experiment and not a survey - therefore there is no chance for bias due to an interviewing process

SECTION 9.2 EXERCISES

p' = sample proportion	p' = x/n
x = number of successes	
n = sample size (number of independent trials)	

9.63 a. Only two possible outcomes, therefore the "opposite" or "not success" is equal to failure. p + q cover the entire outcome list of possibilities for the experiment. p + q = 1

b. p + q = 1 is equivalent to q = 1 - p

c. q = 1 - p = 1 - 0.6 = <u>0.4</u>

d. p' = 1 - q' = 1 - 0.273 = <u>0.727</u>

9.65 a. n = 1,027 people surveyed, a trial is the surveying of each person, success is when they "say that ketchup is their preferred burger condiment", p = P(say ketchup), x = number of surveyed 1,027 people who say ketchup

b. 0.47 (47%). It is a statistic. It is being used to estimate the parameter.

c. $1.96 \cdot \sqrt{(0.47)(0.53)/1027} = 1.96(0.0156) = 0.031$

d. They have the same value.

e. 0.47 ± 0.031 or 0.439 to 0.501

Estimating p - the population proportion

1. point estimate: p' = x/n

2. confidence interval: $\underbrace{p'}_{\text{point estimate}} \pm \underbrace{z(\alpha/2) \cdot \sqrt{p'q'/n}}_{\text{maximum error of estimate}}$

Computer and/or calculator commands to calculate a confidence interval for the population proportion, p, can be found in ES9-p.434

Review: ES9-p350, ST-p229, "The Confidence Interval: A Five-Step Procedure" if necessary.

Explore the relationship betweem a z* value and its corresponding confidence interval using Interactivity 9B.

9.67 Step 1: The proportion of students that support the proposed budget amount

Step 2: a. The sample was randomly selected and each subject's response was independent of those of the others surveyed.

b. $n = 60$; $n > 20$, $np = (60)(22/60) = 22$, $nq = (60)(38/60) = 38$, np and nq both > 5

c. $1 - \alpha = 0.99$

Step 3: $n = 60$, $x = 22$, $p' = x/n = 22/60 = 0.367$

Step 4: a. $z(\alpha/2) = z(0.005) = 2.58$

b. $E = z(\alpha/2)\cdot\sqrt{p'q'/n} = 2.58\sqrt{(0.367)(0.633)/60}$
$= (2.58)(0.0622) = 0.161$

c. $p' \pm E = 0.367 \pm 0.161$

Step 5: <u>0.206 to 0.528</u>, the 0.99 interval for $p = P(\text{favor budget})$

9.69 Step 1: The proportion of managers and professionals who work late five days a week

Step 2: a. The sample was randomly selected and each subject's response was independent of those of the others surveyed.

b. $n = 1742$; $n > 20$, $np = (1742)(0.278) = 484.3$, $nq = (1742)(0.722) = 1257.7$, np and nq both > 5

c. $1 - \alpha = 0.99$

Step 3: $n = 1742$, x not given, $p' = 0.278$ (given)

Step 4: a. $z(\alpha/2) = z(0.005) = 2.58$

b. $E = z(\alpha/2)\cdot\sqrt{p'q'/n} = 2.58\cdot\sqrt{(0.278)(0.722)/1742}$
$= (2.58)(0.0107) = 0.028$

c. $p' \pm E = 0.278 \pm 0.028$

Step 5: <u>0.250 to 0.306</u>, the 0.99 interval for $p = P(\text{working late five days a week})$

9.71 There are many possible reasons why the results could be biased. Here are a few:

1. Many people will not reveal information about credit cards to telephone callers.
2. Many people will not ever admit to being "easy prey" to offers like this.
3. Many people answer questions with answers they think the caller wants to hear hoping to end the call quickly.

9.73 a. $E = 1.96\sqrt{(0.70)(0.30)/1020} = 1.96(0.01435) = 0.028$

$E = 1.96\sqrt{(0.57)(0.43)/1019} = 1.96(0.01551) = 0.030$

$E = 1.96\sqrt{(0.24)(0.76)/1020} = 1.96(0.01337) = 0.026$

b. The variation was caused by the differing values of p. More explicitly, the differing product of pq: 0.2100, 0.2451, and 0.1824.

b. Yes.

c. The "margin of error" (MoE) is typically reported as the maximum error of estimate calculated using p = 0.5 because this yields the maximum value.Rounding up yields a slightly larger error, in turn a wider interval. "Conservative" equates with a less restrictive (narrower) interval.

e. p = 0.5

9.75 a. - e. The distributions do not look normal, they are skewed right. The gaps are caused by working with a discrete distribution, the binomial distribution. Both histograms look the same because they are both showing the same set of data, one as a proportion and the other as a standardized proportion.

f. The distribution is not normal and the normal distribution should not be used to calculate probabilities.

Sample Size Determination Formula for a Population Proportion

$$n = \frac{[z(\alpha/2)]^2 \cdot p^* \cdot q^*}{E^2}$$

Sample Size - A Four-Step Prodecure

Step 1: Use the level of confidence, $1-\alpha$, to find $z(\alpha/2)$
Step 2: Find the maximum error of estimate
Step 3: Determine p^* and $q^* = 1 - p^*$ (if not given, use $p^* = 0.5$)
Step 4: Use formula to find n

9.77 a. Step 1: $1 - \alpha = 0.90$; $z(\alpha/2) = z(0.05) = 1.65$
Step 2: $E = 0.02$
Step 3: $p^* = 0.81$ and $q^* = 0.19$
Step 4: $n = \{[z(\alpha/2)]^2 \cdot p^* \cdot q^*\}/E^2$
$= (1.65^2)(0.81)(0.19)/(0.02^2) = 1047.48 = 1048$

b. Step 1: $1 - \alpha = 0.90$; $z(\alpha/2) = z(0.05) = 1.65$
Step 2: $E = 0.04$
Step 3: $p^* = 0.81$ and $q^* = 0.19$
Step 4: $n = \{[z(\alpha/2)]^2 \cdot p^* \cdot q^*\}/E^2$
$= (1.65^2)(0.81)(0.19)/(0.04^2) = 261.87 = 262$

c. Step 1: $1 - \alpha = 0.98$; $z(\alpha/2) = z(0.01) = 2.33$
Step 2: $E = 0.02$
Step 3: $p^* = 0.81$ and $q^* = 0.19$
Step 4: $n = \{[z(\alpha/2)]^2 \cdot p^* \cdot q^*\}/E^2$
$= (2.33^2)(0.81)(0.19)/(0.02^2) = 2088.77 = 2089$

d. Increasing the maximum error decreases the required sample size. The maximum error is located in the denominator of formula 9.8 and therefore an increase will reduce the resulting value for n.

e. Increasing the level of confidence increases the required sample size. The level of confidence determines the value of z used, and it is located in the numerator of formula 9.8 and therefore an increase in $1-\alpha$ will increase z and will increase the resulting value for n.

Hypotheses are written with the same rules as before. Now replace μ, the population mean, with p, the population proportion.

(ex.: H_o: p = P(driving a convertible) = 0.45 vs.
H_a: p = P(driving a convertible) $\neq$ 0.45,
if driving a convertible is considered the success)

Review: ES9-pp371&373, ST-pp241&242, if necessary.

9.79 a. H_o: p = P(work) = 0.60 ($\leq$) vs. H_a: $p > 0.60$

b. H_o: p = P(win tonight) = 0.50 ($\geq$) vs. H_a: $p < 0.50$

c. H_o: p = P(interested in quitting) = 1/3 ($\leq$) vs. H_a: $p > 1/3$

d. H_o: p = P(believe in spanking) = 0.50 ($\geq$) vs. H_a: $p < 0.50$

e. H_o: p = P(vote for) = 0.50 ($\leq$) vs. H_a: $p > 0.50$

f. H_o: p = P(seriously damaged) = 3/4 ($\geq$) vs. H_a: $p < 3/4$

g. H_o: p = P(H|tossed fairly) = 0.50 vs. H_a: $p \neq 0.50$

h. H_o: p = P(odd|random) = 0.50 vs. H_a: $p \neq 0.50$

Determining the test criteria will also follow the same procedures as before. We are again working with the normal distribution. Review: ES9-pp392&393, ST-p251, if necessary.

Determining the p-value will follow the same procedures as before. We are again working with the normal distribution. Review: ES9-p377, Table 8.6; ST-p245, if necessary.

9.81 a. **P** = $2P(z > 1.48) = 2(0.5000 - 0.4306) = 2(0.0694) = \underline{0.1388}$

b. **P** = $2P(z < -2.26) = 2P(z > 2.26) = 2(0.5000 - 0.4881) = 2(0.0119) = \underline{0.0238}$

c. **P** = $P(z > 0.98) = (0.5000 - 0.3365) = \underline{0.1635}$

d. **P** = $P(z < -1.59) = P(z > 1.59) = (0.5000 - 0.4441) = \underline{0.0559}$

Since $n \le 15$ and x is discrete, Table 2 (Appendix B, ES9-p711) will be used to determine the level of significance, α.
x can be any value, 0 through n, for each experiment.
List all values in numerical sequence.
Draw a vertical line separating the set of values that belong in the critical region and the set of values that belong in the noncritical region.
Add all of the probabilities associated with those numbers in the critical region to find α.

9.83 a. $\alpha = P[x = 12, 13, 14, 15 | B(n=15, p=0.5)]$
$= 0.014 + 0.003 + 2(0+) = \underline{0.017}$

b. $\alpha = P[x = 0, 1 | B(n=12, p=0.3)]$
$= 0.014 + 0.071 = \underline{0.085}$

c. $\alpha = P[x = 0, 1, 2, 3, 9, 10 | B(n=10, p=0.6)]$
$= (0+) + 0.002 + 0.011 + 0.042 + 0.040 + 0.006 = \underline{0.101}$

d. $\alpha = P[x = 4, 5, 6, \ldots, 14 | B(n=14, p=0.05)]$
$= 0.004 + 10(0+) = \underline{0.004}$

List all x values in numerical order, as before. Based on H_a#, add consecutive probabilities until the sum is as close to the given α as possible, without exceeding it. Draw a vertical line at this point, separating the critical and noncritical regions.

If H_a contains $<$, begin adding from $x = 0$ towards $x = n$.
If H_a contains $>$, begin adding from $x = n$ towards $x = 0$.
If H_a contains $\neq$, begin adding simultaneously from $x = 0$ and $x = n$ toward the center.

9.85 a. (1) Correctly fail to reject H_o

b. $\alpha = P[x = 14, 15 | B(n = 15, p = 0.7)]$
$= 0.031 + 0.005 = \underline{0.036}$

c. (4) Commit a type II error

d. **P** $= P[x = 13, 14, 15 | B(n = 15, p = 0.7)]$
$= 0.092 + 0.031 + 0.005 = \underline{0.128}$

Review: "The Probability-Value Hypothesis Test: A Five-Step Procedure"; ES9-p370, ST-p241 and/or "The Classical Hypothesis Test: A Five-Step Procedure"; ES9-pp386&387, ST-p249, if necessary.

Use $z = \dfrac{p' - p}{\sqrt{\dfrac{pq}{n}}}$, for calculating the test statistic.

$p' = x/n$, if not given directly.

Computer and/or calculator commands to perform a hypothesis test for a population proportion can be found in ES9-p442.

Hint for writing the hypotheses for exercise 9.87

Look at the last sentence in the exercise, "If the consumer group ... that fewer than 90% ...?" The words "fewer than" indicate less than (<), therefore the alternative is less than (<). The negation is "NOT less than", which is > or =. Equality (=) is used in the null hypothesis, but stands for greater than or equal to (≥). Remember to use *p* as the population parameter in the hypotheses.

9.87 Step 1: a. The proportion of claims settled within 30 days

b. H_o: $p = P(\text{claim is settled within 30 days}) = 0.90$ (≥) H_a: $p < 0.90$

Step 2: a. independence assumed

b. z; $n = 75$; $n > 20$, $np = (75)(0.90) = 67.5$, $nq = (75)(0.10) = 7.5$, both np and nq > 5

c. $\alpha = 0.05$

Step 3: a. $n = 75$, $x = 55$, $p' = x/n = 55/75 = 0.733$

b. $z = (p' - p)/\sqrt{pq/n}$

$z* = (0.733 - 0.900)/\sqrt{(0.9)(0.1)/75} = -4.82$

Step 4: -- using p-value approach --------------------

a. $\mathbf{P} = P(z < -4.82) = P(z > 4.82)$;

Using Table 3, Appendix B, ES9-p714:

$\mathbf{P} = 0.5000 - 0.499997 = 0.000003$

Using Table 5, Appendix B, ES9-p716:

$\mathbf{P} > 0+$

b. $\mathbf{P} < \alpha$

-- using classical approach -----------------

a. $-z(0.05) = -1.65$

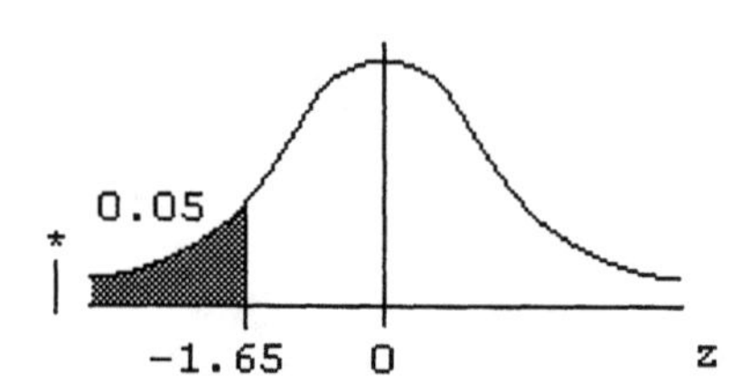

b. z^* falls in the critical region, see Step 4a

Step 5: a. Reject H_o

b. The sample provides sufficient evidence that p is significantly less than 0.90; it appears that less than 90% are settled within 30 days as claimed, at the 0.05 level of significance.

Hint for writing the hypothesis for exercise 9.89
Look at the first sentence of the exercise, "A politician claims she will receive 60% of the vote ..." The words "will receive" indicate <u>equality</u> (=). Since a politician would be interested in a majority, anything <u>equal to or greater than</u> the stated percentage would be acceptable. <u>Greater than or equal to</u> (≥) includes the equality, therefore it belongs in the null hypothesis. Continue to write the null hypothesis with the equal sign (=), but include the greater-than or equal-to sign(≥) in parentheses after it. The negation is "...less than," hence the alternative hypothesis must have a less-than sign (<).

9.89 Step 1: a. The proportion of vote for a politician in an upcoming election

b. H_o: $p = P(\text{vote for}) = 0.60$ [will receive 60% of vote]

H_a: $p < 0.60$ [will receive less than 60%]

Step 2: a. independence assumed

b. z; $n = 100$; $n > 20$, $np = (100)(0.60) = 60$, $nq = (100)(0.40) = 40$, both np and nq > 5

c. $\alpha = 0.05$

Step 3: a. $n = 100$, $x = 50$, $p' = x/n = 50/100 = 0.500$

b. $z = (p' - p)/\sqrt{pq/n}$

$z^* = (0.500 - 0.600)/\sqrt{(0.6)(0.4)/100} = -2.04$

Step 4: -- using p-value approach --------------------

a. **P** = P(z < -2.04) = P(z > 2.04);

Using Table 3, Appendix B, ES9-p714:

P = 0.5000 - 0.4793 = 0.0207

Using Table 5, Appendix B, ES9-p716:

0.0202 < **P** < 0.0228

b. **P** < α

-- using classical approach ------------------

a. -z(0.05) = -1.65

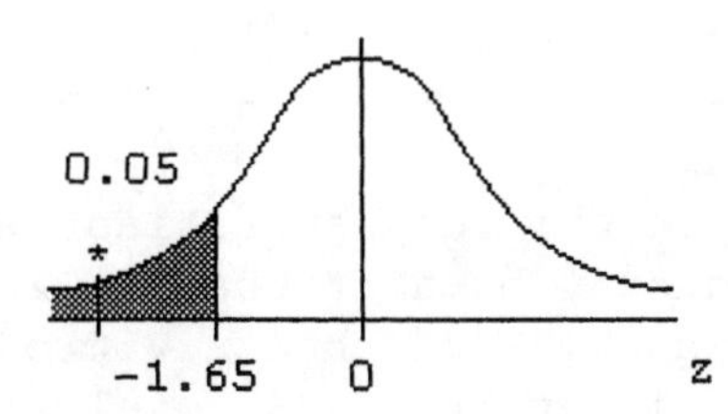

b. z* falls in the critical region, see Step 4a

Step 5: a. Reject H_o

b. The sample provides sufficient evidence that the proportion is significantly less than 0.60, at the 0.05 level; it appears that less than 60% support her.

9.91 Step 1: a. The proportion of women that become pregnant after using an in vitro fertilization procedure

b. H_o: p = P(success) = 0.225 (≤)

H_a: p > 0.225

Step 2: a. independence assumed

b. z; n = 200; n > 20, np = (200)(0.225) = 45, nq = (200)(0.775) = 155, both np and nq > 5

c. α = 0.05

Step 3: a. n = 200, x = 61, p' = x/n = 61/200 = 0.305

b. $z = (p' - p)/\sqrt{pq/n}$

$z^* = (0.305 - 0.225)/\sqrt{(0.225)(0.775)/200} = 2.71$

Step 4: -- using p-value approach --------------------

a. **P** = P(z > 2.71);

Using Table 3, Appendix B, ES9-p714:

P = 0.5000 - 0.4966 = 0.0034

Using Table 5, Appendix B, ES9-p716:

0.0030 < **P** < 0.0035

b. **P** < α

-- using classical approach ------------------

a. $z(0.05) = 1.65$

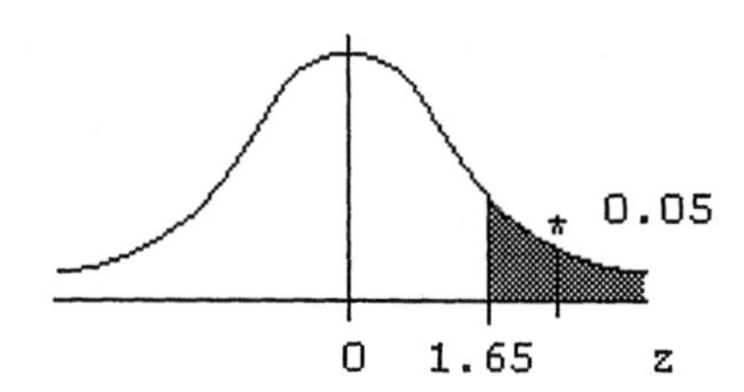

b. z^* falls in the critical region, see Step 4a

Step 5: a. Reject H_o

b. The sample provides sufficient evidence that p is significantly greater than 0.225; it appears that more than 22.5% of the in vitro procedures are successful, at the 0.05 level of significance.

9.93 a. $p' = 324/1000 = 0.324$

$z = (p' - p)/\sqrt{pq/n}$

$z^* = (0.324 - 0.35)/\sqrt{(0.35)(0.65)/1000}$
$= -0.026/0.015083 = -1.72$

b & c.

Step 1: a. The proportion of profession women that fear public speaking

b. H_o: $p = 0.35$ ($\geq$)
H_a: $p < 0.35$

Step 2: a. independence assumed

b. z; $n = 1000$; $n > 20$, $np = (1000)(0.35) = 350$, $nq = (1000)(0.65) = 650$, both np and nq > 5

c. $\alpha = 0.01$

Step 3: a. $n = 1000$, $x = 324$, $p' = x/n = 324/1000 = 0.324$

b. $z^* = -1.72$ (from part a above)

Step 4: -- using p-value approach ------------------

a. $\mathbf{P} = P(z < -1.72) = P(z > 1.72)$;
Using Table 3, Appendix B, ES9-p714:
$\mathbf{P} = 0.5000 - 0.4573 = 0.0427$
Using Table 5, Appendix B, ES9-p716:
$0.0401 < \mathbf{P} < 0.0446$

b. $\mathbf{P} > \alpha$

-- using classical approach ------------------

a. $z(0.01) = -2.33$

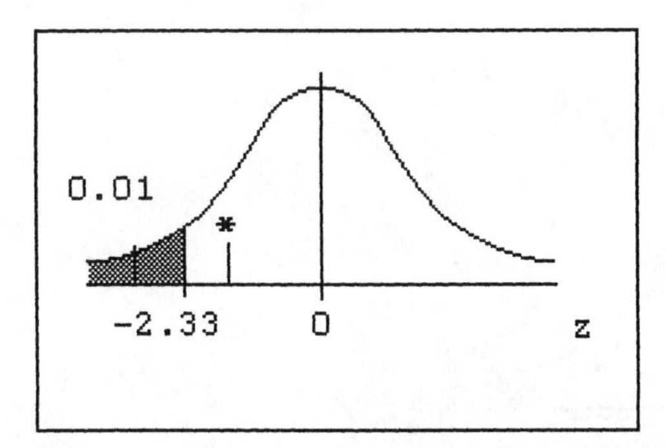

b. z^* falls in the noncritical region, see Step 4a

Step 5: a. Fail to reject H_o
b. At the 0.01 level of significance there is insufficient evidence to show that less than 35% of the country's professional women fear public speaking.

SECTION 9.3 ANSWER NOW EXERCISES

9.94 a. $\chi^2(10,\ 0.01) = 23.2$ b. $\chi^2(12,\ 0.025) = 23.3$

9.95 a. $\chi^2(10,\ 0.95) = 3.94$ b. $\chi^2(22,\ 0.995) = 8.64$

9.96 Chi-Square with 15 DF

x	P(X <= x)
20.2000	0.8356

a. $P(\chi^2 < 20.2 | df = 15) = 0.8356$
b. $P(\chi^2 > 20.2 | df = 15) = 1 - 0.8356 = 0.1644$

9.97 $n = 18,\ s^2 = 785$
$\chi^2{*} = (n-1)s^2/\sigma^2 = (17)(785)/(532) = \underline{25.08}$

9.98 $\mathbf{P} = P(\chi^2 > 25.08 | df = 17)$
Using Table 8, $0.05 < \mathbf{P} < 0.10$
$\alpha = 0.01$; $\mathbf{P} > \alpha$; Fail to reject H_o

9.99 $\chi^2(17, 0.01) = 33.4$ $\chi^2 \geq 33.4$

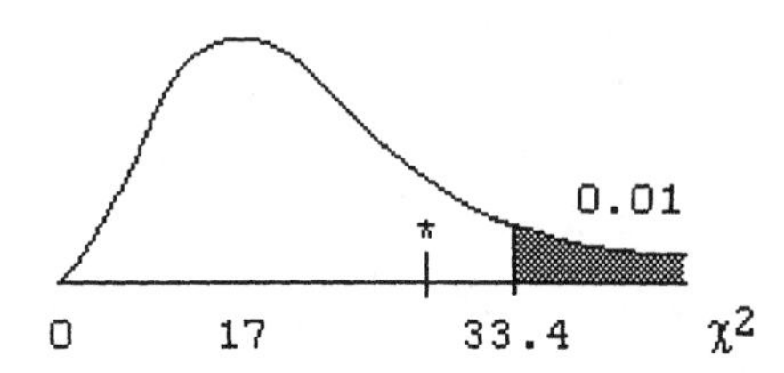

χ^2* falls in the noncritical region; Fail to reject H_o

9.100 $n = 41$, $s^2 = 78.2$
$\chi^2* = (n-1)s^2/\sigma^2 = (40)(78.2)/(52) = \underline{60.15}$

9.101 $\mathbf{P} = 2P[\chi^2 > 60.15 | df = 40]$;
$0.01 < \frac{1}{2}\mathbf{P} < 0.025$; $\underline{0.02 < \mathbf{P} < 0.05}$
$\alpha = 0.05$; $\mathbf{P} < \alpha$; Reject H_o

9.102 $\chi^2(40, 0.975) = 24.4$; $\chi^2(40, 0.025) = 59.3$

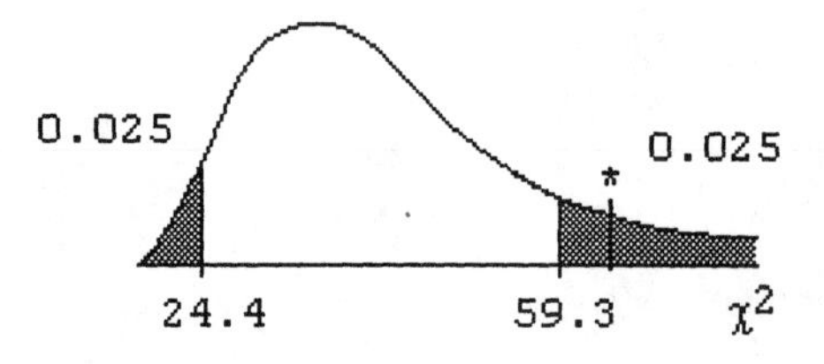

χ^2* falls in the critical region; Reject H_o

9.103 Chi-Square with 14 DF

x	P(X <= x)
6.8700	0.0604

P = $2P(\chi^{2*} > 6.87 | df = 14) = 2(0.0604) = \underline{0.1208}$

SECTION 9.3 EXERCISES

χ^2 Distribution
(used for inferences concerning σ and σ^2)

Key facts about the χ^2 curve:
1) the total area under a χ^2 curve is 1
2) it is skewed to the right (stretched out to the right side, not symmetrical)
3) it is nonnegative, starts at zero and continues out towards $+\infty$
4) a different curve exists for each sample size
5) uses α and degrees of freedom, df, to determine table values
6) degrees of freedom is abbreviated as 'df', where, df = n-1
7) for df > 2, the mean of the distribution is df.

Notation: $\chi^2(df,\alpha) = \chi^2$("degrees of freedom","area to the right")

χ^2 ↑ Table 8; "degrees of freedom" ↑ row id #; "area to the right" ↑ column id #

ex.) Right tail: $\chi^2(13,.025) = 24.7$ (n must have been 14)

Left tail: $\chi^2(13,.975) = 5.01$

Note: For left tail, "area to right" includes both the area in the "middle" and the area of the "right" tail.

Explore the chi-square distribution for various degrees of freedom using Interactivity 9C.

9.105 a. 34.8 b. 28.8 c. 13.4 d. 48.3

e. 12.3 f. 3.25 g. 37.7 h. 10.9

Draw a diagram of a chi square distribution, labeling the given information and shading the specified region(s).
Percentiles: P_k = k^{th} percentile ⇒ k% of the data lies <u>below</u> (to left of) this value

9.107 a. $\chi^2(5, 0.05)$ = <u>11.1</u>

b. $\chi^2(5, 0.05)$ = <u>11.1</u>

c. $\chi^2(5, 0.10)$ = <u>9.24</u>

9.109 1 - (0.01 + 0.05) = <u>0.94</u>

Hypotheses for variability are written with the same rules as before. Now, in place of μ or p, the population standard deviation, σ, or the population variance, σ^2, will be used.
(ex. H_o: $\sigma = 3.7$ vs. H_a: $\sigma \neq 3.7$) Review: ES9-p373, ST-p241, if necessary.

9.111 a. H_o: $\sigma = 24$ ($\leq$) vs. H_a: $\sigma > 24$

b. H_o: $\sigma = 0.5$ ($\leq$) vs. H_a: $\sigma > 0.5$

c. H_o: $\sigma = 10$ vs. H_a: $\sigma \neq 10$

d. H_o: $\sigma^2 = 18$ ($\geq$) vs. H_a: $\sigma^2 < 18$

e. H_o: $\sigma^2 = 0.025$ vs. H_a: $\sigma^2 \neq 0.025$

f. H_o: $\sigma^2 = 34.5$ ($\leq$) vs. H_a: $\sigma^2 > 34.5$

Chi-square test statistic - $\chi^{2*} = (n-1)s^2/\sigma^2$

Hypothesis tests (p-value approach) will be completed in the same format as before. You may want to review: ES9-p370, ST-p241. The only differences are in:

1. the calculated test statistic, which is χ^2, where
   $\chi^2 = (n-1)s^2/\sigma^2$
2. using Table 8 to estimate the p-value
   a) locate df row
   b) locate the calculated χ^2-value between two critical values in the df row, the p-value or ½p-value is in the interval between the two corresponding probabilities of the critical values

Computer and/or calculator probability and cumulative probability commands for values of χ^2 can be found in ES9-p452.

9.113 a. $\mathbf{P} = 2P(\chi^{2*} > 27.8|df = 14)$; $0.01 < ½\mathbf{P} < 0.025$;
$\underline{0.02 < \mathbf{P} < 0.05}$

b. $\mathbf{P} = P(\chi^{2*} > 33.4|df = 17) = \underline{0.01}$

c. $\mathbf{P} = 2P(\chi^{2*} > 37.9|df = 25)$; $0.025 < ½\mathbf{P} < 0.05$;
$\underline{0.05 < \mathbf{P} < 0.10}$

d. $\mathbf{P} = P(\chi^{2*} < 26.3|df = 40)$; $\underline{0.025 < \mathbf{P} < 0.05}$

Hypothesis tests (classical approach) will be completed using the same format as before. You may want to review: ES9-pp386&387, ST-p249. The only differences are :

1. the χ^2 distribution
   a) draw a skewed right distribution (starting at zero)
   b) locate df value as middle value
   c) shade in the critical region(s) based on the alternative hypothesis (H_a).
2. finding critical value(s) from Table 8
   a) degrees of freedom (n-1) is the row id #
   b) α, area to the right, is the column id #
3. the left tail uses $1-\alpha$ or $1-\alpha/2$ for its probability
4. the calculated test statistic is χ^{2*}, where
   $\chi^{2*} = (n-1)s^2/\sigma^2$

Hint for writing the hypotheses for Exercise 9.115
Look at the last sentence in the exercise, "Does this sample ... the population standard deviation is not equal to 8 ...?". "Not equal to" ($\neq$) belongs in the alternative hypothesis. The negation becomes "equal to" and the null hypothesis would be written with an equality sign (=). Remember to use the population standard deviation, σ, as the parameter in the hypotheses.

9.115 Step 1: a. The standard deviation, σ

b. H_o: $\sigma = 8$

H_a: $\sigma \neq 8$

Step 2: a. normality assumed

b. χ^2 c. $\alpha = 0.05$

Step 3: a. $n = 51$, $\overline{x} = 98.2$, $s^2 = 37.5$

b. $\chi^{2*} = (n-1)s^2/\sigma^2 = (50)(37.5)/8^2 = 29.3$

Step 4: -- using p-value approach --------------------

a. $\mathbf{P} = 2P(\chi^2 < 29.3 | df = 50)$; $0.005 < \frac{1}{2}\mathbf{P} < 0.010$;

$0.01 < \mathbf{P} < 0.02$

b. $\mathbf{P} < \alpha$

-- using classical approach ------------------

a. $\chi^2(50, 0.975) = 32.4$, $\chi^2(50, 0.025) = 71.4$

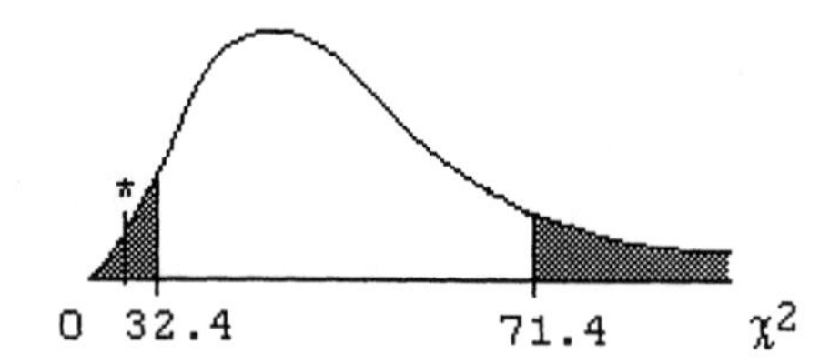

b. $\chi^{2}*$ falls in the critical region, see Step 4a

Step 5: a. Reject H_o

b. There is sufficient reason to conclude that the population standard deviation is not equal to 8, at the 0.05 level of significance.

9.117 a. Allows the use of the chi-square distribution to calculate probabilities.

b. Examine the distribution of the sample data.

c & d.

Step 1: a. The standard deviation of women's shoes obtained by mail-order sales

b. H_o: $\sigma = 0.32$ ($\leq$)

H_a: $\sigma > 0.32$

Step 2: a. normality indicated

b. χ^2 c. $\alpha = 0.01$

Step 3: a. $n = 27$, $s = 0.51$

b. $\chi^{2*} = (n-1)s^2/\sigma^2 = (26)(0.51^2)/(0.32^2) = 66.04$

Step 4: -- using p-value approach --------------------

a. $\mathbf{P} = P(\chi^2 > 66.04|df = 26)$; $\mathbf{P} < 0.005$

b. $\mathbf{P} < \alpha$

-- using classical approach -----------------

a. $\chi^2(26, 0.01) = 45.6$

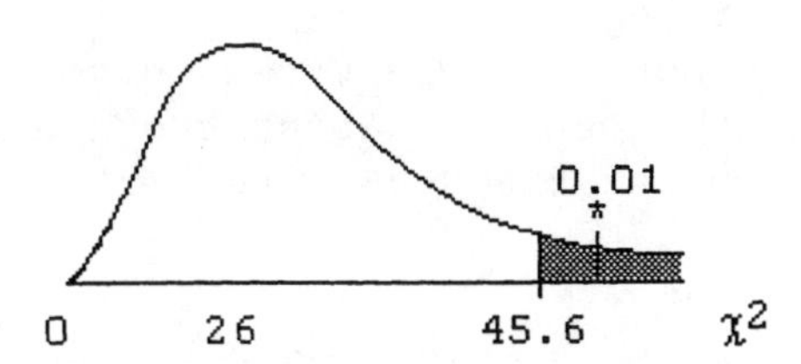

b. χ^{2*} falls in the critical region, see Step 4a

Step 5: a. Reject H_o

b. There is sufficient reason to conclude that the variability in women's mail-order shoes is greater than those supplied by all manufacturers, at the 0.01 level of significance.

9.119 Sample statistics: $n = 15$, $s^2 = 17.4595$

Step 1: a. The standard deviation of miles per gallon for a certain model car

b. H_o: $\sigma = 3.50$ [not different]

H_a: $\sigma \neq 3.50$ [differs]

Step 2: a. normality indicated

b. χ^2 c. $\alpha = 0.05$

Step 3: a. $n = 15$, $s^2 = 17.4595$

b. $\chi^{2*} = (n-1)s^2/\sigma^2 = (14)(17.4595)/3.50^2 = 19.95$

Step 4: -- using p-value approach --------------------

a. $\mathbf{P} = 2 \cdot P(\chi^2 > 19.95|df = 14)$; $0.20 < \mathbf{P} < 0.50$

b. $\mathbf{P} > \alpha$

-- using classical approach ------------------

a. $\chi^2(14, 0.975) = 5.63$, $\chi^2(14, 0.025) = 26.1$

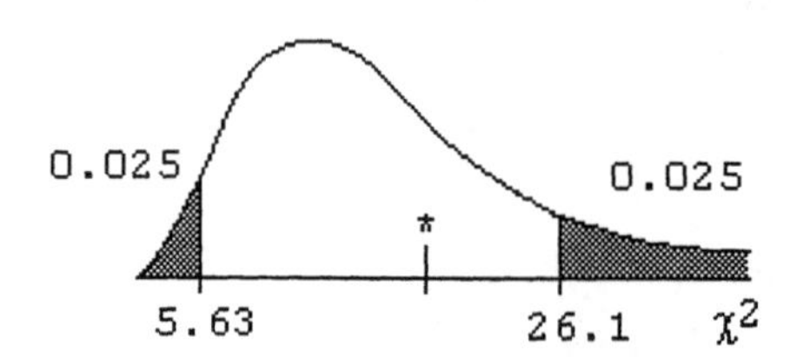

b. χ^{2*} falls in the noncritical region, see Step 4a

Step 5: a. Fail to reject H_o

b. There is not sufficient reason to contradict the manufacturer's claim about the standard deviation, at the 0.05 level of significance.

9.121 a. Results will vary slightly for each student.

b. The mean is very close to the number of degrees of freedom.

c. The mean and median are close, both of which appear to be approximately equal to the degrees of freedom. The mode gets closer to the number of degrees of freedom (also to the mean and median) as the sample size increases.

d. The distributions are no longer skewed to the right. They appear to be symmetrical with the mean ≈ median ≈ mode ≈ df.

RETURN TO CHAPTER CASE STUDY

9.123 a. People in the health profession

b.

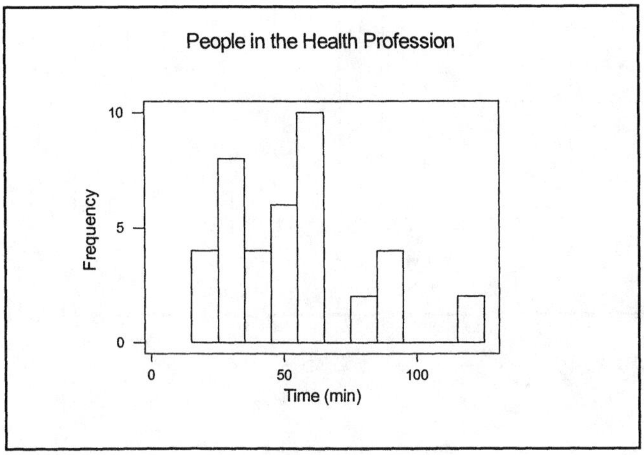

number of observations in Time (min) = 40

9.125 Answers will vary.

CHAPTER EXERCISES

9.127 Step 1: The mean physician fee for cataract removal

Step 2: a. normality indicated

b. t c. $1-\alpha = 0.99$

Step 3: $n = 25$, $\bar{x} = 1550$, $s = 125$

Step 4: a. $\alpha/2 = 0.01/2 = 0.005$; df = 24; $t(24, 0.005) = 2.80$

b. $E = t(df,\alpha/2)\cdot(s/\sqrt{n}) = (2.80)(125/\sqrt{25}) = 70$

c. $\bar{x} \pm E = 1550 \pm 70$

Step 5: $1,480 to $1,620, the 0.99 estimate for μ

9.129 a.

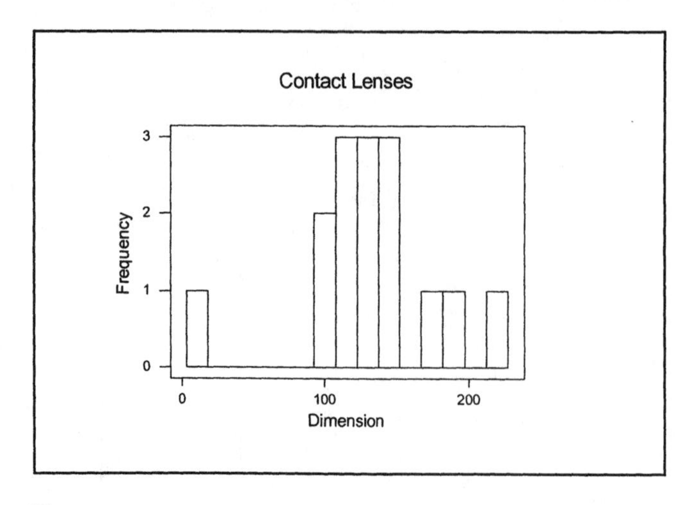

$\overline{x} = 133.0$, $s = 47.3$

b.

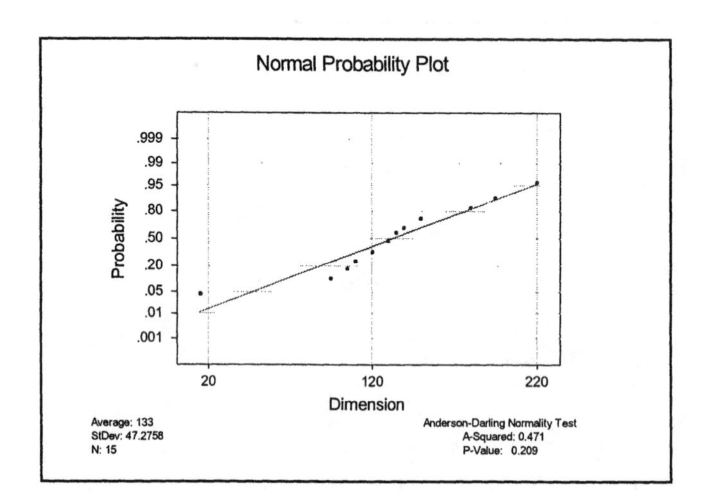

The dimension seems to demonstrate an approximately normal distribution.

c. Step 1: The mean dimension for contact lens molds

Step 2: a. normality assumed, see part b above.

b. t   c. $1-\alpha = 0.95$

Step 3: $n = 15$, $\overline{x} = 133.0$, $s = 47.3$

Step 4: a. $\alpha/2 = 0.05/2 = 0.025$; df $= 14$; $t(14, 0.025) = 2.14$

b. $E = t(df, \alpha/2) \cdot (s/\sqrt{n}) = (2.14)(47.3/\sqrt{15}) = 26.11$

c. $\overline{x} \pm E = 133.0 \pm 26.11$

Step 5: 106.89 to 159.11, the 0.95 estimate for μ

d. With 95% confidence, the true mean dimension of the contact lens is between 106.89 and 159.11 units.

9.131 Step 1: a. The mean maintenance expenditure for a television during the first year following the expiration of the warranty

b. H_o: $\mu = 50$ [no more] ($\leq$)
H_a: $\mu > 50$ [more than]

Step 2: a. normality assumed

b. t   c. $\alpha = 0.01$

Step 3: a. $n = 50$, $\bar{x} = 61.60$, $s = 32.46$

b. $t = (\bar{x} - \mu)/(s/\sqrt{n})$
$t* = (61.60 - 50.00)/(32.46/\sqrt{50}) = 2.53$

Step 4: -- using p-value approach --------------------

a. **P** = $P(t > 2.53|df = 49)$;
Using Table 6, Appendix B, ES9-p717:
$0.005 < \mathbf{P} < 0.01$
Using Table 7, Appendix B, ES9-p718:
$0.006 < \mathbf{P} < 0.008$

b. $\mathbf{P} < \alpha$

-- using classical approach -----------------

a. $t(49, 0.01) = 2.42$

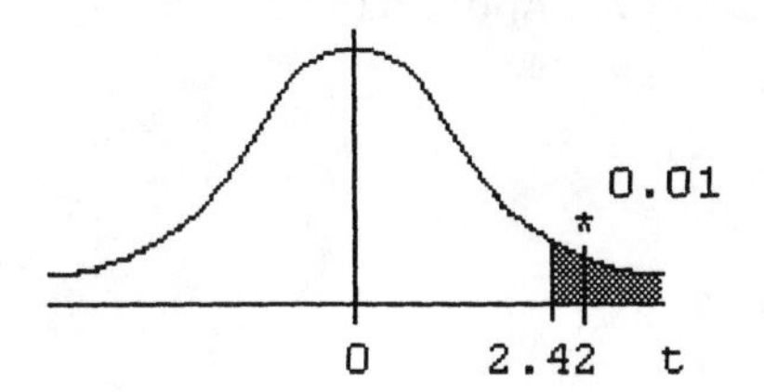

b. t* falls in the critical region, see Step 4a

Step 5: a. Reject H_o

b. The sample does provide sufficient evidence to show the mean expenditure is significantly more than $50, at the 0.01 level.

9.133 Summary of data: $n = 12$, $\Sigma x = 41.3$, $\Sigma x^2 = 146.83$

$\overline{x} = \Sigma x/n = 41.3/12 = \underline{3.44}$

$s^2 = [\Sigma x^2 - (\Sigma x)^2/n]/(n - 1)$

$= [146.83 - (41.3^2/12)]/11 = 0.4263$

$s = \sqrt{s^2} = \sqrt{0.4263} = \underline{0.653}$

Step 1: a. The mean of this year's pollution readings
b. H_o: $\mu = 3.8$ ($\geq$)
H_a: $\mu < 3.8$ [lower]
Step 2: a. normality indicated
b. t c. $\alpha = 0.05$
Step 3: a. $n = 12$, $\overline{x} = 3.44$, $s = 0.653$
b. $t = (\overline{x} - \mu)/(s/\sqrt{n})$
$t^* = (3.44 - 3.8)/(0.653/\sqrt{12}) = -1.91$
Step 4: -- using p-value approach --------------------
a. **P** $= P(t < -1.91 | df = 11) = P(t > 1.91 | df = 11)$;
Using Table 6, Appendix B, ES9-p717:
$0.025 <$ **P** < 0.05
Using Table 7, Appendix B, ES9-p718:
$0.034 <$ **P** < 0.043
b. **P** $< \alpha$

-- using classical approach -----------------
a. $-t(11, 0.05) = -1.80$

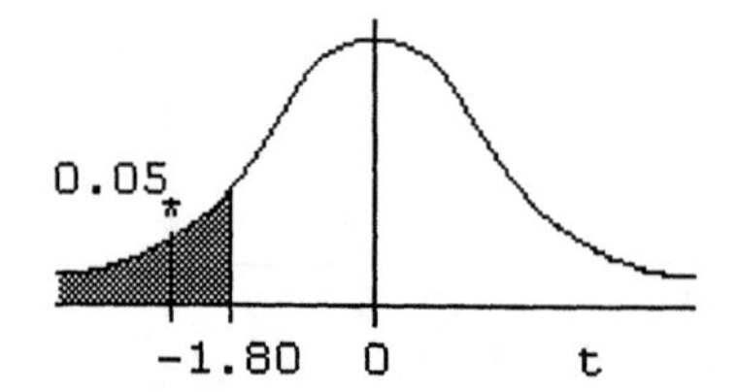

b. t* falls in the critical region, see * Step 4a

Step 5: a. Reject H_o

b. The sample does provide sufficient evidence to justify the contention that the mean of this year's pollution readings is significantly lower than last year's mean, at the 0.05 level.

9.135 a.

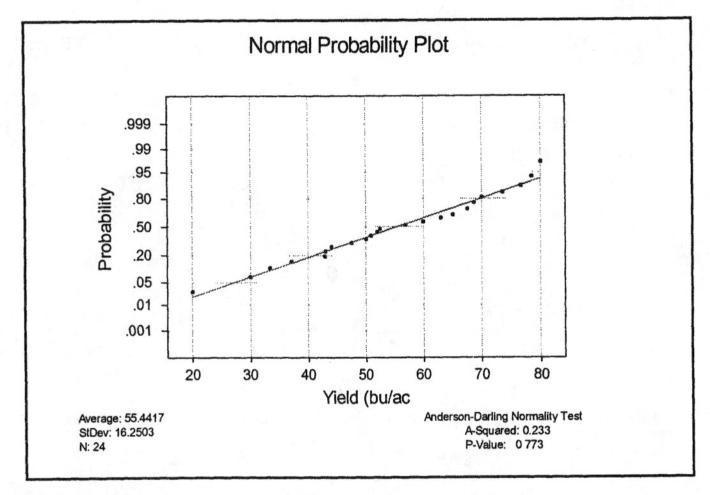

Data appears to have an approximately normal distribution.

b. Step 1: a. The mean oat crop yield for 2001

b. H_o: $\mu = 60$

H_a: $\mu < 60$ [lower]

Step 2: a. normality indicated in step 'a' above

b. t c. $\alpha = 0.05$

Step 3: a. $n = 24$, $\overline{x} = 55.44$, $s = 16.25$

b. $t = (\overline{x} - \mu)/(s/\sqrt{n})$

$t* = (55.44 - 60.0)/(16.25/\sqrt{24}) = -1.37$

Step 4: -- using p-value approach --------------------

a. $\mathbf{P} = P(t < -1.37|df = 23)$;

Using Table 6, Appendix B, ES9-p717:

$0.05 < \mathbf{P} < 0.10$

Using Table 7, Appendix B, ES9-p718:

$0.087 < \mathbf{P} < 0.104$

b. $\mathbf{P} > \alpha$

-- using classical approach ------------------

a. $-t(23, 0.05) = -1.71$; critical region $t \le -1.71$

b. t* falls in the noncritical region

Step 5: a. Fail to reject H_o

b. At the 0.05 level of significance, there is not sufficient evidence to show that the 2001 oat crop mean yield rate is less than 60 bushels per acre.

9.137 a. Half or 0.50

b.

Step 1: The proportion of American who eat at least one beef burger per week

Step 2: a. The sample was randomly selected and each subject's response was independent of those of the others surveyed.

b. $n = 1027$; $n > 20$, $np = (1027)(0.50) = 513.5$, $nq = (1027)(0.50) = 513.5$, np and nq both > 5

c. $1 - \alpha = 0.98$

Step 3: a. $n = 1027$, $p' = 0.50$

Step 4: a. $z(\alpha/2) = z(0.01) = 2.33$

b. $E = z(\alpha/2)\cdot\sqrt{p'q'/n} = 2.33\sqrt{(0.5)(0.5)/1027} = 0.036$

c. $p' \pm E = 0.50 \pm 0.036$

Step 5: 0.464 to 0.536, the 0.98 interval for p

c. 50% ± 3.6%

9.139 Step 1: The proportion who recognize and respect a particular woman golfer

Step 2: a. The sample was randomly selected and each subject's response was independent of those of the others surveyed.

b. $n = 100$; $n > 20$, $np = (100)(16/100) = 16$, $nq = (100)(84/100) = 84$, np and nq both > 5

c. $1 - \alpha = 0.95$

Step 3: $n = 100$, $x = 16$, $p' = x/n = 16/100 = 0.16$

Step 4: a. $z(\alpha/2) = z(0.025) = 1.96$

b. $E = z(\alpha/2)\cdot\sqrt{p'q'/n} = 1.96\sqrt{(0.16)(0.84)/100}$
$= (1.96)(0.03666) = 0.072$

c. $p' \pm E = 0.160 \pm 0.072$

Step 5: 0.088 to 0.232, the 0.95 interval for p = P(recognized)

9.141

p =	0.1	0.2	0.3	0.4	0.5	0.6	0.7	0.8	0.9
q =	0.9	0.8	0.7	0.6	0.5	0.4	0.3	0.2	0.1
pq =	0.09	0.16	0.21	0.24	0.25	0.24	0.21	0.16	0.09

9.143 a. Step 1: $1 - \alpha = 0.95$; $z(\alpha/2) = z(0.025) = 1.96$

Step 2: $E = 0.03$

Step 3: $p^* = 0.37$ and $q^* = 0.63$

Step 4: $n = \{[z(\alpha/2)]^2 \cdot p^* \cdot q^*\}/E^2$
$= ((1.96^2)(0.37)(0.63))/(0.03^2) = 994.97 = \underline{995}$

b. Step 1: $1 - \alpha = 0.95$; $z(\alpha/2) = z(0.025) = 1.96$

Step 2: $E = 0.03$

Step 3: $p^* = 0.37$ and $q^* = 0.63$

Step 4: $n = \{[z(\alpha/2)]^2 \cdot p^* \cdot q^*\}/E^2$
$= ((1.96^2)(0.5)(0.5))/(0.03^2) = 067.11 = \underline{1068}$

c. Yes, the extra knowledge about the variable is of value. It allows the sample size to be approximately 75 smaller, thus costing less to obtain.

9.145 Let x = number of defective in the sample of 50. When H_o is true, we may treat x as a binomial variable with n = 50 and p = 0.005.

α is **P**(of rejecting H_o when it is true):

$\alpha = P[x \geq 2 | B(n = 50, p = 0.005)]$

$= 1.0 - P[x = 0 \text{ or } 1 | B(n = 50, p = 0.005)]$

$= 1.0 - [\binom{50}{0} \cdot (0.005)^0 \cdot (0.995)^{50} + \binom{50}{1} \cdot (0.005)^1 \cdot (0.995)^{49}]$

$= 1.0000 - [0.7783 + 0.1956] = \underline{0.0261}$

9.147 H_o: p = P(7 or 11 picked randomly) = 2/15

H_a: $p \neq 2/15$ [not picked randomly]

$\alpha = 0.05$

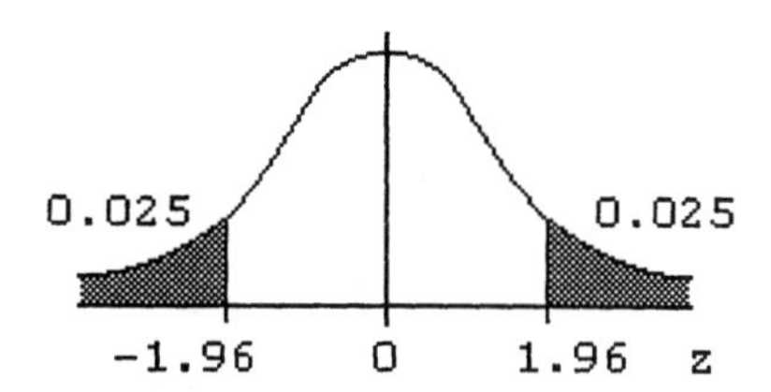

$z = (p' - p)/\sqrt{pq/n}$

$1.96 = [p' - (2/15)]/\sqrt{(2/15)(13/15)/54}$

$p' = (2/15) + (1.96)\sqrt{(2/15)(13/15)/54} = 0.224$

Since $p' = x/n$; then $x = n \cdot p' = 54(0.224) = 12.09 = \underline{13}$

9.149 Step 1: a. The standard deviation on a standard test
b. H_o: $\sigma = 12$
H_a: $\sigma \neq 12$ [different]

Step 2: a. normality assumed
b. χ^2 c. $\alpha = 0.05$

Step 3: a. $n = 40$, $s^2 = 155$
b. $\chi^{2*} = (n-1)s^2/\sigma^2 = (39)(155)/12^2 = 42.0$

Step 4: -- using p-value approach --------------------
a. $\mathbf{P} = 2\mathbf{P}(\chi^2 > 42.0 | df = 39) = 0.6844$
b. $\mathbf{P} > \alpha$
-- using classical approach -----------------
a. $\chi^2(39, 0.975) = 23.7$, $\chi^2(39, 0.025) = 58.1$

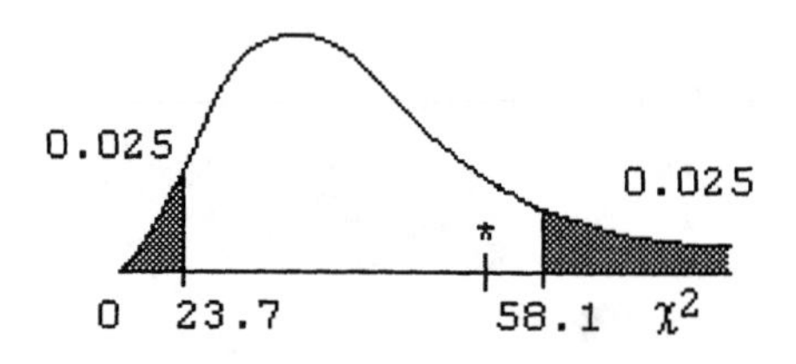

b. χ^{2*} falls in the noncritical region, see Step 4a

Step 5: a. Fail to reject H_o
c. There is not sufficient reason to conclude that the standard deviation is not equal to 12, at the 0.05 level of significance.

9.151 Step 1: a. The standard deviation of produced parts
b. H_o: $\sigma = 0.5$ [in control] (≤)
H_a: $\sigma > 0.5$ [out of control]

Step 2: a. normality indicated
b. χ^2 c. $\alpha = 0.05$

Step 3: a. $n = 30$, $s = 0.87$
b. $\chi^{2*} = (n-1)s^2/\sigma^2 = (29)(0.87^2)/0.5^2 = 87.8$

Step 4: -- using p-value approach --------------------
a. $\mathbf{P} = P(\chi^2 > 87.8 | df = 29)$; $\mathbf{P} < 0.005$
b. $\mathbf{P} < \alpha$
-- using classical approach ------------------
a. $\chi^2(29, 0.05) = 42.6$

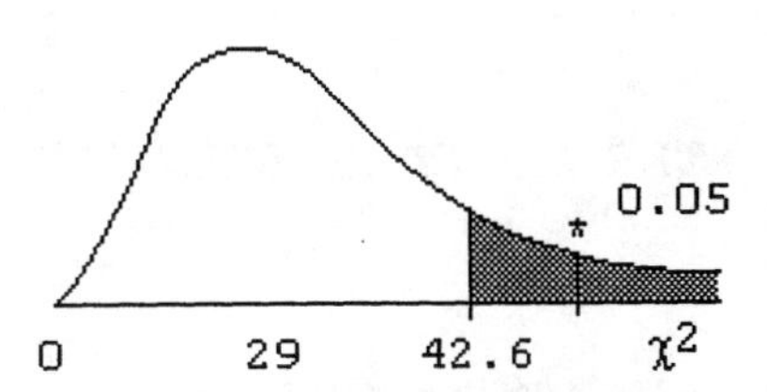

b. χ^{2*} falls in the critical region, see Step 4a

Step 5: a. Reject H_o
b. There is sufficient reason to conclude that the process is out of control with regard to standard deviation, at the 0.05 level of significance.

CHAPTER 10 ∇ INFERENCES INVOLVING TWO POPULATIONS

Chapter Preview

In Chapters 8 and 9, the concepts of confidence intervals and hypothesis tests were introduced. Each of these was demonstrated with respect to a single mean, standard deviation, variance or proportion. In Chapter 10, these concepts will be extended to include two means, two proportions, two standard deviations, or two variances, thereby enabling us to compare two populations. Distinctions will have to be made with respect to dependent and independent samples, in order to select the appropriate testing procedure and test statistic.

A report published by Nellie Mae on the credit card usage of college students is presented in the Case Study for this chapter.

CHAPTER 10 CASE STUDY

10.1 a. Freshmen: 97/200 = 0.485 or 48.5%; Nellie Mae = 54%
Sophomore: 187/200 = 0.935 or 93.5%; Nellie Mae = 92%

b. mounded, but skewed with the right-hand tail being much longer

c.

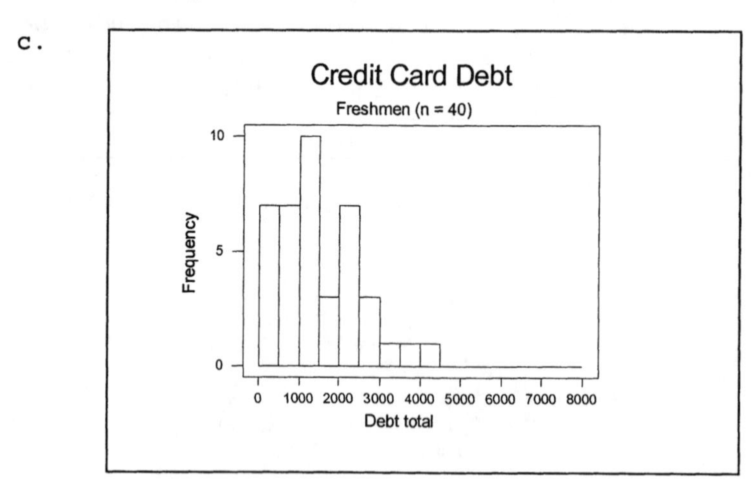

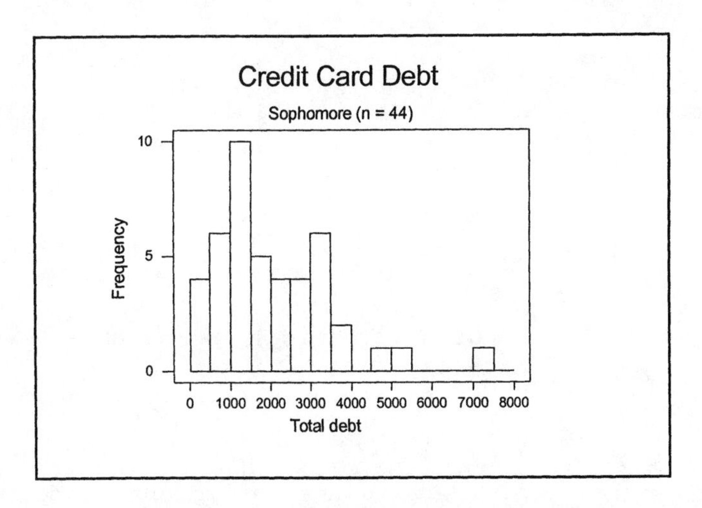

Both distributions are mounded and skewed to the right. The distribution of debts for sophomores is more dispersed.

d. Freshmen: mean = \$1519, standard deviation = \$1036
Sophomore: mean = \$2079, standard deviation = \$1434

e. Step 1: The mean credit card debt for freshmen
Step 2: a. The sampled population appears to be mounded, CLT is satisfied.
b. t c. $1-\alpha = 0.95$
Step 3: $n = 40$, $\overline{x} = 1519$, $s = 1036$
Step 4: a. $\alpha/2 = 0.05/2 = 0.025$; df = 39;
$t(39, 0.025) = 2.03$
b. $E = t(df,\alpha/2)\cdot(s/\sqrt{n}) = 2.03(1036/\sqrt{40}) = 332.5$
c. $\overline{x} \pm E = 1519 \pm 332.5$
Step 5: \$1186.50 to \$1851.50 the 0.95 confidence Interval for μ

f. Step 1: a. The mean credit card debt for sophomores
b. H_o: $\mu = 1825$ ($\leq$)
H_a: $\mu > 1825$ (higher)
Step 2: a. normality indicated
b. t c. $\alpha = 0.05$
Step 3: a. $n = 44$, $\overline{x} = 2079$, $s = 1434$
b. $t = (\overline{x} - \mu)/(s/\sqrt{n})$
$t* = (2079 - 1825)/(1434/\sqrt{44}) = 254/216.2) = 1.17$

Step 4: -- using p-value approach --------------------

a. **P** = P(t > 1.17|df = 43);

Using Table 6, Appendix B, ES9-p717:

0.10 < **P** < 0.25

Using Table 7, Appendix B, ES9-p718:

0.118 < **P** < 0.139

Using a computer: **P** = 0.123

b. **P** > α

-- using classical approach ------------------

a. t(43, 0.05) = 1.68

b. t* falls in the noncritical region

Step 5: a. Fail to reject H_o

b. There is not sufficient evidence to conclude that the sampled population of sophomores have a significantly higher credit card debt than the national average reported by Nellie Mae using = 0.05.

SECTION 10.1 ANSWER NOW EXERCISES

10.2 Identical twins are so much alike that the information obtained from one would not be independent from the information obtained from the other twin.

10.3 a. Divide the class into two groups, males and females. Randomly select a sample from each of the two groups.

b. Randomly select a set of students, obtaining the two heights from each of the selected students.

SECTION 10.1 EXERCISES

INDEPENDENT SAMPLES - Two samples are independent if the selection of one sample from a population has no effect on the selection of the other sample from another population. (They do not have to be different populations.)
ex.: the repair costs for two different brands of VCRs

DEPENDENT SAMPLES (paired samples) - two samples are dependent if the objects or individuals selected for one sample from a population are paired in some meaningful way with the objects or individuals selected for the second sample from the same or another population.
ex.: "before and after" experiments - change in weight for smokers who became nonsmokers

10.5 Dependent samples. The two sets of data were obtained from the same set of 20 people, each person providing one piece of data for each sample.

10.7 Independent samples. The gallon of paint serves as the population of many (probably millions) particles. Each set of 10 specimens forms a separate and independent samples.

10.9 a. Independent samples will result if the two sets are selected in such a way that there is no relationship between the two resulting sets.

b. Dependent samples will result if the 1,000 men and women were husband and wife or if they were brother and sister, or related in some way.

SECTION 10.2 ANSWER NOW EXERCISES

10.11 a.

Pairs	1	2	3	4	5
d=A-B	1	1	0	2	-1

b. $n = 5,\ \Sigma d = 3,\ \Sigma d^2 = 7$

$\bar{d} = \Sigma d/n = 3/5 = \underline{0.6}$

c. $s_d = \sqrt{(\Sigma d^2 - (\Sigma d)^2/n)/(n-1)} = \sqrt{(7 - (3)^2/5)/(4)}$

$= \sqrt{1.3} = \underline{1.14}$

10.12 $t(15,\ 0.025) = \underline{2.13}$; it determines the number of standard errors the confidence interval should extend in either direction of the mean difference. Its value depends on the sample size and level of confidence.

10.13 a. Step 1: The mean difference, μ_d

Step 2: a. normality assumed

b. t c. $1-\alpha = 0.95$

Step 3: $n = 26,\ \bar{d} = 6.3,\ s_d = 5.1$

Step 4: a. $\alpha/2 = 0.05/2 = 0.025$; df = 25;

$t(25,\ 0.025) = 2.06$

b. $E = t(df,\alpha/2)\cdot(s_d/\sqrt{n}) = (2.06)(5.1/\sqrt{26})$

$= (2.06)(1.0002) = 2.06$

c. $\bar{d} \pm E = 6.3 \pm 2.06$

Step 5: $\underline{4.24 \text{ to } 8.36}$, the 0.95 confidence interval for μ_d

b. The same $\bar{d}$ and s_d values were used with a much larger n, resulting in a narrower confidence interval.

10.14 d = Before - After:

Confidence Intervals

Variable	N	Mean	StDev	SEMean	95% C.I.
C3	6	3.0	3.35	1.37	(-0.51,6.51)

$\underline{-0.51 \text{ to } 6.51}$, the 0.95 confidence interval for μ_d

10.15 Data Summary: $n = 5$, $\Sigma d = 90$, $\Sigma d^2 = 2700$

Step 1: a. The mean difference, μ_d
b. H_o: $\mu_d = 0$
H_a: $\mu_d > 0$

Step 2: a. normality indicated
b. t c. $\alpha = 0.05$

Step 3: a. $n = 5$, $\bar{d} = 18$, $s_d = 16.43$
b. $t* = (\bar{d} - \mu_d)/(s_d/\sqrt{n}) = (18 - 0)/(16.43/\sqrt{5}) = 2.45$

Step 4: -- using p-value approach --------------------
a. $\mathbf{P} = P(t > 2.45|df = 4)$;
Using Table 6, Appendix B, ES9-p717:
$0.025 < \mathbf{P} < 0.05$
Using Table 7, Appendix B, ES9-p718:
$0.033 < \mathbf{P} < 0.037$
b. $\mathbf{P} < \alpha$

-- using classical approach ------------------
a. $t(4, 0.05) = 2.13$

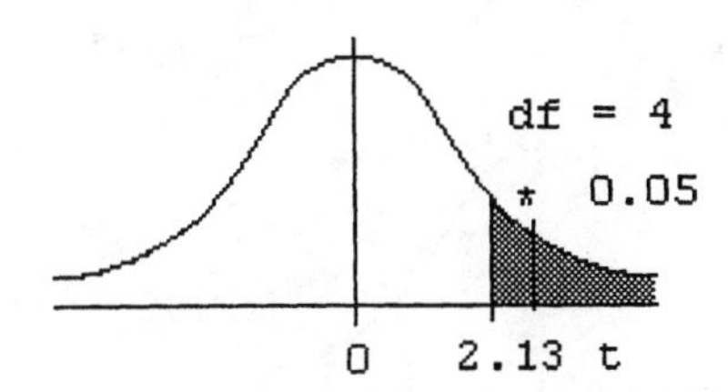

b. t* falls in the critical region, see Step 4a

Step 5: a. Reject H_o
b. At the 0.05 level of significance, there is sufficient evidence that the mean difference is greater than zero.

10.16 $d = M - N$, $\alpha = 0.02$:

T-Test of the Mean
Test of $\mu_d = 0.00$ vs $\mu_d < 0.00$

Variable	N	Mean	StDev	SEMean	T	P-value
C3	6	-2.33	4.41	1.80	-1.30	0.13

$P > \alpha$; fail to reject H_o

10.17 Data Summary: $n = 5$, $\Sigma d = 12$, $\Sigma d^2 = 92$

Step 1: a. The mean difference, μ_d

b. H_o: $\mu_d = 0$

H_a: $\mu_d \neq 0$

Step 2: a. normality indicated

b. t c. $\alpha = 0.01$

Step 3: a. $n = 5$, $\bar{d} = 2.4$, $s_d = 3.97$

b. $t* = (\bar{d} - \mu_d)/(s_d/\sqrt{n}) = (2.4 - 0)/(3.97/\sqrt{5}) = 1.35$

Step 4: -- using p-value approach --------------------

a. $\mathbf{P} = 2P(t > 1.35|df = 4)$;

Using Table 6, Appendix B, ES9-p717:

$0.10 < \frac{1}{2}\mathbf{P} < 0.25$; $0.20 < \mathbf{P} < 0.50$

Using Table 7, Appendix B, ES9-p718:

$0.117 < \frac{1}{2}\mathbf{P} < 0.132$; $0.234 < \mathbf{P} < 0.264$

b. $\mathbf{P} > \alpha$

-- using classical approach ------------------

a. $\pm t(4, 0.005) = \pm 4.60$

b. t* falls in the noncritical region, see Step 4a

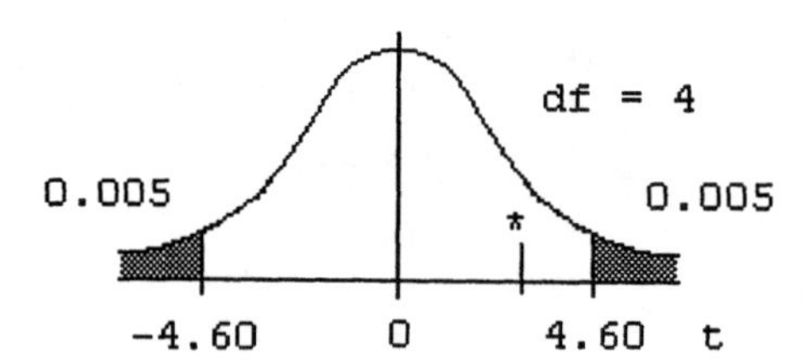

Step 5: a. Fail to reject H_o

b. At the 0.01 level of significance, there is not sufficient evidence that the mean difference is different than zero.

10.18 a. The null hypothesis is, "the average difference is zero."

b. The "t-calculated" and the "t-critical" values are being used to make the decision as in the classical approach.

c. The test is two-tailed and the t-distribution is symmetric, making the number of multiples each statistic is from zero the only information needed. Further, the absence of negative numbers makes the table less confusing to most.

d. The decision was to "fail to reject the null hypothesis" in 12 of them. Actually the calculated t (2.224) for the No. 4 sieve size is less than the critical value (2.228) and therefore leads to a "fail to reject" decision also. But for some reason, they viewed it as too close.
e. The conclusion reached was, "the two methods of sampling are equivalent with respect to Gmb, Gmm, asphalt binder content and gradation."
f. The recommended action was, "the revised Florida method for sampling (FM 1-T 168) be accepted and implemented statewide."

SECTION 10.2 EXERCISES

Estimating μ_d - the population mean difference

1. point estimate: $\bar{d} = \dfrac{\sum d}{n}$

2. confidence interval: $\bar{d} \pm t(df, \alpha/2) \cdot (s_d/\sqrt{n})$

$\bar{d}$ = point estimate; $t(df, \alpha/2)$ = confidence coefficient; $(s_d/\sqrt{n})$ = estimated standard error

confidence coefficient · estimated standard error = maximum error of estimate

Follow the steps outlined in "The Confidence Interval: A Five-Step Procedure" in: ES9-p350, ST-p229.

Computer and/or calculator commands to construct a confidence interval for the mean difference can be found in ES9-pp479&480.

10.19 Data Summary: $n = 8$, $\Sigma d = 8$, $\Sigma d^2 = 48$

a. Point estimate = 1.0
b. Step 1: The mean reduction in diastolic blood pressure

10.21 Sample statistics: $d = A - B$

$n = 8$, $\bar{d} = 3.75$, $s_d = 5.726$

Step 1: The mean difference in weight gain for pigs fed ration A as compared to those fed ration B

Step 2: a. normality indicated

b. t c. $1-\alpha = 0.95$

Step 3: $n = 8$, $\bar{d} = 3.75$, $s_d = 5.726$

Step 4: a. $\alpha/2 = 0.05/2 = 0.025$; df = 7; $t(7, 0.025) = 2.36$

b. $E = t(df,\alpha/2)\cdot(s_d/\sqrt{n}) = (2.36)(5.726/\sqrt{8})$
$= (2.36)(2.0244) = 4.78$

c. $\bar{d} \pm E = 3.75 \pm 4.78$

Step 5: <u>-1.03 to 8.53</u>, the 0.95 interval for μ_d

10.23 Sample statistics: $d = I - II$

$n = 10$, $\bar{d} = 0.8$, $s_d = 1.32$

Step 1: The mean difference between two routes

Step 2: a. normality indicated

b. t c. $1-\alpha = 0.95$

Step 3: $n = 10$, $\bar{d} = 0.8$, $s_d = 1.32$

Step 4: a. $\alpha/2 = 0.05/2 = 0.025$; df = 9;
$t(9, 0.025) = 2.26$

b. $E = t(df,\alpha/2)\cdot(s_d/\sqrt{n}) = (2.26)(1.32/\sqrt{10})$
$= (2.26)(0.4174) = 0.943$

c. $\bar{d} \pm E = 0.8 \pm 0.943$

Step 5: <u>-0.143 to 1.743</u>, the 0.95 confidence interval for μ_d

WRITING HYPOTHESES FOR TEST OF TWO DEPENDENT MEANS

μ_d = population mean difference

<u>null hypothesis</u> - H_o: $\mu_d = 0$
("the mean difference equals 0, that is, there is no difference within the pairs of data")

<u>possible alternative hypotheses</u> -

H_a: $\mu_d > 0$
H_a: $\mu_d < 0$
H_a: $\mu_d \neq 0$,
("the mean difference is significant, that is, there is a difference within the pairs of data")

10.25 a. H_o: $\mu_d = 0$ (≤); H_a: $\mu_d > 0$; d = posttest - pretest

b. H_o: $\mu_d = 0$; H_a: $\mu_d \neq 0$; d = after - before

c. H_o: $\mu_d = 10$ (≥); H_a: $\mu_d < 10$; d = after - before

d. H_o: $\mu_d = 12$ (≥); H_a: $\mu_d < 12$; d = before - after

Reviewing how to determine the test criteria in: ES9-pp392&393, ST-pp251-252, may be helpful. Remember the t-distribution uses Table 6 (Appendix B, ES9-p717), therefore α and degrees of freedom, df = n - 1, are needed.

Hypothesis Tests for Two Dependent Means

In this form of hypothesis test, each data value of the first sample is compared to its corresponding (or paired) data value in the second sample. The differences between these paired data values are calculated, thereby forming a sample of differences or *d* values. It is these differences or *d* values that we wish to use to test the difference between two dependent means.
Review the parts to a hypothesis test (p-value & classical) as outlined in: ES9-pp350&386, ST-pp241&249, if needed. Changes will occur in:

1) the calculated value of the test statistic, t;

$$t = \frac{\bar{d} - \mu_d}{s_d / \sqrt{n}}, \quad \text{where } \bar{d} = \frac{\sum d}{n}, s_d = \sqrt{\frac{\sum d^2 - (\sum d)^2 / n}{n - 1}}$$

and n = # of paired differences

2) a. p-value approach
Use Table 6 (Appendix B, ES9-p717) to <u>estimate</u> the p-value
1) Locate df row.
2) Locate the absolute value of the calculated t-value between two critical values in the df row. ...

3) The p-value is in the interval between the two heading if H_a is one-tailed, or from the *two-tailed* headings if H_a is two-tailed.

OR

Use Table 7 (Appendix B, ES9-p718) to estimate or place bounds on the p-value

1) locate the absolute value of the calculated t-value along the left margin and the df along the top,then read the p-value directly from the table where the row and column intersect

OR

2) locate the absolute value of the calculated t-value and its df between appropriate bounds. From the box formed at the intersection of these row(s) and column(s), use the upper left and lower right values for the bounds on **P**.

b. classical approach
Use Table 6 (Appendix B, ES9-p717) with df = n - 1 and α to find the critical value

3) if H_o is rejected, a significant difference as stated in H_a is indicated if H_o is not rejected, no significant difference is indicated

Computer and/or calculator commands to perform a hypothesis test for μ_d can be found in ES9-p483.

NOTE: To find $\overline{d}$ and s_d, set up a table of corresponding pairs of data. Calculate *d*, the difference (be careful to subtract in the same direction each time). Calculate a d^2 for each pair and find summations, Σd and $\Sigma(d^2)$.
The sample of paired differences are assumed to be selected from an approximately normally distributed population with a mean μ_d and a standard deviation σ_d. Since σ_d is unknown, the calculated t-statistic is found using an estimated standard error of $s_d/\sqrt{n}$.

Hint for writing the hypotheses for exercise 10.27

Look at the second and third sentences of the exercise; "The data ... where d is the amount of corrosion on the coated portion subtracted from the amount of corrosion on the uncoated portion. Does this sample provide sufficient reason to conclude that the coating is beneficial?" For the coating to be beneficial, there would have to be less corrosion on the coated portion. This implies that the differences, "uncoated - coated," if beneficial, would be positive, which in turn implies a greater than zero (> 0). Therefore the alternative hypothesis is a greater than (>). The negation is "NOT beneficial", which indicates < or =. Therefore the null hypothesis should be written with the equal sign (=), but include the less-than or equal-to sign (≤) in parentheses after it.

10.27 Sample statistics: $n = 40$, $\bar{d} = 5.5$, $s_d = 11.34$

Step 1: a. The mean difference between coated and uncoated sections of steel pipe

b. H_o: $\mu_d = 0$

H_a: $\mu_d > 0$ (beneficial)

Step 2: a. normality assumed, CLT with n = 40.

b. t   c. $\alpha = 0.01$

Step 3: a. $n = 40$, $\bar{d} = 5.5$, $s_d = 11.34$

b. $t* = (\bar{d} - \mu_d)/(s_d/\sqrt{n})$

$= (5.5 - 0.0)/(11.34/\sqrt{40}) = 3.067$

Step 4: -- using p-value approach --------------------

a. **P** $= P(t > 3.067|df = 39)$;

Using Table 6, Appendix B, ES9-p717:

P < 0.005

Using Table 7, Appendix B, ES9-p718:

P ≈ 0.002

b. **P** $< \alpha$

-- using classical approach -----------------

a. $t(39, 0.01) = 2.44$

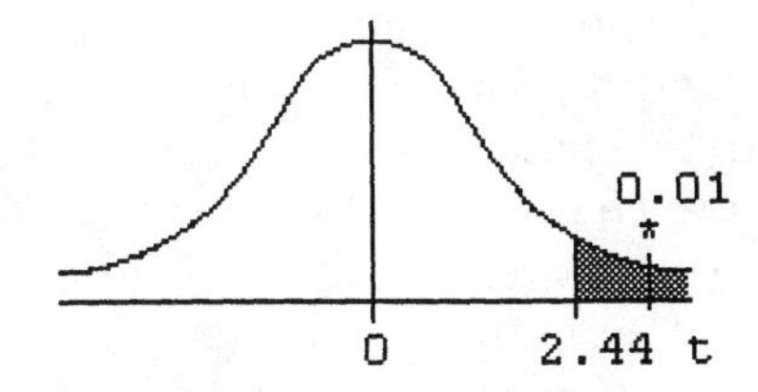

b. t* falls in the critical region, see * Step 4a

Step 5: a. Reject H_o

b. At the 0.01 level of significance, there is a significant benefit to coating the pipe.

Computer and/or calculator commands to perform a hypothesis test for μ_d can be found in ES9-p483. The order of subtraction needs to match the "planned" approach as determined by H_a.

10.29 Verify - answers given in exercise.

10.31 Sample statistics: d = LD - HD

$n = 20, \quad \bar{d} = 1.35, \quad s_d = 1.631$

Step 1: a. The mean difference in the level of pain and discomfort experienced by patients receiving heavier dosages of anesthetic prior to eye surgery

b. H_o: $\mu_d = 0$ (no difference)
H_a: $\mu_d > 0$ (less pain and discomfort)

Step 2: a. normality assumed

b. t c. $\alpha = 0.01$

Step 3: a. $n = 20, \ \bar{d} = 1.35, \ s_d = 1.631$

b. $t* = (\bar{d} - \mu_d)/(s_d/\sqrt{n})$
$= (1.35 - 0.0)/(1.631/\sqrt{20}) = 3.70$

Step 4: -- using p-value approach --------------------

a. $\mathbf{P} = P(t > 3.70|df=19)$;
Using Table 6, Appendix B, ES9-p717:
$\mathbf{P} < 0.005$
Using Table 7, Appendix B, ES9-p718:
$\mathbf{P} = 0.001$

b. $\mathbf{P} < \alpha$

-- using classical approach ------------------

a. critical region: $t \geq 2.54$

b. t* falls in the critical region

Step 5: a. Reject H_o

b. At the 0.01 level of significance, there is sufficient evidence that there is less pain and discomfort.

SECTION 10.3 ANSWER NOW EXERCISES

10.33 $\sqrt{(s_1^2 / n_1) + (s_2^2 / n_2)} = \sqrt{(190 / 12) + (150 / 18)} = \sqrt{24.1667} = \underline{4.92}$

10.34 Case I: df will be between 17 and 40
Case II; df = 17 (smaller df)

10.35 Step 1: The difference between two means, $\mu_1 - \mu_2$
Step 2: a. normality indicated
b. t c. $1-\alpha = 0.90$
Step 3: sample information given in exercise;
$\overline{x}_1 - \overline{x}_2 = 35 - 30 = 5$
Step 4: a. $\alpha/2 = 0.10/2 = 0.05$; df = 14;
$t(14, 0.05) = 1.76$
b. $E = t(df, \alpha/2) \cdot \sqrt{(s_1^2 / n_1) + (s_2^2 / n_2)}$
$= (1.76)\sqrt{(22^2/20) + (16^2/15)}$
$= (1.76)(6.42) = 11.3$
c. $(\overline{x}_1 - \overline{x}_2) \pm E = 5 \pm 11.3$
Step 5: $\underline{-6.3 \text{ to } 16.3}$, the 0.90 confidence interval for $\mu_1 - \mu_2$

10.36 $t* = [(\overline{x}_2 - \overline{x}_1) - (\mu_2 - \mu_1)] / \sqrt{(s_2^2 / n_2) + (s_1^2 / n_1)}$
$= [(43.1 - 38.2) - 0]/\sqrt{(10.6^2/25) + (14.2^2/18)} = \underline{1.24}$

10.37 With the smaller degrees of freedom, df = 9, a higher calculated value is needed making it more difficult to reject H_o. This is due to the lack of reliability with a small sample.

10.38 a. **P** = 2P(t* > 1.3|df = 18);
Using Table 6, Appendix B, ES9-p717:
<u>0.20 < **P** < 0.50</u>
Using Table 7, Appendix B, ES9-p718:
P = 2(0.105) = <u>0.210</u>

b. $\alpha = 0.05$; $\pm t(18, 0.025) = \pm 2.10$

10.39 a. $\sqrt{[((1.589)^2 / 15) + ((3.023)^2 / 13)]} = \sqrt{[0.168328 + 0.702964]}$

$= \sqrt{0.871292} = 0.9334 = 0.933$

b. $(9.6667 - 3.1538)/0.933 = 6.5129/0.933 = 6.9806$

c. 12 to 26
d. It was calculated using a computer program.
e. 12, the number of degrees of freedom associated with the smaller sample
f. $t^* = 6.98$ is significant for a very small level of significance, p-value is less than 0.0005.

SECTION 10.3 EXERCISES

Estimating $(\mu_1 - \mu_2)$ - the difference between two population means,

1. Point Estimate: $\bar{x}_1 - \bar{x}_2$
2. Confidence Interval

$$(\bar{x}_1 - \bar{x}_2) \pm t(df, \alpha/2) \cdot \sqrt{(s_1^2 / n_1) + (s_2^2 / n_2)}$$

point estimate ($(\bar{x}_1 - \bar{x}_2)$); confidence coefficient ($t(df, \alpha/2)$); estimated standard error ($\sqrt{(s_1^2 / n_1) + (s_2^2 / n_2)}$)

maximum error of estimate (confidence coefficient · estimated standard error)

estimate df by using the smaller value of df_1 or df_2

Review "The Confidence Interval: A Five-Step Procedure" in: ES9-p350, ST-p229, if necessary.
Subtract sample means ($\bar{x}_1 - \bar{x}_2$ or $\bar{x}_2 - \bar{x}_1$) in whichever order results in a positive difference.
Also, use appropriate subscripts to designate the source.

10.41 Step 1: The difference between the mean absorption rates for two drugs, $\mu_B - \mu_A$

Step 2: a. normality indicated
b. t c. $1-\alpha = 0.98$

Step 3: $n_A = 36$, $\bar{x}_A = 7.9$, $s_A = 0.11$
$n_B = 36$, $\bar{x}_B = 8.5$, $s_B = 0.10$
$\bar{x}_B - \bar{x}_A = 8.5 - 7.9 = 0.60$

Step 4: a. $\alpha/2 = 0.02/2 = 0.01$; df = 35;
$t(35, 0.01) = 2.44$

b. $E = t(df,\alpha/2)\cdot\sqrt{(s_B^2 / n_B) + (s_A^2 / n_A)}$

$= (2.44)\cdot\sqrt{(0.10^2 / 36) + (0.11^2 / 36)}$

$= (2.44)(0.025) = 0.06$

c. $(\bar{x}_B - \bar{x}_A) \pm E = 0.60 \pm 0.06$

Step 5: 0.54 to 0.66, the 0.98 confidence interval for $\mu_B-\mu_A$

10.43 Step 1: The difference between the average lengths of workweek for Mining and Manufacturing, $\mu_{Min}-\mu_{Man}$

Step 2: a. normality indicated

b. t c. $1-\alpha = 0.95$

Step 3: sample information given in exercise;
$\bar{x}_{Min} - \bar{x}_{Man} = 47.5 - 43.5 = 4.0$

Step 4: a. $\alpha/2 = 0.05/2 = 0.025$; df = 9;
$t(9, 0.025) = 2.26$

b. $E = t(df,\alpha/2)\cdot\sqrt{(s_{Min}^2 / n_{Min}) + (s_{Man}^2 / n_{Man})}$

$= (2.26)\sqrt{(5.5^2/15) + (4.9^2/10)}$

$= (2.26)(2.1018) = 4.75$

c. $(\bar{x}_{Min} - \bar{x}_{Man}) \pm E = 4.0 \pm 4.75$

Step 5: -0.75 to 8.75, the 0.95 confidence interval for $\mu_{Min}-\mu_{Man}$

Computer and/or calculator commands to construct a confidence interval for the difference between two means can be found in ES9-pp498&499.

10.45 Verify - answer given in exercise

10.47 Sugar Beet Sucrose: n = 15, $\bar{x}$ = 16.948, s = 0.862

Sugar Cane Sucrose: n = 12, $\bar{x}$ = 13.925, s = 0.912

Step 1: The difference between the mean sucrose percentages for sugar beet producing counties and sugar cane producing counties, $\mu_1-\mu_2$

Step 2: a. normality assumed

b. t c. $1-\alpha = 0.95$

Step 3: sample information given above;
$\bar{x}_1 - \bar{x}_2 = 16.948 - 13.925 = 3.023$

Step 4: a. $\alpha/2 = 0.05/2 = 0.025$; df = 11;
$t(11, 0.025) = 2.20$

b. $E = t(df,\alpha/2) \cdot \sqrt{(s_1^2 / n_1) + (s_2^2 / n_2)}$

$= (2.20)(\sqrt{((0.862^2 / 15) + (0.912^2 / 12))})$

c. $(\bar{x}_1 - \bar{x}_2) \pm E = 3.023 \pm 0.758$

Step 5: <u>2.265 to 3.781</u>, the 0.95 confidence interval for $\mu_1 - \mu_2$

WRITING HYPOTHESES FOR THE DIFFERENCE BETWEEN TWO MEANS

null hypothesis:

$H_o: \mu_1 = \mu_2$ <u>or</u> $H_o: \mu_1 - \mu_2 = 0$ <u>or</u> $H_o: \mu_1 - \mu_2 = \#$

possible alternative hypotheses:

$H_a: \mu_1 > \mu_2$ <u>or</u> $H_a: \mu_1 - \mu_2 > 0$ <u>or</u> $H_a: \mu_1 - \mu_2 > \#$

$H_a: \mu_1 < \mu_2$ <u>or</u> $H_a: \mu_1 - \mu_2 < 0$ <u>or</u> $H_a: \mu_1 - \mu_2 < \#$

$H_a: \mu_1 \neq \mu_2$ <u>or</u> $H_a: \mu_1 - \mu_2 \neq 0$ <u>or</u> $H_a: \mu_1 - \mu_2 \neq \#$

10.49 a. $H_o: \mu_1 - \mu_2 = 0$ vs. $H_a: \mu_1 - \mu_2 \neq 0$

b. $H_o: \mu_1 - \mu_2 = 0$ ($\leq$) vs. $H_a: \mu_1 - \mu_2 > 0$

c. $H_o: \mu_1 - \mu_2 = 20$ ($\leq$) vs. $H_a: \mu_1 - \mu_2 > 20$

d. $H_o: \mu_A - \mu_B = 50$ ($\geq$) vs. $H_a: \mu_A - \mu_B < 50$
or equivalently
$H_o: \mu_B - \mu_A = 50$ ($\leq$) vs. $H_a: \mu_B - \mu_A > 50$

Review the rules for calculating the p-value in: ES9-p377, ST-p245, if necessary. Remember to use the t-distribution, therefore either Table 6, Table 7 or the computer/calculator will be used to find probabilities.
Review of the use of the tables can be found in: ES9-pp422-424, ST-pp 273&276.

10.51 Table 6, Appendix B, ES9-p717
Table 7, Appendix B, ES9-p718

a. **P** = $P(t > 1.3 | df = 5)$;
Using Table 6: $0.10 <$ **P** < 0.25
Using Table 7: **P** = 0.125

b. **P** = $P(t < -2.8|df = 8) = P(t > 2.8|df = 8)$;
Using Table 6: $0.01 < \mathbf{P} < 0.025$
Using Table 7: **P** = 0.012

c. **P** = $2P(t > 1.8|df = 15)$;
Using Table 6: $0.05 < \mathbf{P} < 0.10$
Using Table 7: **P** = 2(0.046) = 0.092

d. **P** = $2P(t^* > 1.8|df = 25)$;
Using Table 6: $0.05 < \mathbf{P} < 0.10$
Using Table 7: **P** = 2(0.042) = 0.084

Use Table 6 (Appendix B, ES9-p717) with the smaller of df_1 or df_2 and the given α to find the critical value(s). Reviewing how to determine the test criteria in: ES9-p415, ST-p251, as it is applied to the t-distribution may be helpful.

Hypothesis Test for the Difference Between Two Means, Independent Samples

Review the parts to a hypothesis test as outlined in: ES9-pp370&386, ST-pp241&249, if needed. Slight changes will occur in:

1. **the hypotheses**: (see box before exercise 10.49)

2. **the calculated test statistic**

$$t = \frac{(\bar{x}_1 - \bar{x}_2) - (\mu_1 - \mu_2)}{\sqrt{(s_1^2 / n_1) + (s_2^2 / n_2)}}, \text{ using df = smaller of } df_1 \text{ or } df_2$$

3. If H_o is rejected, a significant difference between the means is indicated.
If H_o is not rejected, no significant difference between the means is indicated.

Any subscripts may be used on the hypotheses. Try to use letters that indicate the source. The form H_o: $\mu_2 - \mu_1 = 0$ (versus H_o: $\mu_1 = \mu_2$) is the preferred form since it establishes the order for subtraction that will be needed when calculating the test statistic.

Computer and/or calculator commands to perform a hypothesis test for the difference between two means can be found in ES9-pp498&499.

10.53 $t = (1.2 - 1.05)/\sqrt{((0.32^2/16) + (0.25^2/14))}$

$= 0.15/\sqrt{(0.0064 + 0.00446)} = 0.15/0.104 = 1.44$

p-value $= 2 \cdot P(t > 1.44|df=13) \approx 2(0.09) \approx \underline{0.18}$

10.55 Step 1: a. The difference between the mean scores of students using an electronic study guide to help them learn accounting principles and those not using one

b. $H_o: \mu_1 - \mu_2 = 0$
$H_a: \mu_1 - \mu_2 > 0$

Step 2: a. normality assumed, CLT with $n_1 = 38$ and $n_2 = 36$.

b. t c. $\alpha = 0.01$

Step 3: a. sample information given in exercise

b. $t* = [(\overline{x}_1 - \overline{x}_2) - (\mu_1 - \mu_2)]/\sqrt{(s_E^2/n_E) + (s_C^2/n_C)}$

$= [(79.6 - 72.8)-0]/[\sqrt{(6.9^2/38) + (7.6^2/36)}]$

$= 4.02$

Step 4: -- using p-value approach -------------------

a. **P** $= P(t > 4.02|df = 35)$;
Using Table 6, Appendix B, ES9-p717:
P < 0.005
Using Table 7, Appendix B, ES9-p718:
P $= 0+$

b. **P** $< \alpha$

-- using classical approach -----------------

a. critical region: $t \geq 2.44$

b. t* is in the critical region

Step 5: a. Reject H_o

b. At the 0.05 level of significance, there is sufficient evidence to conclude the experimental group of rats consumed a significantly larger amount of lead.

10.57 Step 1: a. The difference between the mean number of days missed at work by people receiving CSM treatment and those undergoing physical therapy, CSM(1) and Therapy(2), $\mu_2 - \mu_1$

b. $H_o: \mu_2 - \mu_1 = 0$ (no difference)
$H_a: \mu_2 - \mu_1 > 0$ (physical therapy less effective)

Step 2: a. normality assumed

b. t c. $\alpha = 0.01$

Step 3: a. $n_1 = 32$, $\overline{x}_1 = 10.6$, $s_1 = 4.8$
$n_2 = 28$, $\overline{x}_2 = 12.5$, $s_2 = 6.3$

b. $t* = [(\overline{x}_2 - \overline{x}_1) - (\mu_2 - \mu_1)] / \sqrt{(s_2^2/n_2) + (s_1^2/n_1)}$

$= [(12.5 - 10.6) - 0] / [\sqrt{(6.3^2 / 28) + (4.8^2 / 32)}]$
$= 1.30$

Step 4: -- using p-value approach --------------------
a. **P** $= P(t > 1.30 | df = 27)$;
Using Table 6, Appendix B, ES9-p717:
$0.10 < \mathbf{P} < 0.25$
Using Table 7, Appendix B, ES9-p718:
$0.102 < \mathbf{P} < 0.103$

b. $\mathbf{P} > \alpha$

-- using classical approach ------------------
a. critical region: $t \geq 2.47$
b. t* is in the noncritical region

Step 5: a. Fail to reject H_o
b. At the 0.01 level of significance, there is insufficient evidence to show that people treated by chiropractors using CSM miss fewer days of work due to acute lower back pain than people undergoing physical therapy.

10.59 a. Verify - answers given in exercise.

b. $\mathbf{P} = 2P(t > 0.59 | df = 20)$;
$2(0.277 < \frac{1}{2}\mathbf{P} < 0.312)$, $0.554 < \mathbf{P} < 0.624$

c. $\mathbf{P} = 2P(t > 0.59 | df = 12)$;
$2(0.280 < \frac{1}{2}\mathbf{P} < 0.313)$, $0.560 < \mathbf{P} < 0.626$

10.61 Step 1: a. The difference between mean weight gained on two diets, A and B, $\mu_B - \mu_A$
b. H_o: $\mu_B - \mu_A = 0$
H_a: $\mu_B - \mu_A > 0$

Step 2: a. normality assumed
b. t c. $\alpha = 0.05$

Step 3: a. $n_A = 10$, $\overline{x}_A = 10.0$, $s_A^2 = 10.44$
$n_B = 10$, $\overline{x}_B = 14.7$, $s_B^2 = 46.01$

b. $t* = [(\bar{x}_B - \bar{x}_A) - (\mu_B - \mu_A)] / \sqrt{(s_B^2 / n_B) + (s_A^2 / n_A)}$

$= [(14.7-10.0)-0] / [\sqrt{(10.44/10) + (46.01/10)}]$

$= 1.98$

Step 4: -- using p-value approach --------------------

a. $\mathbf{P} = P(t > 1.98 | df = 9)$;

Using Table 6, Appendix B, ES9-p717:

$0.025 < \mathbf{P} < 0.05$

Using Table 7, Appendix B, ES9-p718:

$0.037 < \mathbf{P} < 0.047$

b. $\mathbf{P} < \alpha$

-- using classical approach ------------------

a. critical region: $t \geq 1.83$

b. t* is in the critical region

Step 5: a. Reject H_o

b. At the 0.05 level of significance, there is sufficient evidence to show that Diet B had a significantly higher weight gain.

10.63 Everybody will get different results, but they can all be expected to look very similar to the following.

a. Minitab commands:

Choose: Calc > Random > Normal to generate both distributions into C1 and C2

Choose: Stat > Basic Statistics > Display Descriptive Statistics for the means and standard deviations.

Choose: Graph > Histogram

Use cutpoints as noted for both distributions.

Excel commands:

Choose: Tools > Data Analysis > Random Number Generation > Normal to generate both distributions in columns A and B.

Choose: Tools > Data Analysis > Descriptive Statistics for the means and standard deviations.

Choose: Tools > Data Analysis > Histogram

Use cutpoints as noted for both distributions.

N(100,20)

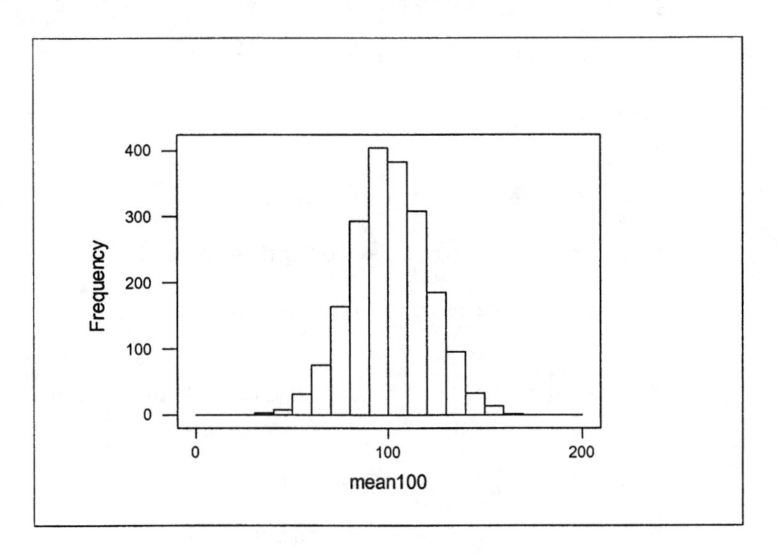

N(120,20)

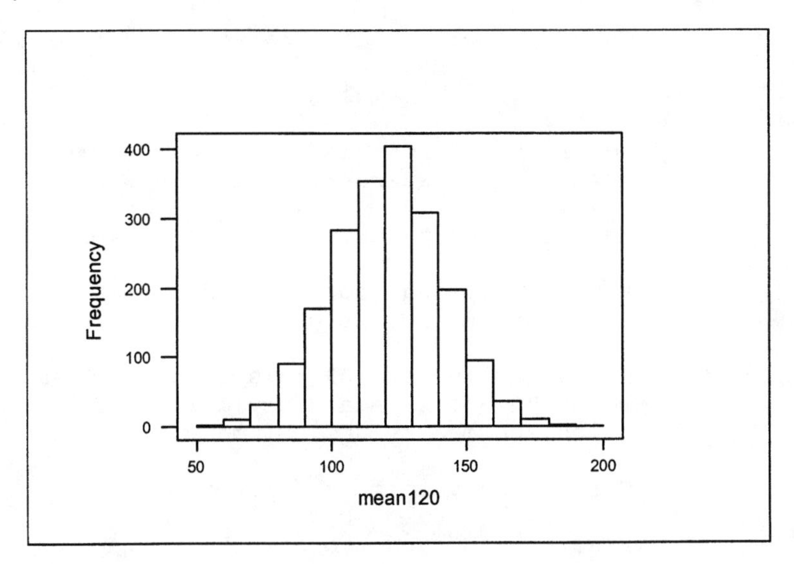

b. The sampling distribution is expected to be normal in shape with a mean of 20 (120-100) and have a standard error of $\sqrt{\frac{20^2}{8}+\frac{20^2}{8}}$ or 10.

c. Minitab commands:

Choose: Calc > Random > Normal

Generate 100 rows into C3-C10 with mean = 100 and standard deviation = 20

Choose: Calc > Row Statistics

Select Mean for C3-C10 and store in C11.

Repeat above for 100 rows into C12-C19 with mean = 120

and standard deviation = 20. Also select Mean for C12-C19 and store in C20.

Choose: Calc > Calculator

Store in C21 the expression: C20 - C11

Excel commands:

Choose: Tools > Data Analysis > Random Number

Generation > Normal

Generate 100 rows into columns C through J with mean 100 and standard deviation = 20.

Choose: Insert function > All > Average

Find the average for C1-J1 and store in K1. Drag down for other averages.

Repeat above for 100 rows into L through S with mean 120 and standard deviation = 20. Also calculate means into column T.

Choose: Edit formula (=)

Find T1 - K1 and store in U1. Drag down for other subtractions.

d. Minitab commands:

Choose: Stat > Basic Statistics > Display Descriptive Statistics for C21

Choose: Graph > Histogram

Use cutpoints -20:60/10 for C21

<u>Excel commands</u>:

Choose: Tools > Data Analysis > Descriptive Statistics for column U.

Choose: Tools > Data Analysis > Histogram

Use classes from -20 to 60 in increments of 10 for column U.

100 values for the difference between two sample means:

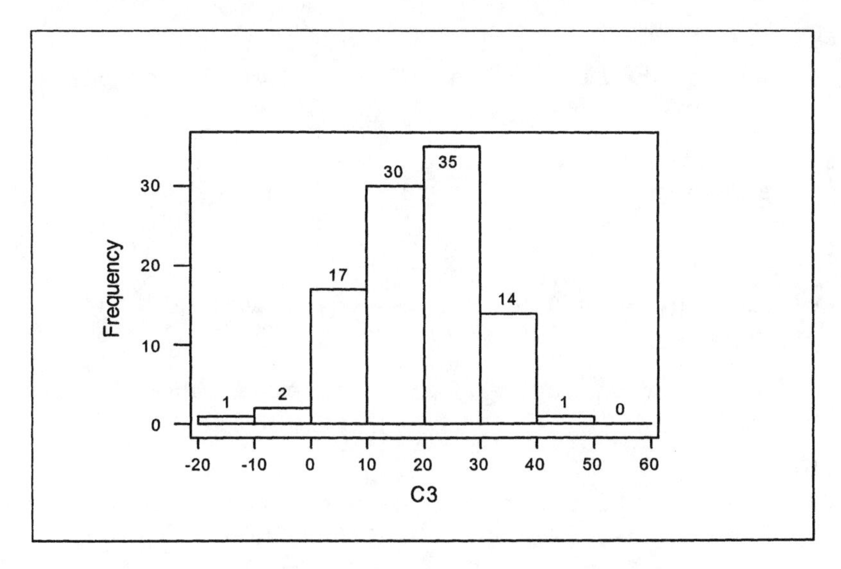

e. For the empirical sampling distribution, the mean is 19.51 and the standard error is 10.71. There is 65%, 96% and 99% of the values within one, two and three standard errors of the expected mean of 20. This seems to agree closely with the empirical rule, thus suggesting a normal distribution occurred.

f. You can expect very similar results to occur on repeated trials.

For Exercise 10.65, adjust the Minitab and/or Excel commands in Exercise 10.63

10.65 Everybody will get different results, but they can all be expected to look very similar to the results found in exercise 10.63. It turns out that the t* statistic is very "robust", meaning "it works quite well even when the assumptions are not met." This it one of the reasons the t-test for the mean and the t-test for the difference between to means are such important tests.

SECTION 10.4 ANSWER NOW EXERCISES

10.67 $x = \underline{75}$, $n = \underline{250}$,
$p' = x/n = 75/250 = \underline{0.30}$, $q' = 1 - p' = 1 - 0.30 = \underline{0.70}$

10.68 a. $n_1p_1 = 40(0.9) = \underline{36}$, $n_1q_1 = 40(0.1) = \underline{4}$
$n_2p_2 = 50(0.9) = \underline{45}$, $n_2q_2 = 50(0.1) = \underline{5}$

b. No, n_1q_1 and n_2q_2 are not larger than 5.

10.69 Step 1: The difference between proportions, $p_A - p_B$
Step 2: a. n's > 20, np's and nq's all > 5
b. z c. $1-\alpha = 0.95$
Step 3: sample information given in exercise
$p'_A = x_A / n_A = 45/125 = 0.36, q'_A = 1 - 0.36 = 0.64$
$p'_B = x_B / n_B = 48/150 = 0.32, q'_B = 1 - 0.32 = 0.68$
$p'_A - p'_B = 0.36 - 0.32 = 0.04$
Step 4: a. $\alpha/2 = 0.05/2 = 0.025$; $z(0.025) = 1.96$
b. $E = z(\alpha/2)\cdot\sqrt{(p_Aq_A)/n_A + (p_Bq_B)/n_B}$
$= (1.96)\cdot\sqrt{(0.36)(0.64)/125 + (0.32)(0.68)/150}$
$= (1.96)(0.057) = 0.11$
c. $(p'_A - p'_B) \pm E = 0.04 \pm 0.11$
Step 5: -0.07 to 0.15, the 0.95 confidence interval for $p_A - p_B$

10.70 $p'_p = (x_E + x_R)/(n_E + n_R) = (15 + 25)/(250 + 275)$
$= 40/525 = \underline{0.076}$

$q'_p = 1 - p'_p = 1 - 0.076 = \underline{0.924}$

10.71 $p'_G = 323/380 = 0.85, \quad p'_H = 332/420 = 0.79$

$p'_p = (x_G + x_H)/(n_G + n_H) = (323 + 332)/(380 + 420)$
$= 655/800 = 0.82$

$q'_p = 1 - p'_p = 1 - 0.82 = 0.18$

$z* = (p'_G - p'_H)/\sqrt{(p'_p)(q'_p)[(1/n_G)+(1/n_H)]}$
$= (0.85 - 0.79)/\sqrt{(0.82)(0.18)[(1/380)+(1/420)]}$
$= 0.06/0.0272 = \underline{2.21}$

10.72 Rewrite the alternative hypothesis for easier understanding:
$H_a: p_R - p_E > 0$

$p'_R = 25/275 = 0.091, \quad p'_E = 15/250 = 0.06$

$z* = (p'_R - p'_E)/\sqrt{(p'_p)(q'_p)[(1/n_R)+(1/n_E)]}$
$= (0.091 - 0.06)/\sqrt{(0.076)(0.924)[(1/275) + (1/250)]}$
$= 0.031/0.0232 = 1.34$

P $= P(z > 1.34) = (0.5000 - 0.4099) = \underline{0.0901}$

10.73 a.

Step 1: a. The difference in proportions of survival rates for transplants from a no heartbeat donor and a heart beating donor, $p_n - p_h$

b. $H_o: p_n - p_h = 0$
$H_a: p_n - p_h > 0$

Step 2: a. n's > 20, np's and nq's > 5

b. z

Step 3: a. $n_n = 125, \; p'_n = 0.79, \quad n_h = 125, \; p'_h = 0.77$

$x_n = (125)(0.79) = 99$
$x_h = (125)(0.77) = 96$
$p'_p = (x_n + x_h)/(n_n + n_h) = (99 + 96)/(125 + 125)$
$= 195/250 = 0.78$
$q'_p = 1 - p'_p = 1 - 0.78 = 0.22$

b. $z^* = (p'_n - p'_h) / \sqrt{(p'_p)(q'_p)[(1/n_n) + (1/n_h)]}$

$z^* = (0.79 - 0.77) / \sqrt{(0.78)(0.22)((1/125) + (1/125))}$

$= 0.02/0.0524 = 0.38$

Step 4: -- using p-value approach --------------------

a. **P** = P(z > 0.38);

Using Table 3, Appendix B, ES9-p714:

P = (0.5000 - 0.1480) = 0.3520

Using Table 5, Appendix B, ES9-p716:

0.3446 < **P** < 0.3632

b. **P** > α

-- using classical approach ------------------

a. Critical region: $z \geq 1.65$

b. z* falls in the noncritical region, see Step 4a

Step 5: a. Fail to reject H_o

b. There is not sufficient evidence to show a difference, at the 0.05 level.

b. size of each sample

c. Yes, it means that one method is as good as the other, and means that there will be many more donated organs available.

SECTION 10.4 EXERCISES

Estimating $(p_1 - p_2)$ - the difference between two population proportions - independent samples (large samples)

1. Point estimate: $p'_1 - p'_2$

2. Confidence interval:

$$(p'_1 - p'_2) \pm z(\alpha/2) \cdot \sqrt{(p'_1 \cdot q'_1 / n_1) + (p'_2 \cdot q'_2 / n_2)}$$

↑ point estimate — ↑ confidence coefficient — ↑ estimated standard error

(confidence coefficient × estimated standard error): Maximum error of the estimate

Computer and/or calculator commands to construct a confidence interval for the difference between two proportions can be found in ES9-p508.

10.75 Step 1: The difference in proportions of nurses who experienced a change in position based on their participation in a program, $p_w - p_n$

Step 2: a. n's > 20, np's and nq's all > 5

b. z c. $1-\alpha = 0.99$

Step 3: $n_w = 341$, $x_w = 87$, $p'_w = 87/341 = 0.255$,

$q'_w = 1 - 0.255 = 0.745$

$n_n = 40$, $x_n = 9$, $p'_n = 9/40 = 0.225$,

$q'_n = 1 - 0.225 = 0.775$

$p'_w - p'_n = 0.255 - 0.225 = 0.03$

Step 4: a. $\alpha/2 = 0.01/2 = 0.005$; $z(0.005) = 2.58$

b. $E = z(\alpha/2)\cdot\sqrt{(p'_w \cdot q'_w / n_w) + (p'_n \cdot q'_n / n_n)}$

$= 2.58\cdot\sqrt{(0.255)(0.745)/341 + (0.225)(0.775)/40}$

$= (2.58)(0.07) = 0.18$

c. $(p'_w - p'_n) \pm E = 0.03 \pm 0.18$

Step 5: <u>-0.15 to 0.21</u>, the 0.99 interval for $p_w - p_n$

10.77 Step 1: The difference in proportions of defectives for two machines, $p_1 - p_2$

Step 2: a. n's > 20, np's and nq's all > 5

b. z c. $1-\alpha = 0.90$

Step 3: $n_1 = 150$, $x_1 = 12$, $p'_1 = 12/150 = 0.08$,

$q'_1 = 1 - 0.08 = 0.92$

$n_2 = 150$, $x_2 = 6$, $p'_2 = 6/150 = 0.04$,

$q'_2 = 1 - 0.04 = 0.96$

$p'_1 - p'_2 = 0.08 - 0.04 = 0.04$

Step 4: a. $\alpha/2 = 0.10/2 = 0.05$; $z(0.05) = 1.65$

b. $E = z(\alpha/2)\cdot\sqrt{(p'_1 \cdot q'_1 / n_1) + (p'_2 \cdot q'_2 / n_2)}$

$= 1.65\cdot\sqrt{(0.08\cdot0.92/150) + (0.04\cdot0.96/150)}$

$= (1.65)(0.027) = 0.04$

c. $(p'_1 - p'_2) \pm E = 0.04 \pm 0.04$

Step 5: <u>0.000 to 0.080</u>, the 0.90 interval for $p_1 - p_2$

WRITING HYPOTHESES FOR THE DIFFERENCE BETWEEN TWO PROPORTIONS

a) null hypothesis:

H_o: $p_1 = p_2$ or $p_1 - p_2 = 0$

b) possible alternative hypotheses:

H_a: $p_1 > p_2$ or H_a: $p_1 - p_2 > 0$

H_a: $p_1 < p_2$ or H_a: $p_1 - p_2 < 0$

H_a: $p_1 \neq p_2$ or H_a: $p_1 - p_2 \neq 0$

10.79 a. H_o: $p_m - p_w = 0$ vs. H_a: $p_m - p_w \neq 0$

b. H_o: $p_b - p_g = 0$ $(\leq)$ vs. H_a: $p_b - p_g > 0$

c. H_o: $p_c - p_{nc} = 0$ $(\leq)$ vs. H_a: $p_c - p_{nc} > 0$

Hypothesis Test for the Difference Between Two Proportions, Independent Samples (Large Samples)

Review parts to a hypothesis test as outlined in: ES9-pp370 & 386, ST-pp241 & 249, if needed. Changes will occur in:

1. **the hypotheses**: (see box before exercise 10.79)

2. **the calculated test statistic**

$$z = \frac{(p_1' - p_2') - (p_1 - p_2)}{\sqrt{p_p' q_p' \left(\frac{1}{n_1} + \frac{1}{n_2}\right)}}, \text{ where } p_1' = \frac{x_1}{n_1}, \; p_2' = \frac{x_2}{n_2},$$

$$p_p' = \frac{x_1 + x_2}{n_1 + n_2} \text{ and } q_p' = 1 - p_p'$$

3. If H_o is rejected, a significant difference in proportions is indicated.
   If H_o is not rejected, no significant difference is indicated.

...

NOTE: The sampling distribution of $p_1' - p_2'$ is approximately normally distributed with a mean $(p_1 - p_2)$ and a standard error of $\sqrt{p_1q_1\ n_1 + p_2q_2/n_2}$, if n_1 and n_2 are sufficiently large. Since H_o is assumed to be true, $p_1 - p_2$ is considered equal to 0. Since p_1 and p_2 are also unknown, the best estimate for $p(p = p_1 = p_2)$ is a pooled estimate p_p'.

Computer and/or calculator commands to perform a hypothesis test for the difference between two proportions can be found in ES9-p512.

Hint for writing the hypotheses for exercise 10.81

The word "compare" implies 2 populations, namely the home PC owners and the work PC owners. The results are given in the form of number of successes and sample size, therefore a difference between 2 proportions is suggested. Look at the second to last sentence of the exercise; "Did the home PC owners experience more problems ...?" The word "more" indicates a greater than. Therefore, the alternative is greater than (>). The negation becomes "not greater than" and the null hypothesis would be written with an inequality sign (≤).

10.81 Step 1: a. The difference in the proportion of home PC owners and work PC owners who experience a computer hardware problem that cannot be solved with help from the manufacturer, $p_h - p_w$

b. $H_o: p_h - p_w = 0 \quad (\leq)$

$H_a: p_h - p_w > 0$

Step 2: a. n's > 20, np's and nq's all > 5

b. z

c. $\alpha = 0.05$

Step 3: a. $n_h = 220,\ x_h = 98,\ p_h' = 98/220 = 0.445$

$n_w = 180,\ x_w = 52,\ p_w' = 52/180 = 0.289$

$p_p' = (x_h + x_w)/(n_h + n_w) = (98+52)/(220+180) = 0.375$

$q_p' = 1 - p_p' = 1.000 - 0.375 = 0.625$

b. $z = [(p_h' - p_w') - (p_h - p_w)] / \sqrt{(p_p')(q_p')[(1/n_h) + (1/n_w)]}$

$z^* = (0.445 - 0.289) / \sqrt{(0.375)(0.625)[(1/220) + (1/180)]}$

$= 0.156/0.0487 = 3.21$

Step 4: -- using p-value approach --------------------

a. **P** = P(z > 3.21);

Using Table 3, Appendix B, ES9-p714:

P = 0.5000 - 0.4993 = 0.0007

Using Table 5, Appendix B, ES9-p716:

0.0006 < **P** < 0.0007

b. **P** < α

-- using classical approach ------------------

a. critical region: z ≥ 1.65

b. z* falls in the critical region

Step 5: a. Reject H_o

b. There is not sufficient evidence to show that home PC owners experienced more problems that could not be solved by the manufacturer, at the 0.05 level.

Hint for writing the hypotheses for exercise 10.83
Look at the last sentence of the exercise; "Is there sufficient evidence to show a difference in the effectiveness of the 2 image campaigns...?" The words "2 image campaigns" imply 2 populations, namely the citizens exposed to a conservative campaign and citizens exposed to a moderate campaign. The results are given in the form of proportions, therefore a difference between 2 proportions is suggested. The word "difference" indicates a not equal to. Therefore, the alternative is not equal to (≠). The negation becomes "equal to" and the null hypothesis would be written with an equality sign (=).

10.83 Step 1: a. The difference in the proportions for the effectiveness of two campaign images, $p_m - p_c$

b. $H_o: p_m - p_c = 0$

$H_a: p_m - p_c \neq 0$

Step 2: a. n's > 20, np's and nq's all > 5

b. z c. α = 0.05

Step 3: a. $n_m = 100,\ p'_m = 0.50,\ \ n_c = 100,\ p'_c = 0.40$

$p'_p = (x_m + x_c)/(n_m + n_c) = (50+40)/(100+100) = 0.45$

$q'_p = 1 - p'_p = 1.000 - 0.45 = 0.55$

b. $z = [(p'_m - p'_c) - (p_m - p_c)] / \sqrt{(p'_p)(q'_p)[(1/n_m) + (1/n_c)]}$

$z^* = (0.50-0.40)/\sqrt{(0.45)(0.55)[(1/100) + (1/100)]}$

$= 0.10/0.0704 = 1.42$

Step 4: -- using p-value approach --------------------

a. $\mathbf{P} = 2P(z > 1.42)$;

Using Table 3, Appendix B, ES9-p714:

$\mathbf{P} = 2(0.5000 - 0.4222) = 2(0.0778) = 0.1556$

Using Table 5, Appendix B, ES9-p716:

$2(0.0735 < \frac{1}{2}\mathbf{P} < 0.0806)$; $0.1470 < \mathbf{P} < 0.1612$

a. $\mathbf{P} > \alpha$

-- using classical approach ------------------

a. Critical region: $z \le -1.96$ and $z \ge 1.96$

b. z^* falls in the noncritical region

Step 5: a. Fail to reject H_o

b. There is not sufficient evidence to show a difference, at the 0.05 level.

10.85 a. $H_a: p_m - p_w > 0$

$p'_p = (x_m + x_w)/(n_m + n_w) = (215+170)/(500 + 500) = 0.385$

$q'_p = 1 - p'_p = 1.000 - 0.385 = 0.615$

$z = [(p'_m - p'_w) - (p_m - p_w)]/\sqrt{(p'_p)(q'_p)[(1/n_m) + (1/n_w)]}$

$z^* = (0.43 - 0.34)/\sqrt{(0.385)(0.615)[(1/500) + (1/500)]}$
$= 0.09/0.0308 = 2.92$

$\mathbf{P} = P(z > 2.92)$;

Using Table 3, Appendix B, ES9-p714:

$\mathbf{P} = 0.5000 - 0.4982 = \underline{0.0018}$

Using Table 5, Appendix B, ES9-p716:

$\underline{0.0016 < \mathbf{P} < .0019}$

b. Reject H_o. The smoking rate for male diabetics is significantly higher than for female diabetics, at the 0.05 level.

SECTION 10.5 ANSWER NOW EXERCISES

10.87 Divide both sides of the original inequality by σ_p^2

10.88 $F(12,24,0.01) = \underline{3.03}$ $F(24,12,0.01) = \underline{3.78}$

10.89 $H_o: \sigma_m/\sigma_p = 1$
$H_a: \sigma_m/\sigma_p > 1$

10.90 $F^* = s_1^2/s_2^2 = (3.2)^2/(2.6)^2 = 1.515 = \underline{1.52}$

10.91 $F(6,9,0.05) = \underline{3.37}$

10.92 $F^* = s_1^2/s_2^2 = 14.44/29.16 = \underline{0.495}$
The smaller variance is in the numerator.

10.93 Multiply it by 2 to cover both sides of the distribution.

10.94 a. $H_o: \sigma_N^2 = \sigma_A^2$ (≤) [police applicants not less variable]

$H_a: \sigma_N^2 > \sigma_A^2$ [police applicants less variable]

b. Have a 0.5% chance of rejecting H_o when it is in fact true - unlikely. That is, sample results are very unlikely if the null hypothesis is true.

c. **P** = 1 - 0.9942= $\underline{0.0058}$

SECTION 10.5 EXERCISES

WRITING HYPOTHESES FOR THE RATIO BETWEEN TWO STANDARD DEVIATIONS OR VARIANCES

null hypothesis:

$H_o: \sigma_1 = \sigma_2$ or $H_o: \sigma_1^2 = \sigma_2^2$ or $H_o: \sigma_1^2/\sigma_2^2 = 1$

possible alternative hypotheses:

$H_a: \sigma_1 > \sigma_2$ or $H_o: \sigma_1^2 > \sigma_2^2$ or $H_o: \sigma_1^2/\sigma_2^2 > 1$

$H_o: \sigma_1 \neq \sigma_2$ or $H_o: \sigma_1^2 \neq \sigma_2^2$ or $H_o: \sigma_1^2/\sigma_2^2 \neq 1$

$H_o: \sigma_2 > \sigma_1$ or $H_o: \sigma_2^2 > \sigma_1^2$ or $H_o: \sigma_2^2/\sigma_1^2 > 1$ **

**Note change for "less than", reverse order.

10.95 a. H_o: $\sigma_A^2 = \sigma_B^2$ vs. H_a: $\sigma_A^2 \neq \sigma_B^2$

b. H_o: $\sigma_I = \sigma_{II}$ vs. H_a: $\sigma_I > \sigma_{II}$

c. H_o: $\sigma_A^2 / \sigma_B^2 = 1$ vs. H_a: $\sigma_A^2 / \sigma_B^2 \neq 1$

d. H_o: $\sigma_C^2 / \sigma_D^2 = 1$ vs. H_a: $\sigma_C^2 / \sigma_D^2 < 1$
or equivalently,
H_o: $\sigma_D^2 / \sigma_C^2 = 1$ vs. H_a: $\sigma_D^2 / \sigma_C^2 > 1$

F-DISTRIBUTION

Key facts about the F-distribution:

1. The total area under the F-distribution is 1.
2. It is zero or positively valued.
3. The shape is skewed right (much like χ^2).
4. A different curve exists for each pair of sample sizes.
5. Critical values are determined based on α and degrees of freedom in the numerator (df_n) and degrees of freedom in the denominator (df_d).
6. Degrees of freedom = df = n - 1.

Notation: $F(df_n, df_d, \alpha)$ =

F(df for numerator, df for denominator, area to the right)

↑ Table 9 — ↑ column id # — ↑ row id # — ↑ Table 9a,b,c

ex: F(10,12,0.05) means df_n = 10(column), df_d = 12(row) and $\alpha = 0.05$ (Table 9a)

Using Table 9a, F(10,12,0.05) = 2.75 (df = n-1 for each sample)

Explore the F-distribution using Interactivities 10A and 10B.

10.97 a. 2.51 b. 2.20 c. 2.91 d. 4.10
e. 2.67 f. 3.77 g. 1.79 h. 2.99

10.99 a. $P(s_1^2 \geq 19s_2^2 \text{ or } s_2^2 \geq 19s_1^2) = P(s_1^2/s_2^2 \geq 19) + P(s_2^2 / s_1^2 \geq 19)$

$= 2P(F \geq 19 | df = 2,2) = 2(0.05) = \underline{0.10}$
(since F(2,2,0.05) = 19)

b. $P(s_1^2 \geq 11s_2^2 \text{ or } s_2^2 \geq 11s_1^2) = P(s_1^2/s_2^2 \geq 11) + P(s_2^2 / s_1^2 \geq 11)$

$= 2P(F \geq 11 | df = 5,5) = 2(0.01) = \underline{0.02}$

(since $F(5,5,0.01) = 11$)

Hypothesis Test for the Ratio Between Two Standard Deviations (or Variances), Independent Samples

Review the parts to a hypothesis test as outlined in: ES9-pp370&386, ST-pp241 & 249, if needed. Slight changes will occur in:

1. **the hypotheses**: (see box before exercise 10.95)
2. **the calculated test statistic**

   $F = s_1^2 / s_2^2$, $df_1 = df_n = n_1 - 1$, $df_2 = df_d = n_2 - 1$
3. If H_o is rejected, a significant difference between the standard deviations (variances) is indicated.
   If H_o is not rejected, no significant difference between the standard deviations (variances) is indicated.

Use subscripts on the sample (or population) variables that identify the source.

Estimate the p-value using Tables 9a, b or c, (Appendix B, ES9-pp720-725).

$P = P(F > F^* | df_n, df_d)$

Locate critical values for the given df_n and df_d on each of the Tables. Compare F* to each and give an interval estimate of P using one or two of the following values: 0.05, 0.025, 0.01. The p-value can be calculated using the cumulative probability commands in ES9-pp517&518.

The p-value can be calculated using the cumulative probability commands in ES9-pp517&518.

Computer and/or calculator commands to perform a hypothesis test between two standard deviations or two variances can be found in ES9-p522&523.

10.101 Step 1: a. The ratio of the variances for two ovens

b. H_o: $\sigma_k^2 = \sigma_m^2$

H_a: $\sigma_k^2 \neq \sigma_m^2$

Step 2: a. normality indicated, independence exists
b. F   c. $\alpha = 0.02$

Step 3: a. $n_m = 16$, $s_m^2 = 2.4$, $n_k = 12$, $s_k^2 = 3.2$
b. $F* = s_k^2/s_m^2 = 3.2/2.4 = 1.33$

Step 4: -- using p-value approach --------------------
a. $\mathbf{P} = 2P(F > 1.33|df = 11,15)$;
Using Tables 9, Appendix B, ES9-pp720-725:
$\mathbf{P} > 2(0.05)$; $\mathbf{P} > 0.10$
b. $\mathbf{P} > \alpha$
-- using classical approach ------------------
a. $F(11, 15, 0.01) = 3.73$

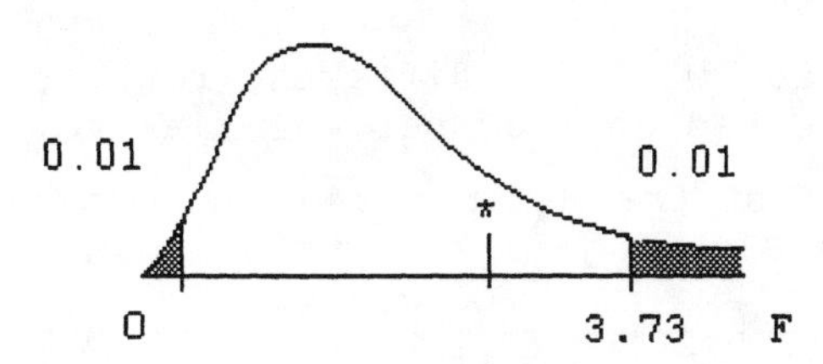

b. F* falls in the noncritical region, see Step 4a

Step 5: a. Fail to reject H_o
b. At the 0.02 level of significance, there is not sufficient evidence to conclude that a difference in variances exist for the two ovens.

10.103 Step 1: a. The ratio of the variances between an older viewing group and a younger viewing group
b. H_o: $\sigma_y^2 = \sigma_o^2$
H_a: $\sigma_y^2 > \sigma_o^2$

Step 2: a. normality assumed, independence exists
b. F   c. $\alpha = 0.05$

Step 3: a. $n_y = 27$, $s_y = 2.9$, $n_o = 28$, $s_o = 1.6$
b. $F* = s_y^2 / s_o^2 = (2.9)^2/(1.6)^2 = 3.29$

Step 4: -- using p-value approach --------------------
a. $\mathbf{P} = P(F > 3.29|df = 26,27)$;
Using Tables 9, Appendix B, ES9-pp720-725
$\mathbf{P} < 0.01$
b. $\mathbf{P} < \alpha$

-- using classical approach ------------------
a. critical region: $F \geq 1.94$
b. F* falls in the critical region

Step 5: a. Reject H_o
At the 0.05 level of significance, there is sufficient evidence that the standard deviation of scores for younger children is larger than the standard deviation for older children.

10.105 $F^* = s_a^2 / s_b^2 = (4.43)^2/(3.50)^2 = \underline{1.60}$

10.107 Everybody will get different results, but they can all be expected to look very similar to the following.

a. Adjust the Minitab and Excel commands from Exercise 10.63a (page 332 of this manual).

N(100,20)

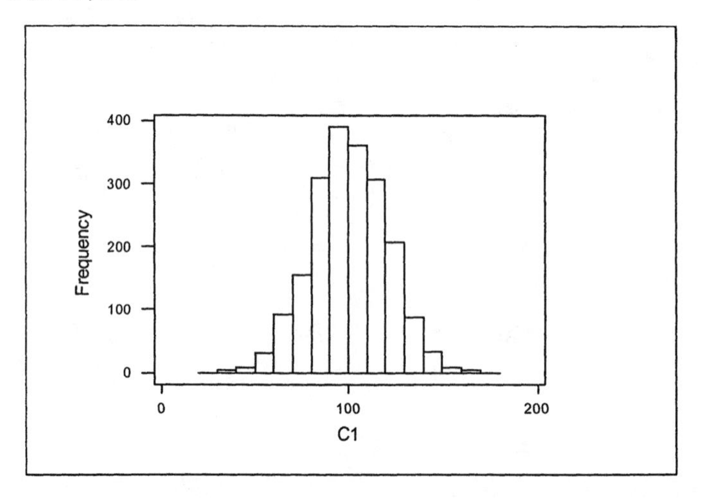

N(120,20)

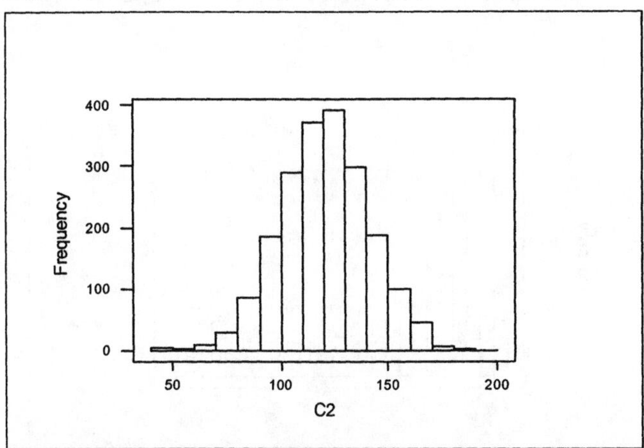

These two very large samples strongly suggest that we are sampling two normal populations.

b. Minitab commands:
Use commands from Exercise 10.63c (page 334) substituting standard deviation for mean in the row statistics and (C20/C11)**2 for the expression in C21.

Excel commands:
Use commands from Exercise 10.63c (page 334) substituting standard deviation for average in the paste function and (T1/K1)**2 for the expression in U1.

c. Use the Minitab and Excel commands from Exercise 10.63d (page 334) using cutpoint classes from 0 to 5 in increments of 0.2 for the column of F's.

An empirical distribution of 100 F* values

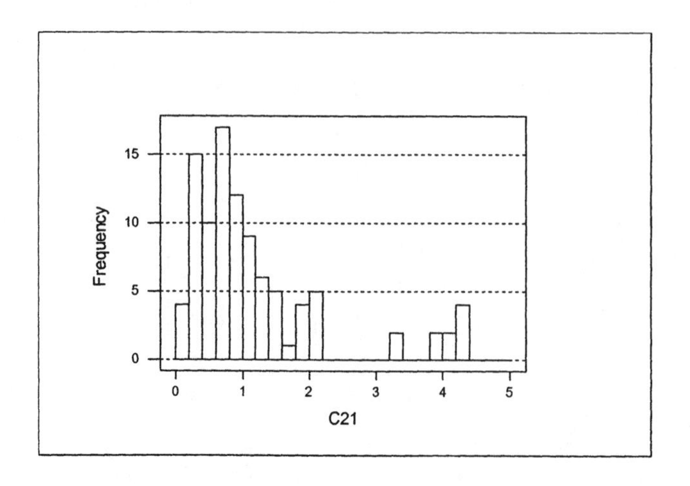

d. Minitab commands:
Use patterned data commands to list from 0 to 5 in increments of 0.2 into C22. Use cumulative probability commands on pages 517 & 518 of the text with 7 degrees of freedom for both the numerator and denominator, C22 as the input column and C23 as the output column. The Plot command, with C23 and C22 and Connect, will draw the graph.

Excel commands:
Use patterned data commands to list from 0 to 5 in increments of 0.2 into column V. Use cumulative probability commands on pages 517 & 518 of the text with 7 degrees of freedom for both the numerator and denominator, column V as the F values and column W as the ouput range. Use Edit formula (=) to calculate column X = 1 - col. W. Use Chart Wizard for the graph of the data in columns X and V.

0.7
0.6
0.5
0.4
0.3
0.2
0.1
0.0
C23
0
1
2
3
4
5
C22

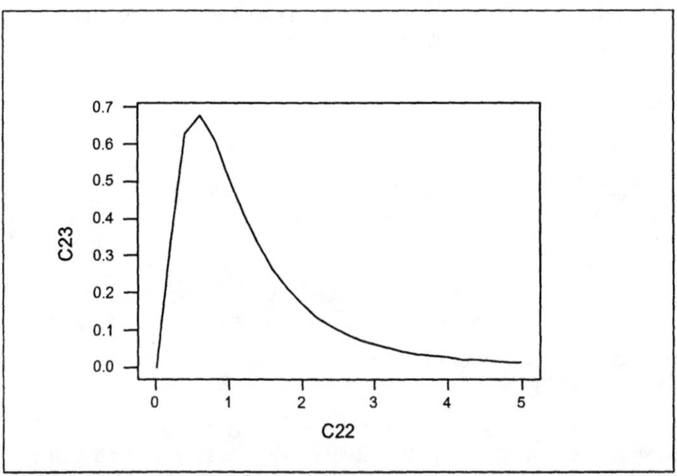

The observed distribution of F*s seems to be very similar to the theoretical distribution. Remember these samples were drawn from populations that were both normal and had the same standard deviations.

RETURN TO CHAPTER CASE STUDY

10.109 a. Freshmen: proportion = 97/200 = 0.485 or 48.5%
Sophomore: proportion = 187/200 = 0.93.5 or 93.5%

b. Point estimate for difference between two proportions: 0.935 - 0.485 = 0.45 or 45%

c. Freshmen: mean = 60752/40 = 1518.8 = $1519
Sophomore: mean = 91473/44 = 2078.9 = $2079

d.

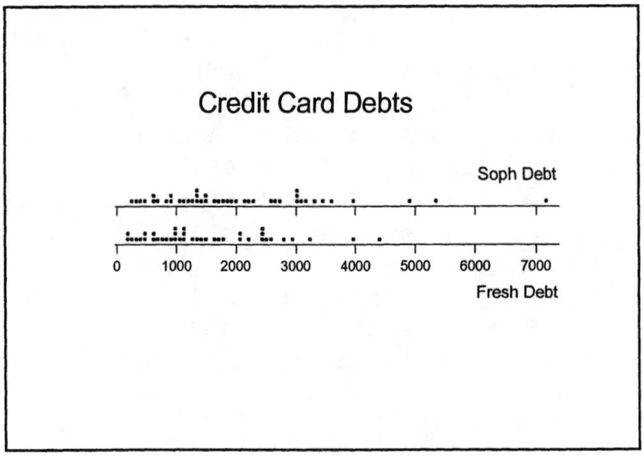

Credit Card Debts
Soph Debt
0
1000
2000
3000
4000
5000
6000
7000
Fresh Debt

e. Point estimate for difference between two means: 2079 - 1519 = \$560.

10.111 Answers will vary.

CHAPTER EXERCISES

10.113 Step 1: The mean difference in IQ scores for oldest and youngest members of a family (d = O - Y)

Step 2: a. normality indicated

b. t   c. $1-\alpha = 0.95$

Step 3: $n = 12$, $\bar{d} = 3.583$, $s_d = 19.58$

Step 4: a. $\alpha/2 = 0.05/2 = 0.025$; df = 11; $t(11, 0.025) = 2.20$

b. $E = t(df,\alpha/2)\cdot(s_d/\sqrt{n}) = (2.20)(19.58/\sqrt{12})$
$= (2.20)(5.65) = 12.435$

c. $\bar{d} \pm E = 3.583 \pm 12.435$

Step 5: <u>-8.85 to 16.02</u>, the 0.95 interval for μ

10.115 Step 1: a. The mean difference in scores for recruits participating in a rifle-shooting competition (d = week later - first day)

b. $H_o: \mu_d = 0$
$H_a: \mu_d > 0$ (improvement)

Step 2: a. normality assumed

b. t   c. $\alpha = 0.05$

Step 3: a. $n = 10$, $\bar{d} = 5.2$, $s_d = 7.406$

b. $t* = (\bar{d} - \mu_d)/(s_d/\sqrt{n})$
$= (5.2 - 0.0)/(7.406/\sqrt{10}) = 2.22$

Step 4: -- using p-value approach --------------------

a. **P** = $P(t > 2.22|df=9)$;
Using Table 6, Appendix B, ES9-p717:
$0.025 <$ **P** < 0.05
Using Table 7, Appendix B, ES9-p728:
$0.022 <$ **P** < 0.029

b. **P** $< \alpha$

-- using classical approach ------------------

a. critical region: $t \geq 1.83$

b. t* falls in the critical region

Step 5: a. Reject H_o

b. There is a significant improvement, at the 0.05 level.

10.117 Step 1: The difference between the mean anxiety scores for males and females, $\mu_f - \mu_m$

Step 2: a. normality assumed, CLT with $n_f = 50$ and $n_m = 50$.

b. t   c. $1-\alpha = 0.95$

Step 3: $n_f = 50$, $\bar{x}_f = 75.7$, $s_f = 13.6$

$n_m = 50$, $\bar{x}_m = 70.5$, $s_m = 13.2$

$\bar{x}_f - \bar{x}_m = 75.7 - 70.5 = 5.2$

Step 4: a. $\alpha/2 = 0.05/2 = 0.0025$; df = 49;

$t(49, 0.025) = 2.02$

b. $E = t(df,\alpha/2)\cdot\sqrt{(s_f^2 / n_f) + (s_m^2 / n_m)}$

$= (2.02)\sqrt{(13.6^2/50) + (13.2^2/50)}$

$= (2.02)(2.68) = 5.41$

c. $(\bar{x}_f - \bar{x}_m) \pm E = 5.2 \pm 5.41$

Step 5: -0.21 to 10.61, the 0.95 interval for $\mu_f - \mu_m$

10.119 Step 1: The difference between mean rifle-shooting scores for two companies, $\mu_B - \mu_A$

Step 2: a. normality assumed

b. t   c. $1-\alpha = 0.95$

Step 3: $n_A = 10$, $\bar{x}_A = 57.0$, $s_A^2 = 209.111$, $s_A = 14.46$

$n_B = 10$, $\bar{x}_B = 62.2$, $s_B^2 = 193.289$, $s_B = 13.90$

$\bar{x}_B - \bar{x}_A = 62.2 - 57.0 = 5.2$

Step 4: a. $\alpha/2 = 0.05/2 = 0.025$; df = 9;

$t(9, 0.025) = 2.26$

b. $E = t(df,\alpha/2)\cdot\sqrt{(s_B^2 / n_B) + (s_A^2 / n_A)}$

$= (2.26)\sqrt{(193.289/10) + (209.111/10)}$

$= (2.26)(6.3435) = 14.34$

c. $(\bar{x}_B - \bar{x}_A) \pm E = 5.2 \pm 14.34$

Step 5: -9.14 to 19.54, the 0.95 confidence interval for $\mu_B - \mu_A$

10.121 Step 1: The difference between means of two methods used in ice fusion, $\mu_A - \mu_B$

Step 2: a. normality assumed

b. t   c. $1-\alpha = 0.95$

Step 3: $n_A = 13$, $\bar{x}_A = 80.021$, $s_A^2 = 0.0005744$, $s_A = 0.02397$

$n_B = 8$, $\bar{x}_B = 79.979$, $s_B^2 = 0.0009839$, $s_B = 0.03137$

$\bar{x}_A - \bar{x}_B = 80.021 - 79.979 = 0.042$

Step 4: a. $\alpha/2 = 0.05/2 = 0.025$; df = 7;
$t(7, 0.025) = 2.36$

b. $E = t(df,\alpha/2)\cdot\sqrt{(s_A^2 / n_A) + (s_B^2 / n_B)}$

$= (2.36)\cdot\sqrt{(0.0005744/13) + (0.0009839/8)}$

$= (2.36)(0.0129) = 0.030$

c. $(\overline{x}_A - \overline{x}_B) \pm E = 0.042 \pm 0.030$

Step 5: <u>0.012 to 0.072</u>, the 0.95 confidence interval for $\mu_A - \mu_B$

10.123 Step 1: a. The difference between the mean 40-yard sprint time recorded by football players on artificial turf and grass, $\mu_2 - \mu_1$

b. H_o: $\mu_2 - \mu_1 = 0$ (no difference)
H_a: $\mu_2 - \mu_1 > 0$ (artif. turf yields a lower time)

Step 2: a. normality assumed, CLT with $n_c = 22$ and $n_h = 22$.
b. t c. $\alpha = 0.05$

Step 3: a. sample information given in exercise

b. $t = [(\overline{x}_2 - \overline{x}_1) - (\mu_2 - \mu_1)] / \sqrt{(s_2^2/n_2) + (s_1^2/n_1)}$

$t^* = [(4.96 - 4.85) - 0] / [\sqrt{(0.42^2/22) + (0.31^2/22)}]$

$= 0.988$

Step 4: -- using p-value approach --------------------

a. **P** = $P(t > 0.988 | df = 21)$;
Using Table 6, Appendix B, ES9-p717:
$0.10 <$ **P** < 0.25
Using Table 7, Appendix B, ES9-p718:
$0.164 <$ **P** < 0.189

b. **P** $> \alpha$

-- using classical approach ------------------

a. critical region: $t \geq 1.72$

b. t^* falls in the noncritical region

Step 5: a. Fail to reject H_o

b. At the 0.05 level of significance, there is insufficient evidence to show that the artificial turf had faster mean sprint times than that of grass.

10.125 a. $n_A = 15$, $\overline{x}_A = \underline{15.53}$, $s_A^2 = \underline{1.98}$, $s_A = \underline{1.41}$

b. $n_B = 15$, $\overline{x}_B = \underline{12.53}$, $s_B^2 = \underline{1.98}$, $s_B = \underline{1.41}$

c. Step 1: a. The difference between mean torques required to remove screws from two different materials, $\mu_A - \mu_B$

b. $H_o: \mu_A - \mu_B = 0$
$H_a: \mu_A - \mu_B \neq 0$

Step 2: a. normality indicated

b. t c. $\alpha = 0.01$

Step 3: a. sample information given above

b. $t* = [(\overline{x}_A - \overline{x}_B) - (\mu_A - \mu_B)] / \sqrt{(s_A^2 / n_A) + (s_B^2 / n_B)}$
$= [(15.53-12.53)-0] / [\sqrt{(1.98/15) + (1.98/15)}]$
$= 5.84$

Step 4: a. **P** $= 2P(t* > 5.84 | df = 14)$;
Using Table 6, Appendix B, ES9-p717:
P < 0.01
Using Table 7, Appendix B, ES9-p718:
$2(\frac{1}{2}\mathbf{P} < 0.001)$, **P** < 0.002

b. **P** $< \alpha$

-- using classical approach ------------------

a. critical region: $t \leq -2.98$, $t \geq 2.98$

b. t* falls in the critical region

Step 5: a. Reject H_o

b. At the 0.01 level of significance, the mean torques are not the same for Material A and Material B.

10.127 a. M: $\overline{x}$ = 74.69, s = 10.19, F: $\overline{x}$ = 79.83, s = 8.80

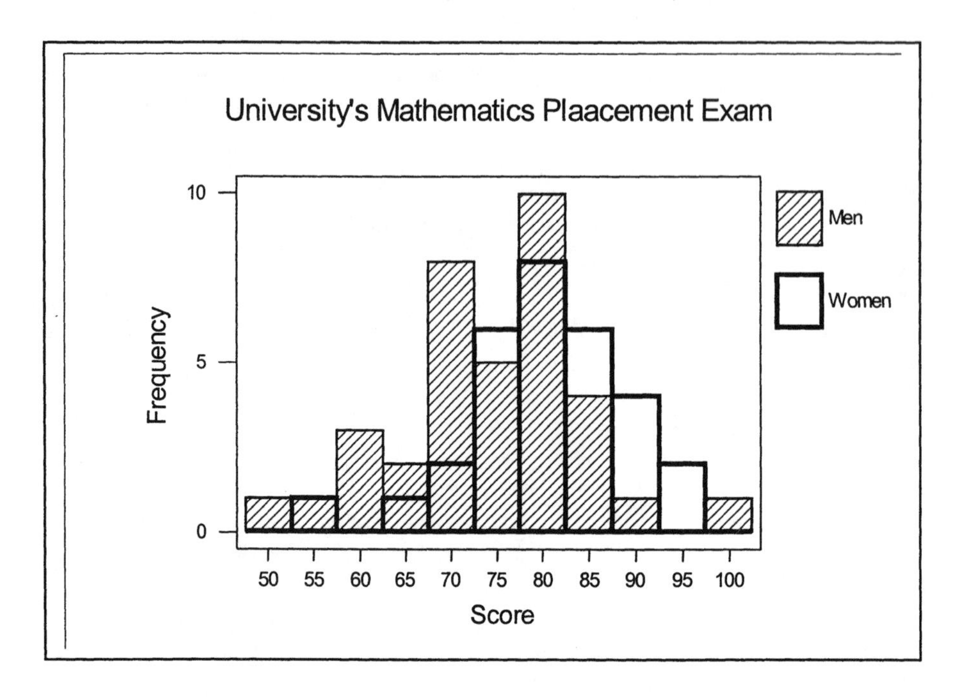

b.

Step 1: a. The mean mathematics placement score for all men

b. H_o: $\mu = 77$

H_a: $\mu \neq 77$

Step 2: a. normality indicated

b. t c. $\alpha = 0.05$

Step 3: a. n = 36, $\overline{x}$ = 74.69, s = 10.19

b. $t = (\overline{x} - \mu)/(s/\sqrt{n})$

$t* = (74.69 - 77.0)/(10.19/\sqrt{36}) = -1.36$

Step 4: -- using p-value approach --------------------

a. **P** = 2P(t > 1.36|df = 35);

Using Table 6, Appendix B, ES9-p717:

0.10 < **P** < 0.20

Using Table 7, Appendix B, ES9-p718:

0.085 < ½**P** < 0.101; 0.170 < **P** < 0.202

Using a computer: **P** = 0.183

b. **P** > α

-- using classical approach -----------------
a. $t \le -2.03$, $t \ge 2.03$
b. t* falls in the noncritical region, see Step 4a

Step 5: a. Fail to reject H_o
b. The sample does provide sufficient evidence that the mean score for men is 77, at the 0.05 level of significance.

Step 1: a. The mean mathematics placement score for all women
b. H_o: $\mu = 77$
H_a: $\mu \neq 77$

Step 2: a. normality indicated
b. t   c. $\alpha = 0.05$

Step 3: a. $n = 30$, $\bar{x} = 79.83$, $s = 8.80$
b. $t = (\bar{x} - \mu)/(s/\sqrt{n})$
$t* = 79.83 - 77.0/(8.80/\sqrt{30} = 1.76$

Step 4: -- using p-value approach -------------------
a. $\mathbf{P} = 2P(t > 1.76|df = 29)$;
Using Table 6, Appendix B, ES9-p717:
$0.05 < \mathbf{P} < 0.10$
Using Table 7, Appendix B, ES9-p718:
$0.041 < \frac{1}{2}\mathbf{P} < 0.050$; $0.082 < \mathbf{P} < 0.100$
Using a computer: $\mathbf{P} = 0.089$
b. $\mathbf{P} > \alpha$
-- using classical approach -----------------
a. $t \le -2.05$, $t \ge 2.05$
b. t* falls in the noncritical region, see Step 4a

Step 5: a. Fail to reject H_o
b. The sample does provide sufficient evidence that the mean score for women is 77, at the 0.05 level of significance.

c. They're both not significantly different than 77. It is easy to jump to the wrong conclusion.

d. Step 1: a. The difference between mean scores for male and female students, $\mu_F - \mu_M$
b. H_o: $\mu_F - \mu_M = 0$
H_a: $\mu_F - \mu_M \neq 0$

Step 2: a. normality indicated

b. t   c. $\alpha = 0.05$

Step 3: a. sample information given above

b. $t* = (79.83 - 74.7)/\sqrt{((8.80^2 / 30) + (10.2^2 / 36))}$
$= 2.19$

Step 4: a. $\mathbf{P} = 2P(t* > 2.19 | df = 29)$;

Using Table 6, Appendix B, ES9-p717:
$0.02 < \mathbf{P} < 0.05$

Using Table 7, Appendix B, ES9-p718:
$2(0.018 < \frac{1}{2}\mathbf{P} < 0.022)$, $0.036 < \mathbf{P} < 0.044$

b. $\mathbf{P} < \alpha$

-- using classical approach ------------------

a. critical region: $t \leq -2.05$, $t \geq 2.05$

b. t* falls in the critical region

Step 5: a. Reject H_o

b. At the 0.05 level of significance, the mean scores for men and women are not equal.

e. & f. No, a significant difference was found. The questions of (b) and (d) are asking different questions. In (b), individually two hypothesis tests are testing if the means are different than 77. In this case, the two sample means are on opposite sides of 77, but not significantly far from 77. Yet the two sample means are themselves far enough apart to be significantly different.

10.129 a. Step 1: The difference in proportions of stocks making a gain, $p_n - p_a$

Step 2: a. n's > 20, np's and nq's all > 5

b. z   c. $1-\alpha = 0.99$

Step 3: $n_n = 100$, $x_n = 32$, $p'_n = 32/100 = 0.32$

$q'_n = 1 - 0.32 = 0.68$

$n_a = 100$, $x_a = 27$, $p'_a = 27/100 = 0.27$

$q'_a = 1 - 0.27 = 0.73$

$p'_n - p'_a = 0.32 - 0.27 = 0.050$

Step 4: a. $\alpha/2 = 0.01/2 = 0.005$; $z(0.005) = 2.58$

b. $E = z(\alpha/2)\cdot\sqrt{(p'_n \cdot q'_n / n_n) + (p'_a \cdot q'_a / n_a)}$

$= 2.58\cdot\sqrt{(0.32 \cdot 0.68 / 100) + (0.27 \cdot 0.73 / 100)}$

$= (2.58)(0.064397) = 0.166$

c. $(p'_n - p'_a) \pm E = 0.050 \pm 0.166$

Step 5: $\underline{-0.116 \text{ to } 0.216}$, the 0.99 interval for $p_n - p_a$

b. No, there is no significant difference at the 0.01 level because the interval estimate contains the value 0.

10.131 Step 1: The difference in proportions requiring service from two manufacturers, $p_1 - p_2$

Step 2: a. n's > 20, np's and nq's all > 5

b. z   c. $1-\alpha = 0.95$

Step 3: sample information given in exercise

$p'_1 - p'_2 = 0.15 - 0.09 = 0.060$

Step 4: a. $\alpha/2 = 0.05/2 = 0.025$; $z(0.025) = 1.96$

b. $E = z(\alpha/2)\cdot\sqrt{(p'_1 \cdot q'_1 / n_1) + (p'_2 \cdot q'_2 / n_2)}$

$= 1.96\cdot\sqrt{(0.15 \cdot 0.85/75) + (0.09 \cdot 0.91/75)}$

$= (1.96)(0.0528) = 0.104$

c. $(p'_1 - p'_2) \pm E = 0.060 \pm 0.104$

Step 5: $\underline{-0.044 \text{ to } 0.164}$, the 0.95 interval for $p_1 - p_2$

10.133 Find the p-value for each sample size for the situation:

$H_o: p_m - p_w = 0$ vs. $H_a: p_m - p_w \neq 0$

a. n = 100:   No

$x_m = 65$, $x_w = 55$

$p'_p = (x_m + x_w)/(n_m + n_w) = (65+55)/(100+100) = 0.600$

$q'_p = 1 - p'_p = 1.000 - 0.600 = 0.400$

$z* = [(p'_m - p'_w) - (p_m - p_w)] / \sqrt{(p'_p)(q'_p)[(1/n_m)+(1/n_w)]}$

$z* = (0.65 - 0.55) / \sqrt{(0.60)(0.40)[(1/100)+(1/100)]}$

$= 0.10/0.069 = 1.44$

P = 2P(z > 1.44);

Using Table 3, Appendix B, ES9-p714:

P = 2(0.5000 - 0.4251) = 2(0.0749) = $\underline{0.1498}$

b. $n = 150$: No

$x_m = 98,\ x_w = 83$

$p'_p = (x_m + x_w)/(n_m + n_w) = (98+82)/(150+150) = 0.600$

$q'_p = 1 - p'_p = 1.000 - 0.600 = 0.400$

$z* = [(p'_m - p'_w) - (p_m - p_w)] / \sqrt{(p'_p)(q'_p)[(1/n_m)+(1/n_w)]}$

$z* = (0.65 - 0.55) / \sqrt{(0.60)(0.40)[(1/150)+(1/150)]}$

$= 0.10/0.0566 = 1.77$

P $= 2P(z > 1.77)$;

Using Table 3, Appendix B, ES9-p714:

P $= 2(0.5000 - 0.4616) = 2(0.0384) = \underline{0.0768}$

c. $n = 200$: Yes

$x_m = 130,\ x_w = 110$

$p'_p = (x_m + x_w)/(n_m + n_w) = (130+110)/(200+200) = 0.600$

$q'_p = 1 - p'_p = 1.000 - 0.600 = 0.400$

$z* = [(p'_m - p'_w) - (p_m - p_w)] / \sqrt{(p'_p)(q'_p)[(1/n_m)+(1/n_w)]}$

$z* = (0.65 - 0.55) / \sqrt{(0.60)(0.40)[(1/200)+(1/200)]}$

$= 0.10/0.049 = 2.04$

P $= 2P(z > 2.04)$;

Using Table 3, Appendix B, ES9-p714:

P $= 2(0.5000 - 0.4793) = 2(0.0207) = \underline{0.0414}$

As the sample size increases, the 65% and 55% become more standard errors apart. If a = 0.05 is used, 655 and 55% would be significantly different provided each of the samples used were of 200 or more people.

10.135 Step 1: a. The difference in the proportion of accountants and lawyers who believe that the new burden-of-proof tax rules will cause an increase in taxpayer wins in court, $p_a - p_l$

b. H_o: $p_a - p_l = 0$

H_a: $p_a - p_l \neq 0$

Step 2: a. n's > 20, np's and nq's all > 5

b. z c. $\alpha = 0.01$

Step 3: a. $n_a = 175$, $x_a = 101$, $p'_a = 101/175 = 0.5771$

$n_l = 165$, $x_l = 84$, $p'_l = 84/165 = 0.5091$

$p'_p = x_a + x_l)/(n_a + n_l) = (101+84)/(175+165) = 0.5441$

$q'_p = 1 - p'_p = 1.000 - 0.5441 = 0.4559$

b. $z = [(p'_a - p'_l) - (p_a - p_l)] / \sqrt{(p'_p)(q'_p)[(1/n_a) + (1/n_l)]}$

$z* = (0.5771-0.5091) / \sqrt{(0.5441)(0.4559)[(1/175) + (1/165)]}$

$= 0.068/0.054 = 1.26$

Step 4: -- using p-value approach --------------------

a. $\mathbf{P} = 2P(z > 1.26)$;

Using Table 3, Appendix B, ES9-p714:

$\mathbf{P} = 2(0.5000 - 0.3962) = 2(0.1038) = 0.2076$

Using Table 5, Appendix B, ES9-p716:

$0.0968 < ½\mathbf{P} < 0.1056$; $0.1936 < \mathbf{P} < 0.2112$

b. $\mathbf{P} > \alpha$

-- using classical approach ------------------

a. critical region: $z \leq -2.58$ and $z \geq 2.58$

b. z* falls in the noncritical region

Step 5: a. Fail to reject H_o

a. There is not sufficient evidence to show that accountants and lawyers differ in their beliefs about the burden of proof, at the 0.01 level.

10.137 Step 1: a. The ratio of the variance of time needed by men to that needed by women to assemble a product

b. H_o: $\sigma_m^2 = \sigma_f^2$

H_a: $\sigma_m^2 > \sigma_f^2$

Step 2: a. normality indicated, independence exists

b. F   c. $\alpha = 0.05$

Step 3: a. $n_m = 15$, $s_m = 4.5$, $n_f = 15$, $s_f = 2.8$

b. $F* = s_m^2 / s_f^2 = (4.5)^2/(2.8)^2 = 2.58$

Step 4: -- using p-value approach --------------------

a. $\mathbf{P} = P(F > 2.58 | df = 14,14)$;

Using Tables 9, Appendix B, ES9-pp720-725:

$0.025 < \mathbf{P} < 0.05$

b. $\mathbf{P} < \alpha$

-- using classical approach ------------------

a. critical region: $F \geq 2.53$

b. F* falls in the critical region

Step 5: a. Reject H_o
b. At the 0.05 level of significance, there is sufficient evidence to conclude that male assembly times are more variable.

10.139 Step 1: a. The difference between the variances in the threads of lug nuts and studs
b. H_o: $\sigma_n^2 = \sigma_s^2$
H_a: $\sigma_n^2 \neq \sigma_s^2$

Step 2: a. normality indicated, independence exists
b. F c. $\alpha = 0.05$

Step 3: a. $n_n = 60$, $s_n^2 = 0.00213$, $n_s = 40$, $s_s^2 = 0.00166$
b. $F* = s_n^2 / s_s^2 = 0.00213/0.00166 = 1.28$

Step 4: -- using p-value approach --------------------
a. **P** = $P(F > 1.28|df = 59,39)$;
Using Tables 9, Appendix B, ES9-pp720-725:
P > 0.10
b. **P** > α
-- using classical approach ------------------
a. critcal region: $F \geq 1.80$
b. F* falls in the noncritical region

Step 5: a. Fail to reject H_o
b. At the 0.05 level of significance, there is no difference in the variances of lug nuts and studs.

10.141 a.

	N	Mean	StDev
Cont	50	0.005459	0.000763
Test	50	0.003507	0.000683

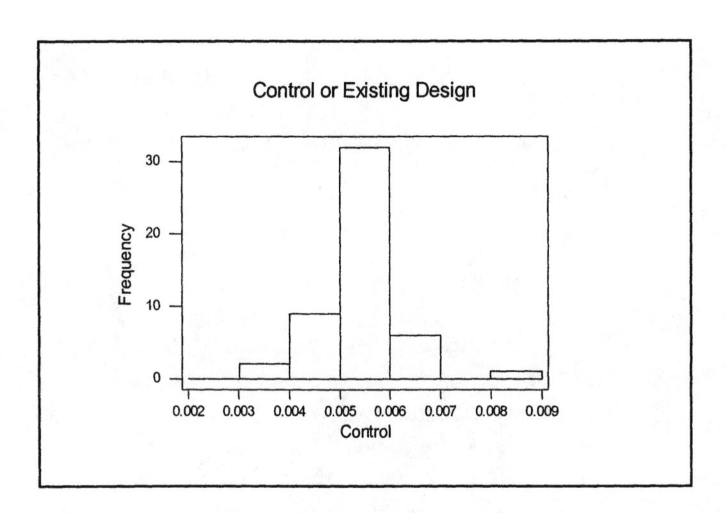

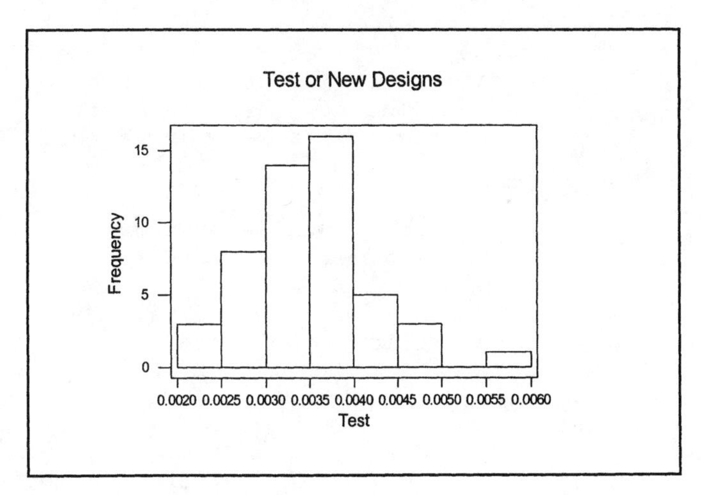

b. Both sets of data have an approximately normal distribution, therefore the assumptions are satisfied.

c. One-tailed - looking for a reduction

d.

Step 1: a. The difference between the variances in two different designs

b. H_o: $\sigma_c^2 = \sigma_t^2$

H_a: $\sigma_c^2 > \sigma_t^2$ (variance for control is greater)

Step 2: a. normality indicated, independence exists

b. F c. $\alpha = 0.05$

Step 3: a. $n_c = 50$, $s_c = 0.000763$, $n_t = 50$, $s_t = 0.000683$

b. $F^* = s_c^2 / s_t^2 = (0.000763^2/0.000683^2) = 1.248$

Step 4: -- using p-value approach --------------------

a. **P** = P(F > 1.25|df = 49,49);

Using Tables 9, Appendix B, ES9-pp720-725:

P > 0.05

Using a computer: **P** = 0.44

b. **P** > α

-- using classical approach -----------------

a. critical region: $F \geq 1.69$

b. F* falls in the noncritical region

Step 5: a. Fail to reject H_o

b. At the 0.05 level of significance, there is no difference in the variances for the two designs. The new design has not significantly reduced the variability.

e. Step 1: a. The difference between mean amount of force for two designs, $\mu_c - \mu_t$

b. H_o: $\mu_c - \mu_t = 0$

H_a: $\mu_c - \mu_t > 0$

Step 2: a. normality indicated

b. t c. $\alpha = 0.05$

Step 3: a. sample data given above

b. t* =

$$[(0.005459-0.003507)-0]/\sqrt{((0.000763^2/50)+(0.000683^2/50))}$$

= 13.48

Step 4: a. **P** = P(t > 13.48|df = 49);

Using Table 6, Appendix B, ES9-p717:

P = +0.000

Using Table 7, Appendix B, ES9-p718:

P = +0.000

Using a computer: **P** = 0.000

b. **P** < α

-- using classical approach ------------------
a. critical region: $t \le -1.68$, $t \ge 1.68$
b. t^* falls in the critical region

Step 5: a. Reject H_o
b. At the 0.05 level of significance, there is sufficient evidence to show that the new design has reduced the mean amount of force

f. The mean force has been reduced, but not the variability.

CHAPTER 11 ∇ APPLICATIONS OF CHI-SQUARE

Chapter Preview

Chapter 11 demonstrates hypothesis tests, as did Chapters 8, 9, and 10. The difference lies in the type of data that is to be analyzed. Enumerative type data, that is, data which can be counted and placed into categories, will be discussed and investigated in three types of tests. Each test will compare actual (observed) results with expected (theoretical) results. One will use the comparisons to determine whether a "claimed" relationship exists, one will determine whether two factors or variables are independent, and the third will determine whether the proportions per variable are the same. Also, the chi square statistic, χ^2, will be reintroduced and utilized in performing these tests.

An article appearing in USA Today showing the ways that people cool their mouths after eating spicy hot food is used in the Case Study for this chapter.

CHAPTER 11 CASE STUDY

11.1 a. and b.

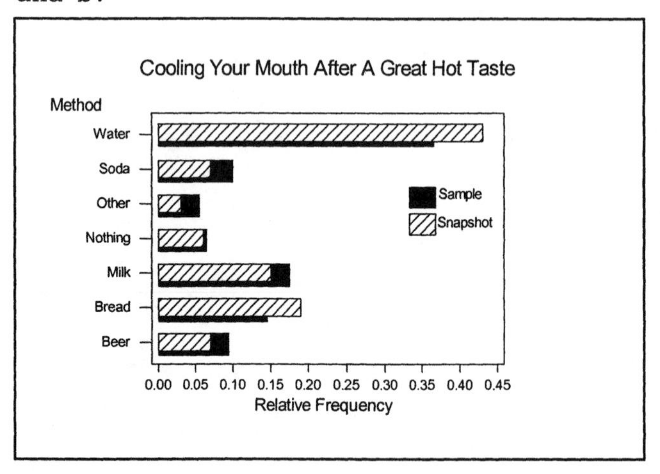

c. The two distributions seem to be fairly similar: water has the greatest response, milk and bread are similar and next most popular, while soda, beer and nothing all have similar shares.

SECTION 11.1 ANSWER NOW EXERCISES

11.2 a. 23.2 b. 23.3 c. 3.94 d. 8.64

SECTION 11.1 EXERCISES

11.3 a. 34.8 b. 28.8 c. 51.8 d. ≈ 69.95
e. $\chi^2(14,\ 0.01) = 29.1$
f. $\chi^2(27,\ 0.975) = 14.6$, $\chi^2(27,\ 0.025) = 43.2$

SECTION 11.2 ANSWER NOW EXERCISES

11.4 9 parts + 3 parts + 3 parts + 1 part = 16 parts total.

11.5 $E = n \cdot p = 556(9/16) = 312.75$

$O - E = 315 - 312.75 = 2.25$

$(O-E)^2/E = (2.25)^2/312.75 = 0.0162$

11.6 a. Each adult surveyed

b. Where lesson learned

c. Mistakes, school, unsure

11.7 Since the percentages sum to more than 1.0, some people use more than one method. In order to be a multinomial experiment, each source most yield exactly one response.

SECTION 11.2 EXERCISES

Characteristics of a Multinomial Experiment

1. There are **n** identical independent trials.
2. The outcome of each trial fits into exactly one of the **k** possible categories or cells.
3. The number of times a trial outcome falls into a particular cell is given by O_i (O - for observed, i for $1 \rightarrow k$).
4. $O_1 + O_2 + O_3 + \ldots + O_k = n$
5. There is a constant probability associated with each of the k cells in such a way that $p_1 + p_2 + \ldots + p_k = 1$.

NOTE: Variable - the characteristic about each item that is of interest.

Various levels of the variable - the k possible outcomes or responses.

Writing Hypotheses for Multinomial Experiments

The null hypothesis is written in a form to show that there is no difference between the experimental (observed) frequencies and the theoretical (expected) frequencies.
The alternative hypothesis is the "opposite" of the null hypothesis. It is written in a form to show that a difference does exist.

ex.: The marital status distribution for New York state is 21%, 64%, 8%, and 7% for the possible categories of single, married, widowed, and divorced.

H_o: P(S) = .21, P(M) = .64, P(W) = .08, P(D) = .07
H_a: The percentages are different than specified in H_o.

ex.: A gambler thinks that a die may be *loaded*.

H_o: P(1)=P(2)=P(3)=P(4)=P(5)=P(6) = 1/6 (not loaded)
H_a: The probabilities are different. (is loaded)

The p-value is estimated using Table 8 (Appendix B, ES9-p719):

a) Locate df row.

b) Locate χ^{2*} between two critical values in the df row; the p-value is in the interval between the two corresponding probabilities at the top of the columns labeled *area to the right*.

OR

The p-value can be calculated using computer and/or calculator commands found in ES9-p452.

11.9 a. $\mathbf{P} = P(\chi^2 > 12.25 | df=3)$;

Using Table 8: $0.005 < \mathbf{P} < 0.01$

Using computer/calculator: $\mathbf{P} = 0.0066$

b. $\mathbf{P} = P(\chi^2 > 5.98 | df=2)$;

Using Table 8: $0.05 < \mathbf{P} < 0.10$

Using computer/calculator: $\mathbf{P} = 0.0503$

Determining the Test Criteria

1. Draw a picture of the χ^2 distribution (skewed right, starting at 0).

2. Locate the critical region (based on α and H_a).
   Since we are testing H_o: "no difference" versus H_a: "difference", all of the α is placed in the right tail to represent a significant or large *difference* between the observed and expected values.

3. Shade in the critical region (the area where you will reject H_o, the right-hand tail)

4. Find the appropriate critical value from Table 8 (Appendix B, ES9-p719), using $\chi^2(df,\alpha)$, where $df = k - 1$. k is equal to the number of cells or categories the data are classified into.

Remember this critical or boundary value divides the area under the χ^2 distribution curve into critical and noncritical regions and is part of the critical region.

Steps for a Hypothesis/Significance Test for Multinomial Experiments

Follow the steps outlined for p-value and classical hypothesis tests in: ES9-pp370 & 386, ST-pp241 & 249.
The variations are noted below.

1. Distribute the sample information (observed frequencies) into the appropriate cells (data may be already categorized).

2. Calculate the expected frequencies using probabilities determined by H_o and the formula:
   $E_i = np_i$, where E_i is the expected frequency for cell i, p_i is the probability for the ith cell

3. Use the observed and expected frequencies from each cell to calculate the test statistic, χ^2.

   Use the formula $$\chi^2 = \sum_{\text{allcells}} \frac{(O - E)^2}{E}$$

4.a) p-value approach:
   Since all of the α is placed in the right tail, the p-value $= P(\chi^2 > \chi^{2*})$.

   The p-value is <u>estimated</u> using Table 8 (Appendix B, ES9-p719):
   a) Locate df row.
   b) Locate χ^{2*} between two critical values in the df row; the p-value is in the interval between the two corresponding probabilities at the top of the columns labeled *area to the right*.

   OR
   The p-value can be calculated using computer and/or calculator commands found in ES9-p452.

   b) Classical approach:
   Follow the steps in determining the test criteria found in: ES9-p543, ST-p369. Then locate χ^{2*} on the χ^2 curve with respect to the critical value.

5. Make a decision and interpret it.
   a) If $P < \alpha$ or the calculated test statistic falls into the critical region, then reject H_o. There is sufficient evidence to indicate that there is a *difference* between the observed and expected frequencies.
   b) If $P > \alpha$ or the calculated test statistic falls into the noncritical region, fail to reject H_o. There is <u>not</u> sufficient evidence to indicate a *difference* between the observed and expected frequencies.

11.11 a. H_o: $P(A) = P(B) = P(C) = P(D) = P(E) = 0.2$

b. χ^2

c. (1) & (2)

Step 1: a. Preference of floor polish, the probability that a particular type is preferred

b. H_o: Equal preference
H_a: preferences not all equal

Step 2: a. Assume that the 100 consumers represent a random sample.

b. χ^2 with df = 4 c. $\alpha = 0.10$

Step 3: a. sample information given in exercise

b. $\chi^2 = \Sigma[(O-E)^2/E]$ (as found on accompanying table) $E = n \cdot p = 100(1/5) = 20$, for all cells

Polish	A	B	C	D	E	Total
Observed	27	17	15	22	19	100
Expected	20	20	20	20	20	100
$(O-E)^2/E$	49/20	9/20	25/20	4/20	1/20	88/20

$\chi^{2}* = 4.40$

Step 4: -- using p-value approach ------------------

a. $\mathbf{P} = P(\chi^2 > 4.40|df=4)$;

Using Table 8: $0.25 < \mathbf{P} < 0.50$

Using computer/calculator: $\mathbf{P} = 0.355$

b. $\mathbf{P} > \alpha$

-- using classical approach ----------------

a. $\chi^2(4, 0.10) = 7.78$

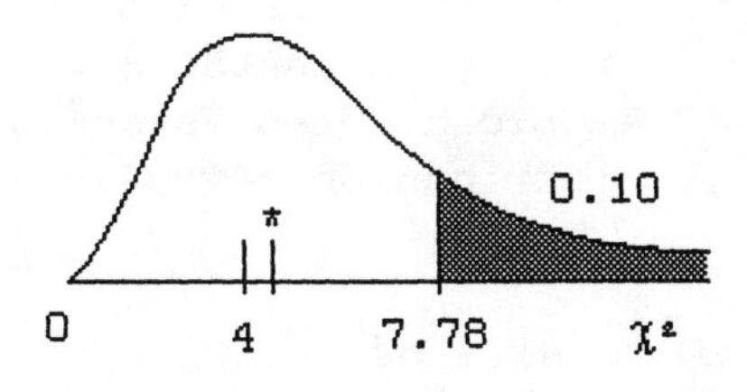

b. $\chi^{2}*$ falls in the noncritical region, see Step 4a

Step 5: a. Fail to reject H_o

b. The preferences of polish are not significantly different from equal proportions.

11.13 Step 1: a. Proportion of roles for African-Americans: Action & Adventure, Comedy, Drama, Horror & Suspense, Romantic Comedy, Other

b. H_o: The distribution of roles are in the following proportions: 13.2%, 31.9%, 23.0%, 12.5%, 8.2%, 11.2%

H_a: Distribution of African/American is different from distribution of all roles

Step 2: a. Assume that the 89 roles represent a random sample.

b. χ^2 with df = 5 c. $\alpha = 0.05$

Step 3: a. sample information given in exercise

b. χ^{2*} =

$$\frac{(9-11.748)^2}{11.748}+\frac{(40-28.391)^2}{28.391}+\frac{(17-20.470)^2}{20.470}+\frac{(11-11.125)^2}{11.125}+\frac{(5-7.298)^2}{7.298}+\frac{(7-9.968)^2}{9.968}$$

= 0.64279 + 4.7468 + 0.58822 + 0.00140+ 0.72360+ .88373
= 7.586

Step 4: -- using p-value approach ------------------

a. $\mathbf{P} = P(\chi^{2*} > 7.586 | df = 5)$;

Using Table 8: $0.10 < \mathbf{P} < 0.25$

Using computer/calculator: $\mathbf{P} = 0.18504$

b. $\mathbf{P} > \alpha$

-- using classical approach -----------------

a. critical region: $\chi^2 \geq 11.1$

b. χ^{2*} falls in the noncritical region

Step 5: a. Fail to reject H_o

b. At the 0.05 level of significance there is no difference in the distribution of African-American roles as compared to the overall distribution of roles in major cinema releases.

11.15 Step 1: a. Results of attempts to obtain help for PC hardware problems using e-mail messages.

b. H_o: There is no difference between service results obtained by your customers and the published study H_a: There is a difference between service results obtained by your customers and the published study

Step 2: a. Assume that the 500 customers represent a random sample.

b. χ^2 with df = 3 c. $\alpha = 0.01$

Step 3: a. sample information given in exercise

b.

$$\chi^{2*} = \frac{(35-70)^2}{70} + \frac{(102-150)^2}{150} + \frac{(125-170)^2}{170} + \frac{(238-110)^2}{110}$$

$$= 17.5 + 15.36 + 11.91 + 148.95 = 193.72$$

Step 4: -- using p-value approach ------------------

a. $\mathbf{P} = P(\chi^{2*} > 193.72 | df = 3)$;

Using Table 8: $\mathbf{P} < 0.005$

Using computer/calculator: $\mathbf{P} = 0+$

b. $\mathbf{P} < \alpha$

-- using classical approach ----------------

a. critical region: $\chi^2 \geq 11.3$

b. χ^{2*} falls in the critical region

Step 5: a. Reject H_o

b. At the 0.01 level of significance there is a difference in the e-mail service results obtained by your customers and the help received by e-mail users in the published study.

One of the properties of random numbers is that they occur with equal probabilities. Since there are 10 values (0-9), each integer will have the same probability 1/10 = 0.1.

11.17 Step 1: a. The probability of a single-digit integer being generated.

b. H_o: $P(0) = P(1) = P(2) = \ldots = P(9) = 0.1$

H_a: The probabilities are not all equal

Step 2: a. Assume that the 100 numbers generated represent a random sample.

b. χ^2 with df = 9 c. $\alpha = 0.05$

Step 3: a. sample information given in exercise
b. $\chi^2 = \Sigma[(O-E)^2/E]$ (as found on accompanying table)

$E = n \cdot p = 100(0.1) = 10$, for all cells

Integer	0	1	2	3	4	5	6	7	8	9	Total
Observed	11	8	7	7	10	10	8	11	14	14	100
Expected	10	10	10	10	10	10	10	10	10	10	100
$(O-E)^2/E$	.1	.4	.9	.9	.0	.0	.4	.1	1.6	1.6	6.0

$\chi^{2*} = 6.00$

Step 4: -- using p-value approach ------------------
a. $\mathbf{P} = P(\chi^2 > 6.00 | df=9)$;
Using Table 8: $0.50 < \mathbf{P} < 0.75$
Using computer/calculator: $\mathbf{P} = 0.740$
b. $\mathbf{P} > \alpha$
-- using classical approach ----------------
a. critical region: $\chi^2 \geq 16.9$
b. χ^{2*} falls in the noncritical region

Step 5: a. Fail to reject H_o
a. The integers generated do not show sufficient reason
to reject the uniform hypothesis, at the 0.05 level of significance.

11.19 a. The calculated value of chi-square will be near zero when the observed values are nearly the same as the expected values. Fail to reject null hypothesis.

b. "Little brother did not roll the die!" He too knew what to expect, so he just reported the expected values, and now you want your money back.

The calculated chi-square will be 'zero.'

The left-hand tail might be interpreted as "the data and the theory expressed by the null hypothesis are too much alike." Either the experimenter, like little brother, knew what was expected and just produced data that agreed or the maker of the hypothesis saw the data and then made up the theory to match the data.

c. Hopefully the experimenter trusts himself to carry out the experiment randomly without influence that contaminates the data.

SECTION 11.3 ANSWER NOW EXERCISES

11.21 E = (40)(50)/200 = 2000/200 = 10

11.22 a. 113 b. 122 c. 300 d. 68.23

11.23 a.

	West	N.Cent.	South	N.East
Baked	55%	46%	47%	41%
Other	45%	54%	53%	59%

b. The information compares several distributions, a distribution for each region of the United States.

c. The distribution of "baked" or "other" are being compared for the 4 different regions.

SECTION 11.3 EXERCISES

Contingency Tables

A contingency table is a table consisting of rows and columns used to summarize and cross-classify data according to two variables. Each row represents the categories for one of the variables, and each column represents the categories for the other variable. The intersections of these rows and columns produce cells. The data will be in a form where two varieties of hypothesis tests are possible. These are:

1. Tests of independence
   - to determine if one variable is independent of the other variable, and
2. Tests of homogeneity
   - to determine if the proportion distribution for one of the variables is the same for each of the categories of the second variable.

Writing Hypotheses for Tests of Independence and/or Homogeneity

The null hypothesis is written in a form to show that there is no difference between the experimental (observed) frequencies and the theoretical (expected) frequencies. The alternative hypothesis is the opposite of the null hypothesis. It is written in a form to show that a difference does exist.

Therefore, in tests of independence:

H_o: One variable is independent of the other variable and in tests of homogeneity:

H_o: The proportions for one variable are distributed the same for all categories of the second variable.

11.25 The *test of independence* has one sample of data that is being cross-tabulated according to the categories of two separate variables; the *test of homogeneity* has multiple samples being compared side-by-side and together these samples form the entire sample used in the contingency table.

The computer or calculator can perform a hypothesis test for independence or homogeneity. Since these tests are completed in the same fashion, the same command may be used. See ES9-p560&561 for more information.

11.27 Step 1: a. The proportions of community size that married men were reared in: P(under 10,000 for present size categories), P(10,000 to 49,999 for present size categories),P(50,000 or over for present size categories).

b. H_o: Size of community of present residence is independent of the size of community reared in.
H_a: Size of community of present residence is not independent of the size of community reared in.

Step 2: a. Given a random sample.
b. χ^2 with df = 4
c. $\alpha = 0.01$

Step 3: a. sample information given in exercise
b. $\chi^2 = \Sigma[(O-E)^2/E]$ (as found on accompanying table)

Expected values:

	less than 10,000	10,000 - 49,999	50,000 or over
less than 10,000	14.36	37.16	62.47
10,000 - 49,999	19.15	48.55	83.30
50,000 or over	29.48	76.28	128.23

χ^{2*} = 6.471 + 1.654 + 4.886 +
0.069 + 4.214 = 2.124 +
2.439 + 6.508 + 7.384 = 35.749

Step 4: -- using p-value approach ------------------
a. **P** = P(χ^2 > 35.749|df=4);
Using Table 8: **P** < 0.005
Using computer/calculator: **P** = 0.0+
b. **P** < α

-- using classical approach ----------------
a. $\chi^2(4,0.01) = 13.3$

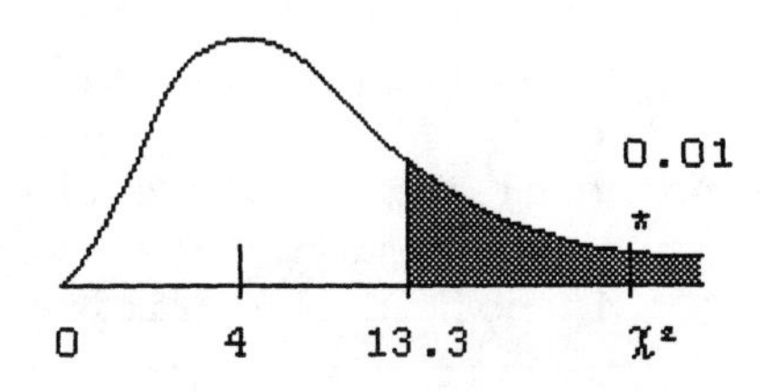

b. χ^{2*} falls in the critical region

Step 5: a. Reject H_o
b. The sample evidence does present significant evidence to show the lack of independence.

11.29 Step 1: a. The proportions supervisor expectation responses: P(true for length of employment category), P(not true for length of employment category).

b. H_o: Response is independent of years of service.
H_a: Response is not independent of years of service.

Step 2: a. Assume a random sample.

b. χ^2 with df = 3

c. $\alpha = 0.10$

Step 3: a. sample information given in exercise

b. $\chi^2 = \Sigma[(O-E)^2/E]$ (as found on accompanying table)

Expected values:

	True	Not true
Less than 1 yr	21.94	9.06
1 to 3 years	19.82	8.18
3 to 10 years	26.18	10.82
10 or more years	24.06	9.94

$$\chi^2* = 0.707 + 1.712 + 0.002 + 0.004 + 0.126 + 0.305 + 0.156 + 0.378 = 3.390$$

Step 4: -- using p-value approach ------------------

a. **P** = $P(\chi^2 > 3.390|df=3)$;

Using Table 8: $0.25 < \mathbf{P} < 0.50$

Using computer/calculator: **P** = 0.335

b. **P** $> \alpha$

-- using classical approach -----------------

a. critical region: $\chi^2 \geq 6.25$

b. χ^2* is not in the critical region

Step 5: a. Fail to reject H_o

b. The evidence is not sufficient to conclude a lack of independence between the length of employment and response.

11.31 Step 1: a. The proportions: P(nondefective for 5 days of the week), P(defective for 5 days of the week)

b. H_o: The number of defective items is independent of the day of the week

H_a: The number of defectives is not independent of day

Step 2: a. Assume a random sample.

b. χ^2 with df = 4 c. $\alpha = 0.05$

Step 3: a. sample information given in exercise

b. $\chi^2 = \Sigma[(O-E)^2/E]$ (as found on accompanying table)

Expected values:

	Mon	Tue	Wed	Thu	Fri	Total
Nondefective	91	91	91	91	91	455
Defective	9	9	9	9	9	45
						500

$\chi^2* = \Sigma[(O-E)^2/E] = 0.396 + 0.011 + 0.176 + 0.176 + 0.011 + 4.000 + 0.111 + 1.778 + 1.778 + 0.111$

$\chi^2* = 8.548$

Step 4: -- using p-value approach ------------------

a. $\mathbf{P} = P(\chi^2 > 8.548|df=4)$;

Using Table 8: $0.05 < \mathbf{P} < 0.10$

Using computer: $\mathbf{P} = 0.074$

b. $\mathbf{P} > \alpha$

-- using classical approach ----------------

a. critical region: $\chi^2 \geq 9.49$

b. χ^2* is in the noncritical region

Step 5: a. Fail to reject H_o

b. There is not sufficient evidence to show that The number of defectives is not the same for all days of the week, at the 0.05 level of significance.

11.33 Step 1: a. The proportions of students: P(A grade for professors), P(B grade for professors), P(C grade for professors), P(Other grade for professors).

b. H_o: The proportion of students receiving each grade is the same for each professor.
H_a: The proportion of students receiving each grade is not the same for all the professors.

Step 2: a. Assume a random sample.
b. χ^2 with df = 6 c. $\alpha = 0.01$

Step 3: a. sample information given in exercise
b. $\chi^2 = \Sigma[(O-E)^2/E]$ (as found on accompanying table)

	No.1	No.2	No.3	Total
A	12 15.52	11 18.97	27 15.52	50
B	16 21.72	29 26.55	25 21.72	70
C	35 24.83	30 30.34	15 24.83	80
Other	27 27.93	40 34.14	23 27.93	90
Total	90	110	90	290

Chi-Sq = 0.797 + 3.346 + 8.497 + 1.508 + 0.226 + 0.494 + 4.168 + 0.004 + 3.890 + 0.031 + 1.007 + 0.871 = 24.838

$\chi^2* = 24.84$

Step 4: -- using p-value approach -------------------
a. **P** = $P(\chi^2 > 24.84|df=6)$;
Using Table 8: **P** < 0.005
Using computer: **P** = 0.000
b. **P** < α
-- using classical approach -----------------
a. critical region: $\chi^2 \geq 16.8$
b. χ^2* is in the critical region

Step 5: a. Reject H_o
b. The distribution of grades is not the same for all professors, at the 0.01 level of significance.

c. Professor #3 gives A's in higher proportion and C's in lower proportions than expected if all graded the same. This can be supported by the value of chi-square that comes from those two cells.

11.35

Step 1: a. The proportion of high school and middle school students who took a gun to school.
b. H_o: The proportion of students who took a gun to school is the same for both groups.
H_a: The proportion of students who took a gun to school is not the same for both groups.

Step 2: a. Assume a random sample.
b. χ^2 with df = 1 c. $\alpha = 0.01$

Step 3: a. sample information given in exercise
b. $\chi^2 = \Sigma[(O-E)^2/E]$

Took a Gun to School.	**At least once**	**Never**	**Total**
Middle school	663 (804.74)	5,633 (5,491.26)	6,296
High school	1,265 (1,123.26)	7,523 (7,664.74)	8,788
Total	1,928	13,156	15,084

Chi-Sq = 24.965 + 3.659 + 17.885 + 2.621
= 49.130

Step 4: -- using p-value approach ------------------
a. **P** = $P(\chi^2 > 49.130|df=1)$;
Using Table 8: **P** < 0.005
Using computer: **P** = 0+
b. **P** < α
-- using classical approach ----------------
a. critical region: $\chi^2 \geq 6.63$
b. χ^{2*} is in the critical region

Step 5: a. Reject H_o
b. The two groups do not have the same proportions, at the 0.01 level of significance.

Step 1: a. The proportion of high school and middle school students who have hit a person because of anger.
b. H_o: The proportion of students who hit someone because they are angry is the same for both groups.
H_a: The proportion of students who hit someone because they are angry is not the same for both groups.

Step 2: a. Assume a random sample.
b. χ^2 with df = 1 c. $\alpha = 0.01$

Step 3: a. sample information given in exercise
b. $\chi^2 = \Sigma[(O-E)^2/E]$

Hit a person because of anger.	**At least once**	**Never**	**Total**
Middle school	4,379 (4,316.07)	1,899 (1,961.93)	6,278
High school	5,954 (6,016.93)	2,798 (2,735.07)	8,752
Total	10,333	4,697	15,030

$$\text{Chi-Sq} = 0.917 + 2.018 + 0.658 + 1.448 = \underline{5.042}$$

Step 4: -- using p-value approach ------------------
a. $\mathbf{P} = P(\chi^2 > 5.042 | df=1)$;
Using Table 8: $0.01 < \mathbf{P} < 0.025$
Using computer: $\mathbf{P} = 0.025$
b. $\mathbf{P} > \alpha$
-- using classical approach ----------------
a. critical region: $\chi^2 \geq 6.63$
b. χ^{2*} is in the noncritical region

Step 5: a. Fail to reject H_o
b. The two groups do have the same proportions, at the 0.01 level of significance.

RETURN TO CHAPTER CASE STUDY

11.37 a. The name of their preferred way to "cool" their mouth after eating a delicious spicy favorite.

b. Population: US adults professing to love eating hot spicy food. Variable: method of cooling the heat.

c. Same bargraph as draw for exercise 11.1 (a) and (b).

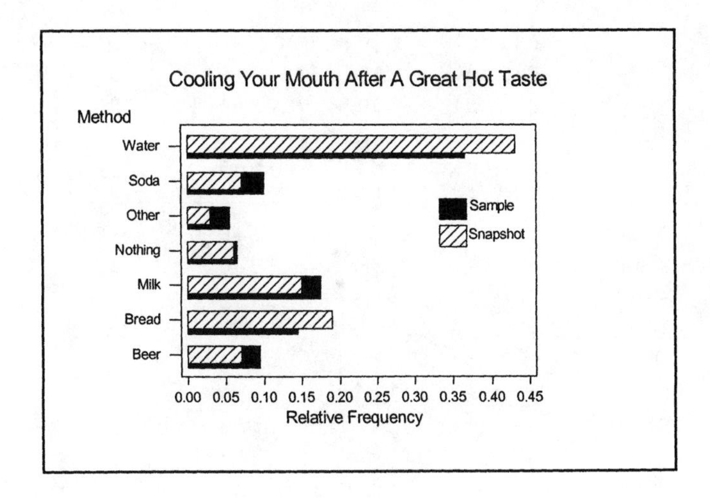

d. The two distributions seem to be fairly similar: the responses appear to breakup into four levels: water has the greatest response, then milk and bread are similar and next most popular, while soda, beer and nothing all have draw similar shares at about 5 or percent, with all others then being grouped together.

11.39 - Answers will vary.

CHAPTER EXERCISES

11.41 Step 1: a. The proportions: P(1st type), P(2nd type), P(3rd type)

b. H_o: 1:3:4 proportions

H_a: proportions are other than 1:3:4

Step 2: a. Assume that the 80 hybrids represent a random sample.

b. χ^2 with df = 2   c. $\alpha = 0.05$

Step 3: a. sample information given in exercise

b. $\chi^2 = \Sigma[(O-E)^2/E]$ (as found on accompanying table)

E(1st) = n·p = 800[1/(1+3+4)] = 100
E(2nd) = 800[3/(1+3+4)] = 300
E(3rd) = 800[4/(1+3+4)] = 400

	1st	2nd	3rd	Total
Observed	80	340	380	800
Expected	100	300	400	800
$(O-E)^2/E$	4.00	5.33	1.00	10.33

$\chi^2* = 10.33$

Step 4: -- using p-value approach ------------------

a. **P** = $P(\chi^2 > 10.33|df=2)$;

Using Table 8: $0.005 <$ **P** < 0.01

Using computer: **P** = 0.006

b. **P** $< \alpha$

-- using classical approach ----------------

a. $\chi^2(2,0.05) = 5.99$ $\chi^2 \geq 5.99$

b. χ^2* is in the critical region.

Step 5: a. Reject H_o

b. We have sufficient evidence to show that the ratiois not the hypothesized 1:3:4 ratio, at the 0.05 level of significance.

11.43 Step 1: a. The proportions: P(parent Home), P(campus), P(off-campus), P(own home), P(other)

b. H_o: P(parent home) = 0.46, P(campus) = 0.26, P(off-campus) = 0.18, P(own home) = 0.09, P(other) = 0.02

H_a: The percentages are different than listed

Step 2: a. Assume that the 1000 individuals represent a random sample.

b. χ^2 with df = 4 c. $\alpha = 0.05$

Step 3: a. sample information given in exercise

b. $\chi^2 = \Sigma[(O-E)^2/E]$ (as found on accompanying table)

E(parent) = n·p = 1000(0.46) = 460,
E(campus) = 1000(0.26) = 260,
E(off-campus) = 1000(0.18) = 180,
E(own home) = 1000(0.09) = 90,
E(other) = 1000(0.02) = 20

	Par.	Cam.	Off.	Own.	Oth.	Total
Observed	484	230	168	96	22	1000
Expected	458	258	178	88	18	1000
$(O-E)^2/E$	1.476	3.039	0.562	0.727	0.889	6.693

$\chi^2* = 6.693$

Step 4: -- using p-value approach ------------------

a. **P** = $P(\chi^2 > 6.693 | df=4)$;
Using Table 8: 0.10 < **P** < 0.25
Using computer: **P** = 0.153

b. **P** > α

-- using classical approach ----------------

a. $\chi^2(4, 0.05) = 9.49$ $\chi^2 \geq 9.49$

b. χ^2* is in the noncritical region.

Step 5: a. Fail to reject H_o

c. We do not have sufficient evidence to show that the sample distribution is different than the newspaper distribution, at the 0.05 level of significance.

11.45 Step 1: a. The largest number of holes played by golfers in one day.

b. H_o: P(18) = 0.05, P(19-27) = 0.12, P(28-36) = 0.28, P(37-45) = 0.20, P(46-54) = 0.18, P(55 or more) = 0.17

H_a: The percentages are different than listed

Step 2: a. Assume that the 200 individuals represent a random sample.

b. χ^2 with df = 5 c. $\alpha = 0.01$

Step 3: a. sample information given in exercise

b. $\chi^2 = \Sigma[(O-E)^2/E]$

$$\chi^{2*} = \frac{(12-10)^2}{10} + \frac{(35-24)^2}{24} + \frac{(60-56)^2}{56} + \frac{(44-40)^2}{40} + \frac{(35-36)^2}{36} + \frac{(14-34)^2}{34}$$

$$= 0.40 + 5.04 + 0.29 + 0.40 + 0.03 + 11.76 = 17.92$$

$\chi^{2*} = 17.92$

Step 4: -- using p-value approach ------------------

a. **P** = $P(\chi^2 > 17.92|df=5)$;

Using Table 8: **P** < 0.005

Using computer: **P** = 0.003

b. **P** < α

-- using classical approach ----------------

a. $\chi^2(5,0.01) = 15.1$ $\chi^2 \geq 15.1$

b. χ^{2*} is in the critical region.

Step 5: a. Reject H_o

b. We do have sufficient evidence to show that the sample distribution is different than the Golf magazine distribution, at the 0.05 level of significance.

11.47 a. Step 1: a. The distribution of colors in a bag of M&M's.

b. H_o: P(brown) = 0.30, P(red & yellow) = 0.20, P(blue, green & orange) = 0.10

H_a: Distributions are different

Step 2: a. Assume a random sample.

b. χ^2 with df = 5 c. $\alpha = 0.05$

Step 3: a. sample information given in exercise

b. $\chi^2 = \Sigma[(O-E)^2/E]$ (as found on accompanying table)

Expected Probability	red 0.20	green 0.10	blue 0.10	orange 0.10	yellow 0.20	brown 0.30
Expected Frequency	11.6	5.8	5.8	5.8	11.6	17.4
Observed Frequency	15	9	3	3	9	19

$\chi^2* = 0.99655 + 1.76552 + 1.35172 + 1.35172 + 0.58276 + 0.14713 = \underline{6.1954}$

Step 4: -- using p-value approach ------------------

a. **P** = $P(\chi^2 > 6.1954 | df=5)$;

Using Table 8: **P** > 0.25

Using computer/calculator: **P** = 0.2877

b. **P** > α

-- using classical approach ----------------

a. critical region: $\chi^2 \geq 12.6$

b. χ^2* falls in the critical region

Step 5: a. Fail to reject H_o

b. There is not significant evidence to show the distribution is different than the target distribution.

b. Step 1: a. The distribution of colors in a bag of M&M's.

b. H_o: P(brown) = 0.30, P(red & yellow) = 0.20, P(blue, green & orange) = 0.10

H_a: Distributions are different

Step 2: a. Assume a random sample.

b. χ^2 with df = 5 c. $\alpha = 0.05$

Step 3: a. sample information given in exercise

b. $\chi^2 = \Sigma[(O-E)^2/E]$ (as found on accompanying table)

Expected Probability	red 0.20	green 0.10	blue 0.10	orange 0.10	yellow 0.20	brown 0.30
Expected Frequency	23.4	11.7	11.7	11.7	23.4	35.1
Observed Frequency	24	26	22	6	12	27

$\chi^2* = 0.0154 + 17.4778 + 9.0675 + 2.7769 + 5.5538 + 1.8692 = \underline{36.761}$

Step 4: -- using p-value approach ------------------
a. $\mathbf{P} = P(\chi^2 > 36.761|df=5)$;
Using Table 8: $\mathbf{P} < 0.005$
Using computer/calculator: $\mathbf{P} = 0+$
b. $\mathbf{P} > \alpha$
-- using classical approach ----------------
a. critical region: $\chi^2 \geq 12.6$
b. χ^{2*} falls in the critical region

Step 5: a. Reject H_o
b. There is significant evidence to show the distribution is different than the target distribution.

c. Step 1: a. The distribution of colors in a bag of M&M's.
b. H_o: P(brown) = 0.30, P(red & yellow) = 0.20, P(blue, green & orange) = 0.10
H_a: Distributions are different
Step 2: a. Assume a random sample.
b. χ^2 with df = 5 c. $\alpha = 0.05$
Step 3: a. sample information given in exercise
b. $\chi^2 = \Sigma[(O-E)^2/E]$ (as found on accompanying table)

Expected Probability	red 0.20	green 0.10	blue 0.10	orange 0.10	yellow 0.20	brown 0.30
Expected Frequency	342.6	171.3	171.3	171.3	342.6	513.9
Observed Frequency	288	222	217	199	413	374

$$\chi^{2*} = 08.7016 + 15.0058 + 12.1920 + 4.4792 + 14.4663 + 38.0853 = \underline{92.93}$$

Step 4: -- using p-value approach ------------------
a. $\mathbf{P} = P(\chi^2 > 92.93|df=5)$;
Using Table 8: $\mathbf{P} < 0.005$
Using computer/calculator: $\mathbf{P} = 0+$
b. $\mathbf{P} > \alpha$
-- using classical approach ----------------
a. critical region: $\chi^2 \geq 12.6$
b. χ^{2*} falls in the critical region

Step 5: a. Reject H_o

b. There is significant evidence to show the distribution is different than the target distribution.

d. The sample sizes were too small. If more samples were taken, the probabilities would begin to approach the target probabilities.

11.49 Step 1: a. The proportions: P(yes for different school locations), P(no for different school locations).

b. H_o: The response and the school's location are independent.

H_a: The response and the school's location are not independent.

Step 2: a. Assume a random sample.

b. χ^2 with df = 4 c. $\alpha = 0.05$

Step 3: a. sample information given in exercise

b. $\chi^2 = \Sigma[(O-E)^2/E]$

Frequencies [expected frequencies]

	Urban	Suburban	Rural	Totals
Yes	57 [57.86]	27 [31.48]	47 [41.66]	131
No	23 [22.53]	16 [12.25]	12 [16.22]	51
Unsure	45 [44.61]	25 [24.27]	31 [32.12]	101
Totals	125	68	90	283

$\chi^2* = \Sigma[(O - E)^2/E] = 0.013 + 0.637 + 0.684 + 0.010 + 1.145 + 1.098 + 0.003 + 0.022 + 0.039 = 3.651$

Step 4: -- using p-value approach ------------------

a. **P** = $P(\chi^2 > 3.651|df=4)$;

Using Table 8: 0.25 < **P** < 0.50

Using computer: **P** = 0.456

b. **P** > α

-- using classical approach ----------------

a. Critical region: $\chi^2 \geq 9.49$

b. χ^2* falls in the noncritical region

Step 5: a. Fail to reject H_o
b. The evidence is not sufficient to show a significant relationship between student response and school's location, at the 0.05 level of significance.

11.51 Step 1: a. The proportion of each political preference for each age group and the proportion of each age group who answer with each political preference.
b. H_o: Political preference is independent of age.
H_a: Political preference is not independent of age.

Step 2: a. Assume a random sample.
b. χ^2 with df = 4 c. $\alpha = 0.01$

Step 3: a. sample information given in exercise
b. $\chi^2 = \Sigma[(O-E)^2/E]$ (as found on accompanying table)

Expected values:

	20-35	36-50	Over 50
Conservative	28.00	30.00	22.00
Moderate	73.50	78.75	57.75
Liberal	38.50	41.25	30.25

$$\chi^2{*} = 2.286 + 3.333 + 0.182 + 0.575 + 0.496 + 2.815 + 0.058 + 6.402 + 7.192 = 23.339$$

Step 4: -- using p-value approach ------------------
a. $\mathbf{P} = P(\chi^2 > 23.339|df=4)$;
Using Table 8: $\mathbf{P} < 0.005$
Using computer/calculator: $\mathbf{P} = 0.000+$
b. $\mathbf{P} < \alpha$
-- using classical approach ----------------
a. $\chi^2(4,0.01) = 13.3$ $\chi^2 \geq 13.3$
b. $\chi^2{*}$ falls in the critical region

Step 5: a. Reject H_o
b. The political preference is not independent of the age group to which the person belongs, at the 0.01 level of significance.

11.53 Step 1: a. The proportions of popcorn that popped and did not pop: P(Brand A), P(Brand B), P(Brand C), P(Brand D)

b. H_o: Proportion of popcorn that popped is the Same for all brands.

H_a: The proportions are not the same for all brands

Step 2: a. Assume a random sample.

b. χ^2 with df = 3 c. $\alpha = 0.05$

Step 3: a. sample information given in exercise

b. $\chi^2 = \Sigma[(O-E)^2/E]$

Expected values:

	A	B	C	D	Total
Popped	88	88	88	88	352
Not popped	12	12	12	12	48
Totals	100	100	100	100	400

$\chi^2* = \Sigma[(O-E)^2/E] = 0.333 + 1.333 + 0.083 + 0.750 + 0.045 + 0.182 + 0.011 + 0.102$

$\chi^2* = 2.839$

Step 4: -- using p-value approach -------------------

a. $\mathbf{P} = P(\chi^2 > 2.839|df=3)$;

Using Table 8: $0.25 < \mathbf{P} < 0.50$

Using computer: $\mathbf{P} = 0.417$

b. $\mathbf{P} > \alpha$

-- using classical approach -----------------

a. $\chi^2(3,0.05) = 7.82$ $\chi^2 \geq 7.82$

b. χ^2* falls in the noncritical region

Step 5: a. Fail to reject H_o

a. There is not sufficient evidence to show that The proportions of popped corn are not the same for all brands, at the 0.05 level of significance.

11.55 Step 1: a. The proportions for cholesterol levels for each gender of nurses.

b. H_o: Cholesterol level is independent of gender.

H_a: Cholesterol level is not independent of gender.

Step 2: a. Given a random sample.
b. χ^2 with df = 1 c. $\alpha = 0.05$

Step 3: a. sample information given in exercise
b. $\chi^2 = \Sigma[(O-E)^2/E]$ (as found on accompanying table)

Expected counts are printed below observed counts

	High	Not High	Total
1	14	20	34
	10.54	23.46	
2	17	49	66
	20.46	45.54	
Total	31	69	100

Chi-Sq = 1.136 + 0.510 + 0.585 + 0.263 = 2.494

$\chi^{2}* = 2.494$

Step 4: -- using p-value approach ------------------
a. **P** = $P(\chi^2 > 2.494 | df=1)$;
Using Table 8: $0.10 < \mathbf{P} < 0.25$
Using computer/calculator: **P** = 0.114
b. **P** > α
-- using classical approach ----------------
a. $\chi^2(1, 0.05) = 3.84$ $\chi^2 \geq 3.84$
b. $\chi^{2}*$ falls in the noncritical region

Step 5: a. Fail to reject H_o
b. The nurses' cholesterol level is independent of gender, at the 0.05 level of significance.

11.57 Conditions for rolling a balanced die 600 times:
The critical value is $\chi^2(5, 0.05) = 11.1$ (6 possible outcomes)

With 600 rolls, the expected value for each cell is 100.

Many combinations of observed frequencies are possible to cause us to reject the equally likely hypothesis. The combinations will have to have a calculated χ^2 value greater than 11.1 or a p-value less than 0.05. Two possibilities are presented.

1. If each observed frequency is different from the expected by the same amount, then 11.1/6 = 1.85 is the amount of chi-square that would come from each cell.

$(O-E)^2/E = (O - 100)^2/100 = 1.85$

$(O - 100)^2 = 185$

$O - 100 = \pm 13.6$

$O = 86$ or 114

That is, if three of the observed frequency values are 86 and the other three are 114, the faces of the die will be declared not to be equally likely.

Row	P	OBS	EXP	CHI-SQ
1	0.166667	86	100.000	1.96005
2	0.166667	86	100.000	1.96005
3	0.166667	86	100.000	1.96005
4	0.166667	114	100.000	1.95994
5	0.166667	114	100.000	1.95994
6	0.166667	114	100.000	1.95994

Row	SUM(P)	SUN(OBS)	SUM(EXP)	CHI-SQ*
1	1.00000	600	600.001	11.7600

DF 5

p-value 0.038

Note χ^2 value and p-value

2. Now suppose just one is different and the other five all occur with the same frequency:
Remember, the total observed most be 600. Therefore, for every five one outcome is different from the expected, the other five each must be different by one to balance. If the five are each different from the expected by x, then the one that is very different is off by 5x. The sum of 5 - x's squared and 5x squared is $30x^2$. Thus,

$30x^2 = 11.1$

$x^2 = 11.1/30 = 0.37$

$(O-E)^2/E = (O-100)^2/100 = 0.37$

$(O-100)^2 = 37$
$O - 100 = \pm 6.08$ (round-up)

O = either 93 or 107 for the five cells, and

O for the other cell must be off by 5(7) or 35; it is either 65 or 135.

Row	P	OBS	EXP	CHI-SQ
1	0.166667	93	100.000	0.4900
2	0.166667	93	100.000	0.4900
3	0.166667	93	100.000	0.4900
4	0.166667	93	100.000	0.4900
5	0.166667	93	100.000	0.4900
6	0.166667	135	100.000	12.2498

Row	SUM(P)	SUN(OBS)	SUM(EXP)	CHI-SQ*
1	1.00000	600	600.001	14.7000

DF 5

p-value 0.012

Note χ^2 and p-value.

CHAPTER 12 ∇ ANALYSIS OF VARIANCE

Chapter Preview

In Chapters 8 and 9, hypothesis tests were demonstrated for testing a single mean. Hypothesis tests between two means was subsequently demonstrated in Chapter 10. To continue in this fashion, Chapter 12 introduces the concept of the analysis of variance technique (ANOVA) so that a hypothesis test for the equality of several means can be completed. The F-distribution will be utilized in this test, since we will be comparing the measures of variation (variance) among the different sets of data and the measure of variation (variance) within the sets the data.

An article, featured in USA Today on time spent reading the newspaper, is used in this chapter's Case Study.

CHAPTER 12 CASE STUDY

12.1 a.

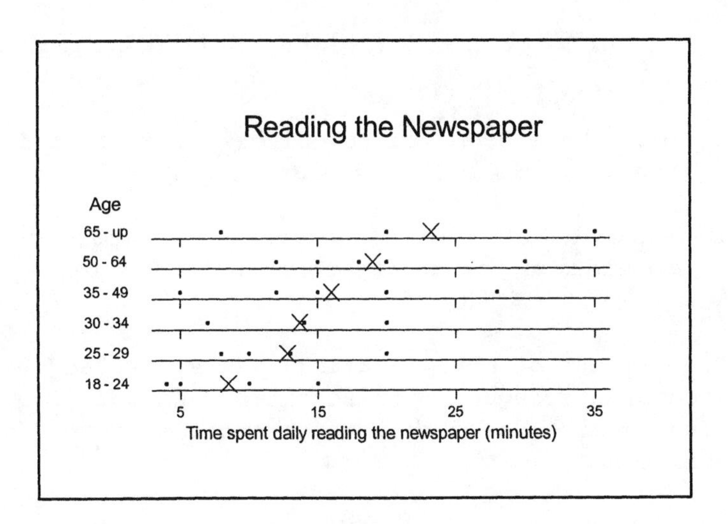

b. The mean amount of time spent reading the newspaper increases steadily from 8.5 minutes daily for the youngest group to 23 minutes daily for the oldest group.

SECTION 12.1 ANSWER NOW EXERCISES

12.2 Units produced per hour at each temperature level

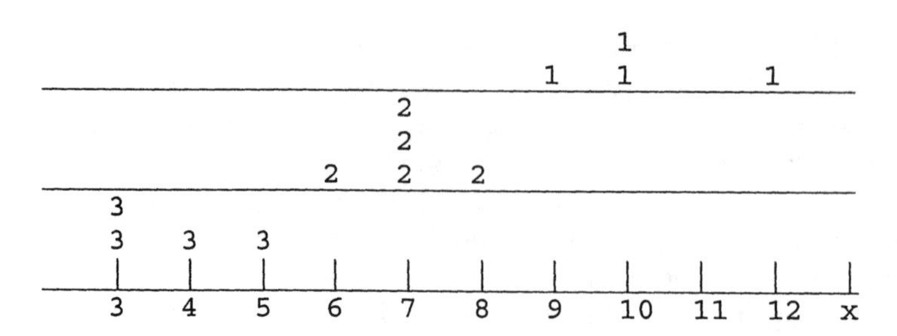

Yes, there appears to be a difference between the three sets.

12.3 $H_o: \mu_1 = \mu_2 = \mu_3$

H_a: means are not all the same

SECTION 12.2 ANSWER NOW EXERCISES

12.4 a. Yes, there seems to be quite a bite of variation between the values shown on the Snapshot.

b. The male and female categories encompass the entire population; as do the other 5 categories.

12.5 The 'amount of money donated' was separated into categories according to age for presentation and analyzing. ANOVA 'subdivides' the data into categories of the factor, age in this case.

SECTION 12.3 ANSWER NOW EXERCISES

12.6 a. 0 b. 2 c. 4 d. 31 e. 393

SECTION 12.3 EXERCISES

WRITING HYPOTHESES FOR THE DIFFERENCE AMONG SEVERAL MEANS

null hypothesis: H_o: $\mu_1 = \mu_2 = \mu_3 = \ldots = \mu_n$

or

H_o: The mean values for all n levels of the experiment are the same.
[factor has no effect]

alternative hypothesis: H_a: The means are not all equal.
[factor has an effect]

or

H_a: At least one mean value is different from the others.

Use subscripts on the population means that correspond to the different levels or sources of the experiment.

12.7 a. H_o: $\mu_1 = \mu_2 = \mu_3 = \mu_4 = \mu_5$ vs. H_a: Means not all equal
[mean scores are all same]

b. H_o: $\mu_1 = \mu_2 = \mu_3 = \mu_4$ vs. H_a: Means not all equal
[mean scores are all same]

c. H_o: $\mu_1 = \mu_2 = \mu_3 = \mu_4$ vs. H_a: Means not all equal
[factor has no effect] [has an effect]

d. H_o: $\mu_1 = \mu_2 = \mu_3$ vs. H_a: Means not all equal
[no effect] [has an effect]

Use Tables 9a, b or c (Appendix B, ES9-pp720-725) depending on α. Locate the critical value using the degrees of freedom for the numerator (factor) and the degrees of freedom for the denominator(error). Reviewing how to determining the test criteria in: ES9-p517, ST-p346, as it is applied to the F-distribution may be helpful.

12.9 a. b.

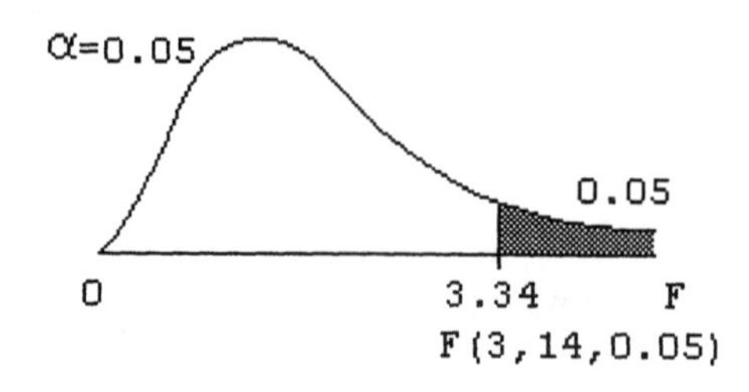

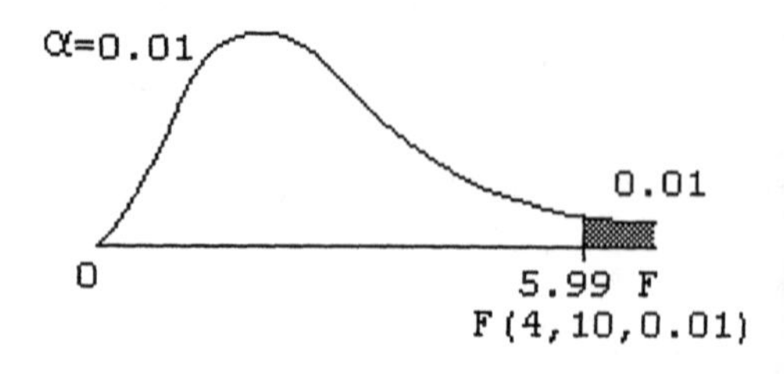

c.

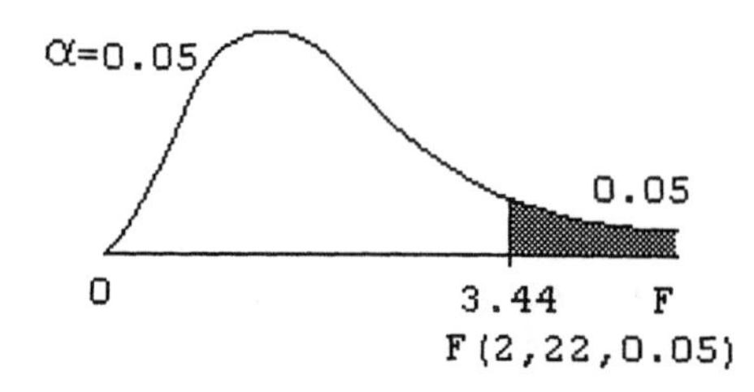

12.11 a. 0.04 of the probability distribution associated with F and a true null hypothesis is more extreme than F*. That is, area under the curve and to the right of F*.

b. Reject the null hypothesis; since the p-value is less than the previously set value for alpha.

c. Fail to reject the null hypothesis; since the p-value is greater than the previously set value for alpha.

12.13 a. The test factor has no effect on the mean at the tested levels

b. The test factor does have an effect on the mean at the tested levels

c. For the p-value approach, $P = P(F > F^*)$ must be $\leq \alpha$. For the classical approach, the calculated value of F must fall in the critical region; that is, the variance between levels of the factor must be significantly larger than variance within the levels.

d. The tested factor has a significant effect on the variable.

e. For the p-value approach, $P = P(F > F^*)$ must be $> \alpha$. For the classical approach, the calculated value of F must fall in the non-critical region; that is, the variance between levels of the factor must not be significantly larger than variance within the levels.

f. The tested factor does not have a significant effect on the variable.

Hypothesis Test for the Difference Among Several Means, Independent Samples

Review the parts to a hypothesis test as outlined in: ES9-pp370 & 386, ST-pp241 & 249, if needed. Slight changes will occur in:

1. **the hypotheses**: (see box before exercise 12.7)
2. **the calculated test statistic**

$$F = \frac{MS(\text{factor})}{MS(\text{error})}$$

3. If H_o is rejected, a significant difference among the means is indicated, the various levels of the factor do have an effect.

   If H_o is not rejected, no significant difference among the means is indicated, the various levels of the factors do not have an effect.

FORMULAS FOR ANOVA TABLE

$$SS(\text{total}) = \sum(x^2) - \frac{(\sum x)^2}{n}$$

...

$$SS(factor) = \left(\frac{C_1^2}{k_1} + \frac{C_2^2}{k_2} + \frac{C_3^2}{k_3} + \ldots\right) - \frac{(\sum x)^2}{n}$$

, where C_i = column total

k_i = number of data values in the ith column

(check: $n = \Sigma k_i$)

$$SS(error) = \sum x^2 - \left(\frac{C_1^2}{k_1} + \frac{C_2^2}{k_2} + \frac{C_3^2}{k_3} + \ldots\right) \quad \text{or} \quad = SS(total) - SS(factor)$$

$df_{total} = n-1$, $df_{factor} = c-1$, $df_{error} = n - c$

$df_{total} = df_{factor} + df_{error}$

ANOVA Table

Source	df	SS	MS
Factor			
Error			
Total			

MS(factor) = SS(factor)/df(factor)

MS(error) = SS(error)/df(error)

12.15 a. 120 [df(total) +1]

b. 3 [df(group) + 1]

c. Verify -- answers given in exercise.

d. Yes. The p-value is very small.

e. No. The p-value is too large.

Computer and/or calculator commands to ANOVA hypothesis test can be found in ES9-p587 & 588. Explanation of the output is found in ES9-p588.

12.17 Step 1: a. The mean level of work for a new worker, the mean level of work for worker A, the mean level of work for worker B.

b. H_o: The mean values for workers are all equal.
H_a: The mean values for workers are not all equal.

Step 2: a. Assume the data were randomly collected and are independent, and the effects due to chance and untested factors are normally distributed.

b. F

c. $\alpha = 0.05$

Step 3: a. $n = 15$, $C_1 = 46$, $C_2 = 58$, $C_3 = 57$, $T = 161$, $\Sigma x^2 = 1771$

Source	df	SS	MS	F*
Work	2	17.73	8.87	4.22
Error	12	25.20	2.10	
Total	14	42.93		

$F* = 8.87/2.10 = 4.22$

Step 4: -- using p-value approach ---------------

a. $\mathbf{P} = P(F > 4.22 | df_n = 2, df_d = 12)$;
Using Table 9: $0.025 < \mathbf{P} < 0.05$
Using computer: $\mathbf{P} = 0.041$

b. $\mathbf{P} < \alpha$

-- using classical approach -------------

a. critical region: $F \geq 3.89$

b. F* falls in the critical region

Step 5: a. Reject H_o.

b. There is significant difference between the workers with regards to mean amount of work produced.

12.19 Step 1: a. The mean ratings obtained by the restaurants in the three categories.

b. H_o: $\mu_F = \mu_D = \mu_S$ (no difference in ratings)
H_a: The means of the ratings obtained by the restaurants in the three categories are not all equal.

Step 2: a. Assume the data were randomly collected and are independent, and the effects due to chance and untested factors are normally distributed.

b. F

c. $\alpha = 0.05$

Step 3: a. $n = 18$, $C_1 = 126$, $C_2 = 102$, $C_3 = 113$, $T = 341$, $\Sigma x^2 = 6587$

Source	df	SS	MS	F*
Rating cat.	2	48.11	24.055	4.58
Error	15	78.83	5.255	
Total	17	126.94		

$F^* = 24.055/5.255 = 4.58$

Step 4: -- using p-value approach ---------------

a. $\mathbf{P} = P(F > 4.58 | df_n = 2, df_d = 15)$;

Using Table 9: $0.025 < \mathbf{P} < 0.05$

Using computer: $\mathbf{P} = 0.028$

b. $\mathbf{P} < \alpha$

-- using classical approach -------------

a. critical region: $F \geq 3.68$

b. F* is in the critical region

Step 5: a. Reject H_o.

b. The data shows significant evidence that would give reason to reject the null hypothesis that the means of the three categories of ratings given to the restaurants are not equal.

12.21 Step 1: a. The means of the won/loss percentages obtained by the teams playing on the road in the three divisions.

b. H_o: $\mu_E = \mu_C = \mu_W$

H_a: The means of the won/loss percentages obtained by the teams playing on the road in the three divisions are not all equal.

Step 2: a. Assume the data were randomly collected and are independent, and the effects due to chance and untested factors are normally distributed.

b. F

c. $\alpha = 0.05$

Step 3: a. $n = 28$, $C_1 = 497.5$, $C_2 = 423.4$, $C_3 = 378.9$, $T = 1299.8$, $\Sigma x^2 = 61924.96$

Source	df	SS	MS	F*
Division	2	284.46	142.23	2.732
Error	25	1301.93	52.07	
Total	27	1586.39		

Step 4: -- using p-value approach --------------

a. $\mathbf{P} = P(F > 2.732 | df_n = 2, df_d = 25)$;

Using Table 9: $\mathbf{P} > 0.05$

Using computer: $\mathbf{P} = 0.0845$

b. $\mathbf{P} > \alpha$

-- using classical approach ------------

a. $F(2,25,0.05) = 3.39 \quad F \geq 3.39$

b. F* is not in the critical region

Step 5: a. Fail to reject H_o.

b. The data shows no significant evidence that would give reason to reject the null hypothesis that the means of the won/loss percentages obtained by the teams playing on the road in the three divisions are equal.

12.23 Calories:

Step 1: a. The means of the calories contained in the four snack categories are not all equal.

b. H_o: $\mu_1 = \mu_2 = \mu_3 = \mu_4$

H_a: The means of the calories contained in the four snack categories are not all equal.

Step 2: a. Assume the data were randomly collected and are independent, and the effects due to chance and untested factors are normally distributed.

b. F

c. $\alpha = 0.01$

Step 3: a. $n = 50$, $k_1 = 12$, $C_1 = 1029$, $k_2 = 10$, $C_2 = 557$, $k_3 = 12$, $C_3 = 1033$, $k_4 = 16$, $C_4 = 1342$, $T = 3961$, $\sum x^2 = 338265$

Source	df	SS	MS	F*
Category	3	6955.56	2318.52	6.088
Error	46	17519.02	380.85	
Total	49	24474.58		

Step 4: -- using p-value approach --------------

a. $\mathbf{P} = P(F > 6.088 | df_n = 3, df_d = 46)$;

Using Table 9: $\mathbf{P} < 0.01$

Using computer: $\mathbf{P} = 0.0014$

b. $\mathbf{P} < \alpha$

-- using classical approach ------------

a. $F(3,46,0.01) \approx 4.26 \quad F \geq 4.26$

b. F* is in the critical region

Step 5: a. Reject H_o.

b. The data shows significant evidence that would give reason to reject the null hypothesis that the calories contained by the snacks in each category are not equal.

Fat content:

Step 1: a. The mean age of three test groups: the mean age for the TTS group, the mean age for the Antivert group, the mean age for the placebo group.

b. H_o: $\mu_1 = \mu_2 = \mu_3 = \mu_4$

H_a: The means of the grams of fat contained in the four snack categories are not all equal.

Step 2: a. Assume the data were randomly collected and are independent, and the effects due to chance and untested factors are normally distributed.

b. F

c. $\alpha = 0.01$

Step 3: a. $n = 50$, $k_1 = 12$, $C_1 = 25.3$, $k_2 = 10$, $C_2 = 12.4$, $k_3 = 12$, $C_3 = 28.2$, $k_4 = 16$, $C_4 = 12.1$, $T = 78.0$, $\Sigma x^2 = 341.4$

Source	df	SS	MS	F*
Category	3	22.46	7.49	1.745
Error	46	197.26	4.29	
Total	49	219.72		

Step 4: -- using p-value approach --------------

a. $\mathbf{P} = P(F > 1.745 | df_n = 3, df_d = 46)$;

Using Table 9: $\mathbf{P} > 0.05$

Using computer: $\mathbf{P} = 0.171$

b. $\mathbf{P} > \alpha$

-- using classical approach ------------

a. $F(3,46,0.01) \approx 4.26$ $F \geq 4.26$

b. F* is not in the critical region

Step 5: a. Fail to reject H_o.

b. The data shows no significant evidence that would give reason to reject the null hypothesis that the grams of fat contained by the snacks in each category are not equal.

12.25 a. Step 1: a. The mean hourly wages paid per month to production or nonsupervisory workers.

b. H_o: The mean hourly wage is the same for each of the twelve months.

H_a: The mean hourly wage is not the same for each of the twelve months.

Step 2: a. Assume the data were randomly collected and are independent, and the effects due to chance and untested factors are normally distributed.

b. F

c. $\alpha = 0.05$

Step 3: a.

Source	df	SS	MS	F*
Factor	11	0.79	0.07	0.04
Error	117	230.58	1.97	
Total	128	231.37		

Step 4: -- using p-value approach --------------

a. $\mathbf{P} = P(F > 0.04 | df_n = 11, df_d = 117)$;

Using Table 9: $\mathbf{P} > 0.05$

Using computer: $\mathbf{P} = 1.000$

b. $\mathbf{P} > \alpha$

-- using classical approach ------------

a. $F(11,117,0.05) = 1.88$

b. F* is not in the critical region

Step 5: a. Fail to reject H_o.

b. There is not sufficient evidence to show that at least one monthly mean is significantly different from the others at the 0.05 level of significance.

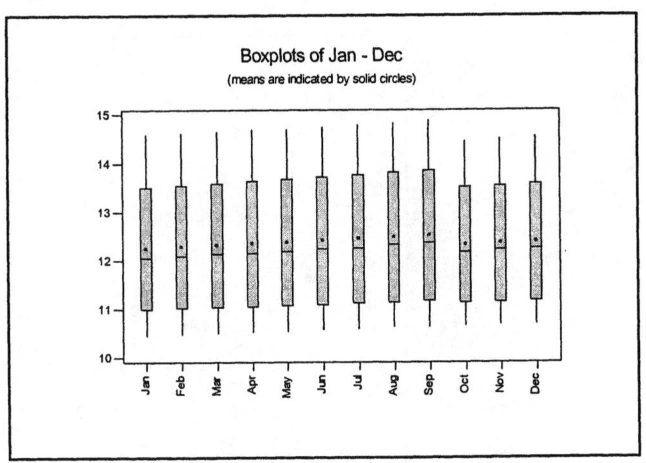

b. Step 1: a. The mean hourly wage paid each year from 1992 - 2002 to production or nonsupervisory workders.

b. H_o: The mean hourly wage is the same for each of the years.

H_a: The mean hourly wage is not the same for each of the years.

Step 2: a. Assume the data were randomly collected and are independent, and the effects due to chance and untested factors are normally distributed.

b. F

c. $\alpha = 0.05$

Step 3: a.

Source	df	SS	MS	F*
Factor	10	229.4794	22.9479	1434.01
Error	118	1.8883	0.0160	
Total	128	231.3677		

Step 4: -- using p-value approach ---------------

a. **P** = $P(F > 1434.01 | df_n = 10, df_d = 118)$;

Using Table 9: **P** < 0.01

Using computer: **P** = 0.000

b. **P** < α

-- using classical approach -------------

a. $F(10,118,0.05) = 1.92$

b. F* is in the critical region

Step 5: a. Reject H_o.

b. There is sufficient evidence to show that at least one mean is significantly different from the others at the 0.05 level of significance.

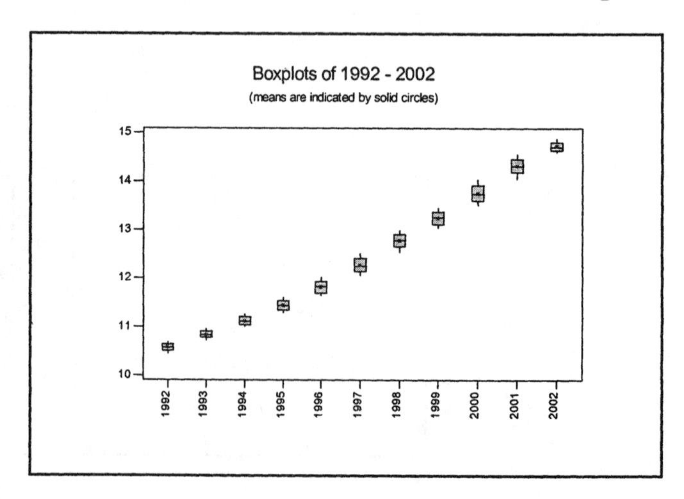

RETURN TO CHAPTER CASE STUDY

12.26 a.

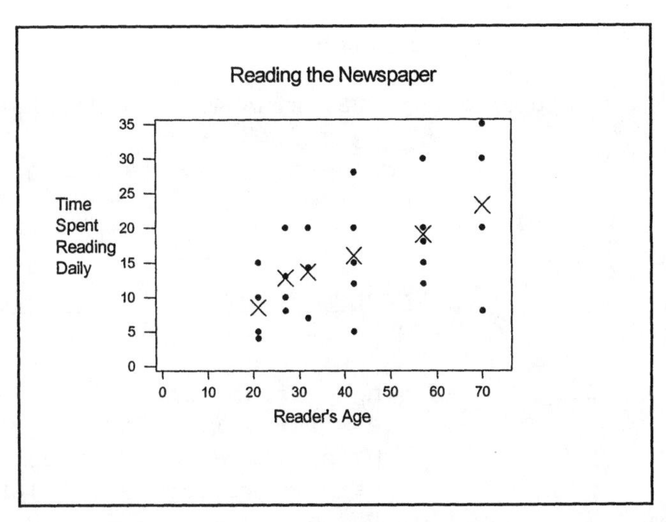

b. Yes, the length of time spent reading the newspaper daily definitely appears to lengthen as the reader ages. However, there is quite a bit of variability within each age group, maybe more than between the age groups. Therefore, it is hard to judge the significance of the pattern without additional statistics.

c. The answer to both questions is, yes age appears to have an effect. They are not identically the same question, but realistically they are the same: if the individual amounts increase, then the average will increase, and also, if the mean increases the data has to increase with it.

12.27 a. Step 1: a. The mean time spent reading newspaper for the six age groups.

b. H_o: $\mu_1 = \mu_2 = \mu_3 = \mu_4 = \mu_5 = \mu_6$
or The mean time spent reading is the same for all age groups or age has no effect on the mean daily amount of time spend reading the newspaper.
H_a: The means are not all equal.
or At least one mean is different, or, age does have an effect on the mean daily amount of time spent reading the newspaper.

Step 2: a. Assume the data were randomly collected and are independent, and the effects due to chance and untested factors are normally distributed.

b. F

c. $\alpha = 0.05$

Step 3: a. $n = 25$, $k_1 = 4$, $C_1 = 34$, $k_2 = 4$, $C_2 = 51$, $k_3 = 3$, $C_3 = 41$, $k_4 = 5$, $C_4 = 80$, $k_5 = 5$, $C_5 = 95$, $k_6 = 4$, $C_6 = 93$, $T = 394$, $\Sigma x^2 = 7904$

Source	df	SS	MS	F*
Age	5	537.4	107.5	1.76
Error	19	1157.2	60.9	
Total	24	1694.6		

Step 4: -- using p-value approach ---------------

a. $\mathbf{P} = P(F > 1.76 | df_n = 5, df_d = 19)$;
Using Table 9: $\mathbf{P} > 0.05$
Using computer: $\mathbf{P} = 0.168$

b. $P > \alpha$

-- using classical approach -------------

a. $F(5,19,0.05) = 2.74$

b. F* is not in the critical region

Step 5: a. Fail to reject H_o.

b. The data does not show significant evidence to conclude the mean amount of time spent reading newspaper yesterday is effected by age, at the 0.05 level of significance.

b. The amount of variability within the age groups is too large, relatively speaking, for the amount of variability between age groups to be significant.

CHAPTER EXERCISES

12.29 a. H_o: The mean amount of salt is the same in all tested brands of peanut butter.
H_a: The mean amount of salt is not the same in all tested brands of peanut butter.

b. Assumptions: samples were randomly selected

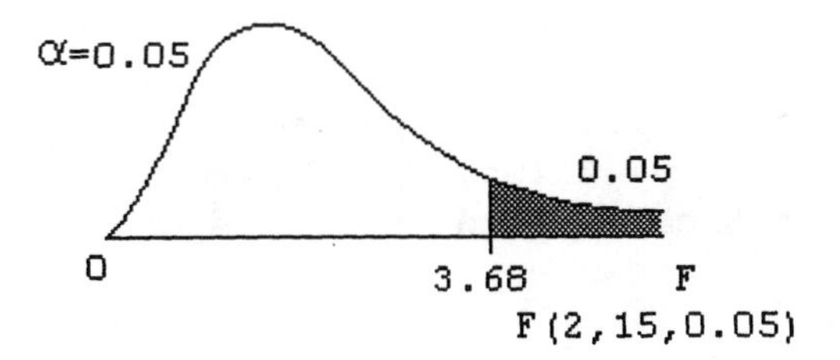

c. Fail to reject H_o. There is no significant difference in the mean amounts of salt in the tested brands.

d. Since the p-value is quite large (much larger than 0.05), it tells us the sample data is quite likely to have occurred under the assumed conditions and a true null hypothesis. Therefore, we 'fail to reject H_o.'

12.31 Step 1: a. The mean lengths of time that pain relief are provided; the mean length of time for drug A, the mean length of time for drug B, the mean length of time for drug C, the mean length of time for drug D.

b. H_o: The mean amount of relief time is the same for all four drugs.
H_a: The mean amount of relief time is not the same for all four drugs.

Step 2: a. Assume the data were randomly collected and are independent, and the effects due to chance and untested factors are normally distributed.

b. F

c. $\alpha = 0.05$

Step 3: a. $n = 16$, $C_A = 20$, $C_B = 20$, $C_C = 50$, $C_D = 10$, $T = 100$, $\Sigma x^2 = 768$

Source	df	SS	MS	F*
Drug	3	108.33	36.11	12.50
Error	12	34.67	2.89	
Total	15	143.00		

Step 4: -- using p-value approach ---------------
a. $\mathbf{P} = P(F > 12.50|df_n = 3, df_d = 12)$;
Using Table 9: $\mathbf{P} < 0.01$
Using computer: $\mathbf{P} = 0.001$
b. $\mathbf{P} < \alpha$
-- using classical approach -------------
a. critical value: $F(3,12,0.05) = 3.49$
critical region: $F \geq 3.49$
b. F* is in the critical region

Step 5: b. Reject H_o.
c. There is a significant difference between the mean amount of time one gets relief from these four drugs.

12.33 Step 1: a. The mean amounts of soft drink dispensed: the mean amount for machine A, the mean amount for machine B, the mean amount for machine C, the mean amount for machine D, the mean amount for machine E.
b. H_o: The mean amounts dispensed by the machines are all equal.
H_a: The mean amounts dispensed by the machines are not all equal.
Step 2: a. Assume the data were randomly collected and are independent, and the effects due to chance and untested factors are normally distributed.
b. F
c. $\alpha = 0.01$
Step 3: a. $n = 18$, $C_A = 16.5$, $C_B = 20.6$, $C_C = 16.9$, $C_D = 19.1$,
$C_E = 21.8$, $T = 94.9$, $\Sigma x^2 = 523.49$

Source	df	SS	MS	F*
Machine	4	20.998	5.2495	31.6
Error	13	2.158	0.166	
Total	17	23.156		

Step 4: -- using p-value approach ---------------
a. $\mathbf{P} = P(F > 31.6|df_n = 4, df_d = 13)$;
Using Table 9: $\mathbf{P} < 0.01$
Using computer: $\mathbf{P} = 0.000$
b. $\mathbf{P} < \alpha$

-- using classical approach ------------

a. critical value: $F(4,13,0.01) = 5.21$
   critical region: $F \geq 5.21$
b. F* is in the critical region

Step 5: a. Reject H_o.
b. There is a significant difference between the machines with regards to mean amount of soft drink dispensed.

12.35 Step 1: a. The mean points scored by teams representing each division.
b. $H_o: \mu_E = \mu_C = \mu_W$
   H_a: The mean points scored by teams representing each division are not all equal.

Step 2: a. Assume the data were randomly collected and are independent, and the effects due to chance and untested factors are normally distributed.
b. F
c. $\alpha = 0.05$

Step 3: a. $n = 31$, $k_E = 10$, $C_E = 3135$, $k_C = 11$, $C_C = 3408$, $k_W = 10$, $C_W = 3481$, $T = 10024$, $\Sigma x^2 = 3342358$

Source	df	SS	MS	F*
Division	2	9110	4555	1.39
Error	28	91939	3284	
Total	30	101049		

Step 4: -- using p-value approach --------------

a. $\mathbf{P} = P(F > 1.39 | df_n = 2, df_d = 28)$;
   Using Table 9: $\mathbf{P} > 0.05$
   Using computer: $\mathbf{P} = 0.266$
b. $\mathbf{P} > \alpha$

-- using classical approach ------------

a. $F(2,28,0.05) \approx 3.36$
b. F* is not in the critical region

Step 5: a. Fail to reject H_o.
b. The data show no evidence that would give reason to reject the null hypothesis that the points scored by the teams in each division are equal, at the 0.05 level of significance.

Step 1: a. The mean points scored by the opponents of the teams representing each division.

b. H_o: $\mu_E = \mu_C = \mu_W$

H_a: The mean points scored by the opponents of the teams representing each division are not all equal.

Step 2: a. Assume the data were randomly collected and are independent, and the effects due to chance and untested factors are normally distributed.

b. F

c. $\alpha = 0.05$

Step 3: a. $n = 31$, $k_E = 10$, $C_E = 3276$, $k_C = 11$, $C_C = 3342$, $k_W = 10$, $C_W = 3406$, $T = 10024$, $\Sigma x^2 = 3374470$

Source	df	SS	MS	F*
Division	2	7353	3676	0.82
Error	28	125808	4493	
Total	30	133161		

Step 4: -- using p-value approach ---------------

a. **P** = $P(F > 0.82 \mid df_n = 2, df_d = 28)$;

Using Table 9: **P** > 0.05

Using computer: **P** = 0.451

b. **P** > α

-- using classical approach -------------

a. $F(2,28,0.05) \approx 3.36$

b. F* is not in the critical region

Step 5: a. Fail to reject H_o.

b. The data show no evidence that would give reason to reject the null hypothesis that the points scored by the opponents of the teams in each division are equal, at the 0.05 level of significance.

12.37 Step 1: a. The mean family income for three different counties in Nebraska: the mean family income for Lancaster, the mean family income for Hall, the mean family income for Sarpy.

b. H_o: The mean family income is the same for all three counties.

H_a: The mean family income is not the same for all three counties.

Step 2: a. Assume the data were randomly collected and are independent, and the effects due to chance and untested factors are normally distributed.
b. F
c. $\alpha = 0.05$

Step 3: a.

Source	df	SS	MS	F
Counties	2	1217.5	608.7	28.74
Error	24	508.4	21.2	
Total	26	1725.8		

Step 4: -- using p-value approach ---------------
a. $\mathbf{P} = P(F > 28.74 | df_n = 2, df_d = 24)$;
Using Table 9: $\mathbf{P} < 0.01$
Using computer: $\mathbf{P} = 0.000+$
b. $\mathbf{P} < \alpha$
-- using classical approach -------------
a. critical value: $F(2,24,0.05) = 3.40$
b. F^* is in the critical region

Step 5: a. Reject H_o.
b. The mean family incomes differ for the three counties, at the 0.05 level of significance.

12.39 Step 1: a. The mean durability for different brands of golf balls: the mean durability for Brand A, the mean durability for Brand B, the mean durability for Brand C, the mean durability for Brand D, the mean durability for Brand E, the mean durability for Brand F.
b. H_o: The six different brands of golf balls withstood the durability test equally well, as measured by the mean number of hits before failure.
H_a: The six different brands of golf balls do not withstand the durability test equally well.

Step 2: a. Assume the data were randomly collected and are independent, and the effects due to chance and untested factors are normally distributed.
b. F
c. $\alpha = 0.05$

Step 3: a.

Source	df	SS	MS	F*
Brand	5	75047	15009.4	5.30
Error	36	101899	2830.5	
Total	41	176946		

Step 4: -- using p-value approach ---------------

a. $\mathbf{P} = P(F > 5.30 | df_n = 5, df_d = 36)$;

Using Table 9: $\mathbf{P} < 0.01$

Using computer: $\mathbf{P} = 0.001$

b. $\mathbf{P} < \alpha$

-- using classical approach -------------

a. critical value: $F(5,36,0.05) = 2.48$

critical region: $F \geq 2.48$

b. F* is in the critical region

Step 5: a. Reject H_o.

b. There is a significant difference between the mean number of hits before failure for the six brands of golf balls tested, at the 0.05 level of significance.

12.41 a.

Step 1: a. The mean petal width for three species of irises.

b. H_o: The mean petal width is the same for each specie of iris

H_a: The mean petal width is not the same for each specie of iris.

Step 2: a. Assume the data were randomly collected and are independent, and the effects due to chance and untested factors are normally distributed.

b. F

c. $\alpha = 0.05$

Step 3: a.

Source	df	SS	MS	F
Specie	2	1671.56	835.78	118.06
Error	27	191.14	7.08	
Total	29	1862.70		

Step 4: -- using p-value approach ---------------

a. $\mathbf{P} = P(F > 118.06 | df_n = 2, df_d = 27)$;

Using Table 9: $\mathbf{P} < 0.01$

Using computer: $\mathbf{P} = 0.000$

b. $\mathbf{P} < \alpha$

-- using classical approach ------------
a. critical value: F(2,27,0.05)   3.37
b. F* is in the critical region

Step 5: a. Reject H_o.
b. There is sufficient evidence to show that at least one specie's petal width is significantly different from the others, at the 0.05 level of significance.

b.

Step 1: a. The mean sepal width for three species of irises.
b. H_o: The mean sepal width is the same for each specie of iris
H_a: The mean sepal width is not the same for each specie of iris.

Step 2: a. Assume the data were randomly collected and are independent, and the effects due to chance and untested factors are normally distributed.
b. F
c. $\alpha = 0.05$

Step 3: a.

Source	df	SS	MS	F
Specie	2	197.1	98.6	7.78
Error	27	342.2	12.7	
Total	29	539.4		

Step 4: -- using p-value approach --------------
a. $\mathbf{P} = P(F > 7.78 | df_n = 2, df_d = 27)$;
Using Table 9: $\mathbf{P} < 0.01$
Using computer: $\mathbf{P} = 0.002$
b. $\mathbf{P} < \alpha$
-- using classical approach ------------
a. critical value: F(2,27,0.05)   3.37
b. F* is in the critical region

Step 5: a. Reject H_o.
b. There is sufficient evidence to show that at least one specie's sepal width is significantly different from the others, at the 0.05 level of significance.

c. Type 0 has the shortest PW and the longest SW. Type 1 has the longest PW and the middle SW. Type 2 has the middle PW and the shortest SW.

12.43 Using formula (12.4):

$SS(error) = 612 - [(14^2/3)+(16^2/3)+(36^2/3)]$
$= 612 - 582.6667 = \underline{29.3333}$

Using formula in exercise:

$s_1^2 = [(8^2+4^2+2^2)-(14^2/3)]/2 = 9.3333$
$s_2^2 = [(6^2+6^2+4^2)-(16^2/3)]/2 = 1.3333$
$s_3^2 = [(10^2+12^2+14^2)-(36^2/3)]/2 = 4.0000$

$SS(error) = 2(9.3333) + 2(1.3333) + 2(4.0000)$
$= \underline{29.3333}$

12.45 $C_1 = 8(97.99) = 783.92$, $C_2 = 8(97.68) = 781.44$,
$C_3 = 8(96.29) = 770.32$, $C_4 = 8(93.92) = 751.36$;
$k_1 = k_2 = k_3 = k_4 = 8$, $n = 32$

$\Sigma x = C_1+C_2+C_3+C_4 = 783.92+781.44+770.32+751.36 = 3087.04$

$SS(factor) = [\Sigma(C_i^2/k_i)] - [(\Sigma x)^2/n]$
$[(783.92^2/8)+(781.44^2/8)+(770.32^2/8)+(751.36^2/8)] -$
$[3087.04^2/32]$

$SS(factor) = \underline{82.4752}$

CHAPTER 13 ∇ LINEAR CORRELATION AND REGRESSION ANALYSIS

Chapter Preview

Chapter 3 introduced the concepts of correlation and regression analysis for bivariate data. In Chapter 13, we will look at these concepts in a more detailed manner using confidence intervals and hypothesis tests. These inference tests will be utilized on the correlation coefficient, the slope of the regression line and the regression line.

Wheat planted and harvasted in the United States during 2002 is the focus of this chapter's Case Study.

CHAPTER 13 CASE STUDY

13.1 a. $r = 0.989$; $\hat{y} = -1.57 + 0.880x$

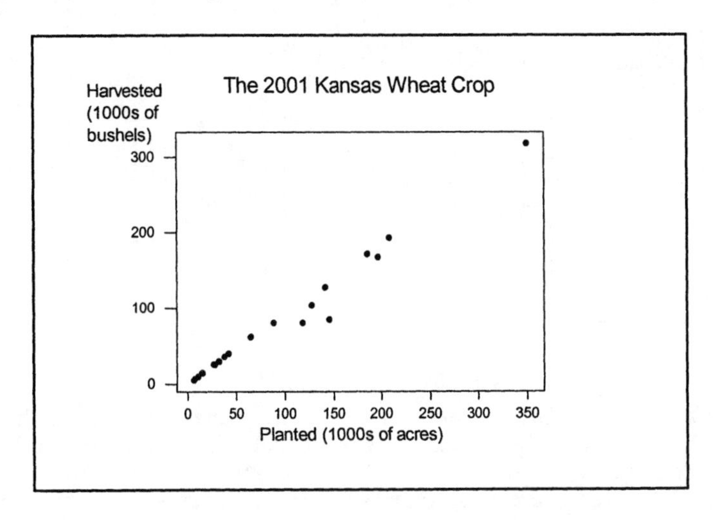

b.

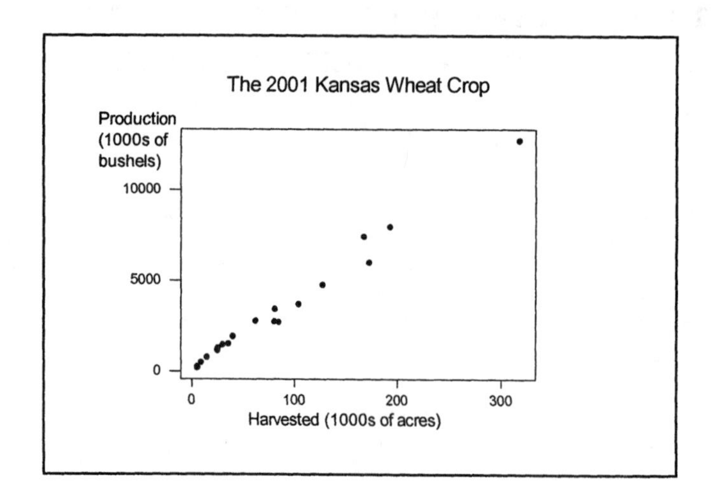

c.

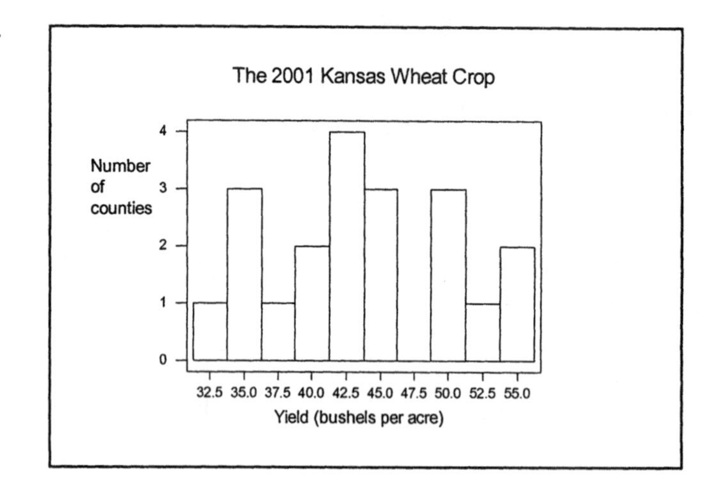

d. The acres planted and acres harvested are highly linearly correlated. The total production and the acres harvested are highly linearly correlated. The yield rate has an approximately normal distribution.

SECTION 13.1 EXERCISES

Formulas for the covariance and correlation coefficient of bivariate data

Covariance of x and y - measure of linear dependency between x and y

$$\text{covar}(x,y) = \frac{\sum (x - \bar{x})(y - \bar{y})}{n - 1}$$

$$s_x = \sqrt{SS(x)/(n-1)}\ ,\ \text{where } SS(x) = \Sigma x^2 - [(\Sigma x)^2/n]$$

$$s_y = \sqrt{SS(y)/(n-1)}\ ,\ \text{where } SS(y) = \Sigma y^2 - [(\Sigma y)^2/n]$$

$$r = \frac{\text{cov ar}(x, y)}{s_x \cdot s_y} \text{ or}$$

$$r = \frac{SS(xy)}{\sqrt{SS(x) \cdot SS(y)}},\ \text{where } SS(xy) = \Sigma xy - [(\Sigma x)(\Sigma y)/n]$$

13.3 a.

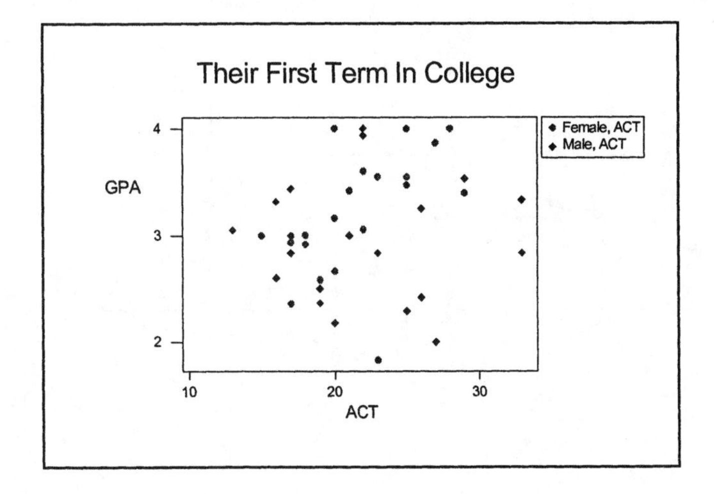

b. Patterns are somewhat similar in that the points pretty much cover the full area of the diagram. The females, with one exception, are all located in the top part of the diagram.

c.

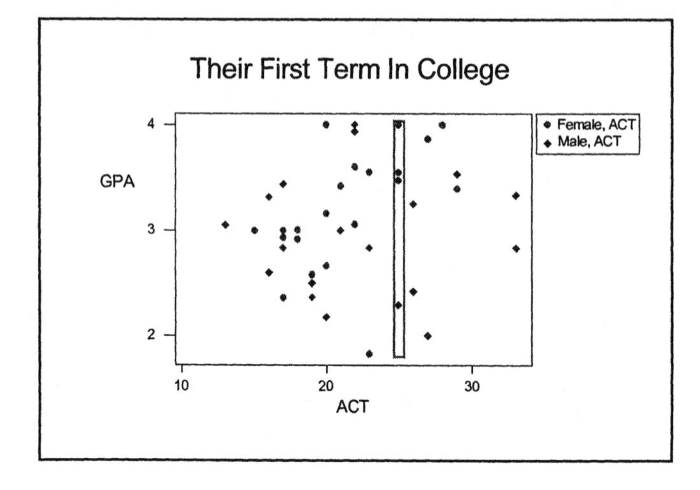

13.5 a.

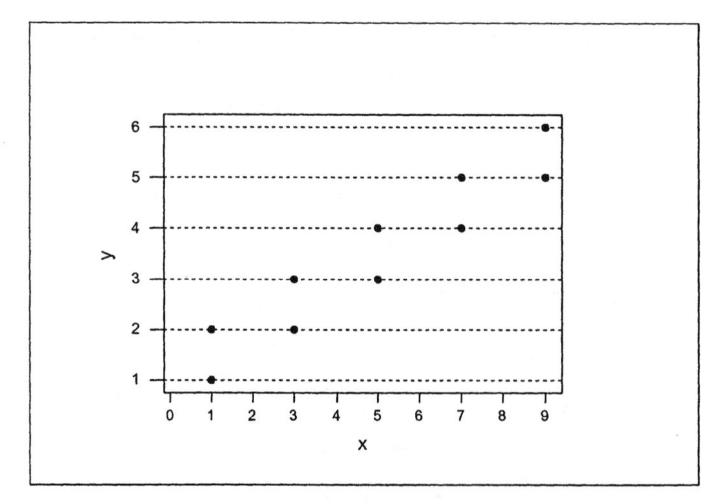

b.

x	y	$x-\overline{x}$	$y-\overline{y}$	$(x-\overline{x})(y-\overline{y})$
1	1	-4	-2.5	+10
1	2	-4	-1.5	+ 6
3	2	-2	-1.5	+ 3
3	3	-2	-0.5	+ 1
5	3	0	-0.5	0
5	4	0	+0.5	0
7	4	+2	+0.5	+ 1
7	5	+2	+1.5	+ 3
9	5	+4	+1.5	+ 6
9	6	+4	+2.5	+10
50	35	0	0.0	40

$\overline{x} = 50/10 = 5.0$ and $\overline{y} = 35/10 = 3.5$

$\text{covar}(x,y) = [\Sigma(x-\overline{x})(y-\overline{y})]/(n-1) = 40/9 = \underline{4.44}$

Summary of data: $n = 10$, $\Sigma x = 50$, $\Sigma y = 35$,
$\Sigma x^2 = 330$, $\Sigma xy = 215$, $\Sigma y^2 = 145$

c. $s_x = \sqrt{[330 - (50^2/10)]/9} = \sqrt{8.889} = \underline{2.981}$

$s_y = \sqrt{[145 - (35^2/10)]/9} = \sqrt{2.50} = \underline{1.581}$

d. $r = 4.444/[(2.981)(1.581)] = \underline{0.943}$

e. $SS(x) = 330 - (50^2/10) = 80$
$SS(y) = 145 - (35^2/10) = 22.5$
$SS(xy) = 215 - [(50)(35)/10] = 40$

$r = 40/\sqrt{(80)(22.5)} = \underline{0.943}$

13.7 Verify -- answers given in exercise.

Computer and/or calculator commands to calculate r, the correlation coefficient can be found in ES9-p148.

13.9 $n = 31$, $\Sigma x = 10024$, $\Sigma y = 10024$, $\Sigma x^2 = 3342358$, $\Sigma xy = 3214143$

$\Sigma y^2 = 3374470$

$SS(x) = 101049$, $SS(y) = 133161$, $SS(xy) = -27165.9$

a. $r = SS(xy)/\sqrt{SS(x) \cdot SS(y)} = -27165.9/\sqrt{(101049)(133161)} = \underline{-0.234}$

b. slight negative relationship

c.

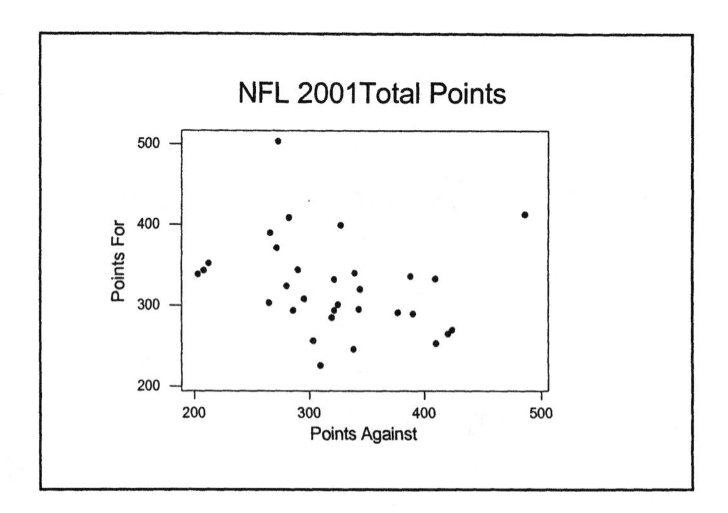

Downward trend shows that as 'points against' increase, 'points for' decrease. Points are somewhat wide spread, causing a fairly weak correlation coefficient.

13.11 a.

x	y	$x-\bar{x}$	$y-\bar{y}$	$(x-\bar{x})(y-\bar{y})$
20	10	-50	-20	1000
30	50	-40	20	-800
60	30	-10	0	0
80	20	10	-10	-100
110	60	40	30	1200
120	10	50	-20	-1000
420	180	0	0	300

$\bar{x} = 420/6 = 70$, $\bar{y} = 180/6 = 30$
covar$(x,y) = 300/5 = \underline{60}$

b. $s_x = 40.99$, $s_y = 20.98$
c. $r = 60/(40.99)(20.98) = 0.0698 = \underline{0.07}$
d. The value for r is the same.

SECTION 13.2 ANSWER NOW EXERCISES

13.13 -0.02 to 0.63

13.14 a. $0.05 < \mathbf{P} < 0.10$ b. $0.025 < \mathbf{P} < 0.05$

13.15 a. ± 0.444
b. -0.378, if left tail critical region; 0.378, if right tail

13.16 a. The linear correlation coefficient, $r = 0.58$ is significant for all levels of $\alpha > 0.008$.
b. 0.464
c. r is significant at the $\alpha = 0.01$ level

SECTION 13.2 EXERCISES

13.17

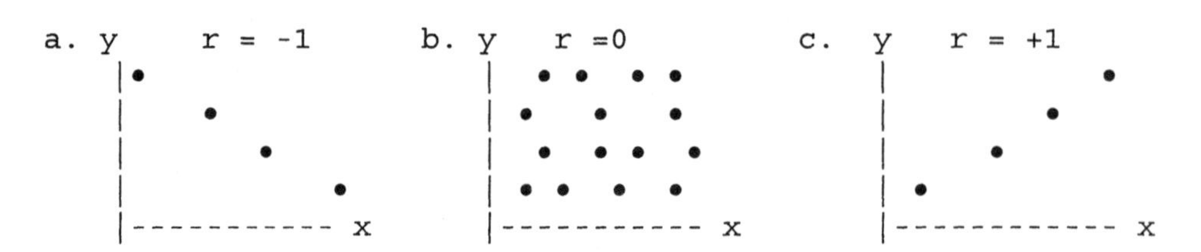

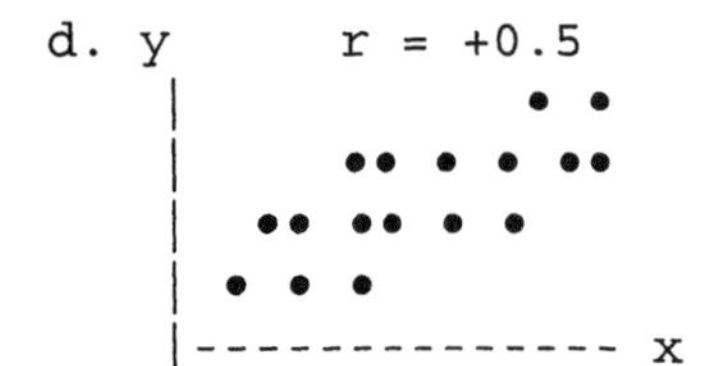

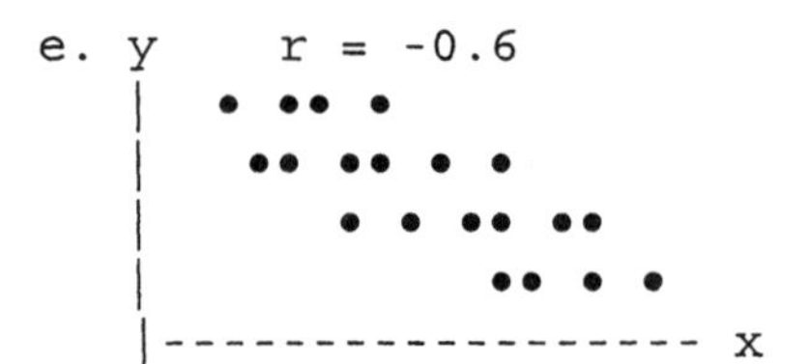

> Estimating ρ - the population correlation coefficient
> 1. point estimate: r
> 2. confidence interval: use Table 10 in Appendix B, ES9-p726, in the text to determine a 95% confidence interval. Locate the r value on the horizontal axis. Follow a vertical line up until the corresponding sample size band intersects. This value is the lower limit. To find the upper limit, locate the r value on the vertical axis. Follow a horizontal line until it intersects the corresponding sample size.

> Reminder: $r = SS(xy)/\sqrt{SS(x)\cdot SS(y)}$ where: $SS(x) = \Sigma x^2 - [(\Sigma x)^2/n]$
> $SS(y) = \Sigma y^2 - [(\Sigma y)^2/n]$
> $SS(xy) = \Sigma xy - [(\Sigma x)(\Sigma y)/n]$

13.19 Summary of data: $n = 10$, $\Sigma x = 746$, $\Sigma y = 736$,
$\Sigma x^2 = 57{,}496$, $\Sigma xy = 56{,}574$, $\Sigma y^2 = 55{,}826$

$SS(x) = 57496 - (746^2/10) = 1844.4$
$SS(y) = 55826 - (736^2/10) = 1656.4$
$SS(xy) = 56574 - [(746)(736)/10] = 1668.4$

$r = SS(xy)/\sqrt{SS(x)\cdot SS(y)} = 1668.4/\sqrt{(1844.4)(1656.4)} = \underline{0.955}$

From Table 10: $\underline{0.78 \text{ to } 0.98}$, the 0.95 interval for ρ

13.21 a. Summary of data: $n = 7$, $\Sigma x = 1189$, $\Sigma y = 5482$,

$\Sigma x^2 = 205787$, $\Sigma xy = 928950$, $\Sigma y^2 = 4333726$

$SS(x) = 205787 - (1189^2/7) = 3826.857$
$SS(y) = 4333726 - (5482^2/7) = 40536.857$
$SS(xy) = 928950 - [(1189)(5482)/7] = -2206.857$

$r = SS(xy)/\sqrt{SS(x) \cdot SS(y)}$
$= -2206.857/\sqrt{(3826.857)(40536.857)} = \underline{-0.177}$

b. From Table 10: -0.78 to 0.62

c. With 95% confidence, the population correlation coefficient is between -0.78 and 0.62.

d. The interval is very wide, largely due to the small sample size.

> Hypotheses for the correlation coefficient are written with the same rules before. Now in place of μ or σ, the population correlation coefficient, ρ, will be used. The standard form is using 0 as the test value, unless some other information is given in the exercise.
> 0 indicates that there is no linear relationship.
> (ex. H_o: $\rho = 0$ vs. H_a: $\rho \neq 0$)

13.23 a. H_o: $\rho = 0$ vs. H_a: $\rho > 0$

b. H_o: $\rho = 0$ vs. H_a: $\rho \neq 0$

c. H_o: $\rho = 0$ vs. H_a: $\rho < 0$

d. H_o: $\rho = 0$ vs. H_a: $\rho > 0$

Test criteria

1. Draw a bell-shaped distribution locating 0 at the center, -1 at the far left and +1 at the far right.
2. Shade in the critical region(s) based on the alternative hypothesis (H_a).
3. Find the critical value(s) from Table 11, Appendix B, ES9-p727:
   a. degrees of freedom (n - 2) is the row id #
   b. α, is the column id #;
      1) use the given α for a two-tailed test
      2) use 2α for a one-tailed test
   c. all values given in the table are positive. Negate the value if the critical region or part of the critical region is to the left of 0.

Hypothesis tests will be completed using the same format as before. You may want to review: ES9-pp370 & 386, ST-pp241 & 249. The only differences are:

1. **writing hypotheses**: (see box before ex. 13.23)
2. **using Table 11**: (see box before ex. 13.24 for the classical approach and ES9-p615 for the p-value approach)
3. **the calculated test statistic**: r*, the sample correlation coefficient

13.25 Step 1: a. The linear correlation coefficient for the population, ρ.

b. $H_o: \rho = 0.0$

$H_a: \rho \neq 0.0$

Step 2: a. random sample, assume normality for y at each x

b. r, df = n - 2 = 20 - 2 = 18

c. $\alpha = 0.10$

Step 3: a. n = 20, r = 0.43

b. r* = 0.43

Step 4: -- using p-value approach -------------------

a. $\mathbf{P} = P(r < 0.43) + P(r > 0.43)$

$= 2P(r > 0.43 | df = 18)$;

Using Table 11, ES9-p727: $0.05 < \mathbf{P} < 0.10$

b. $\mathbf{P} < \alpha$

-- using classical approach ------------------

a. $\pm r(18, 0.10) = \pm 0.378$

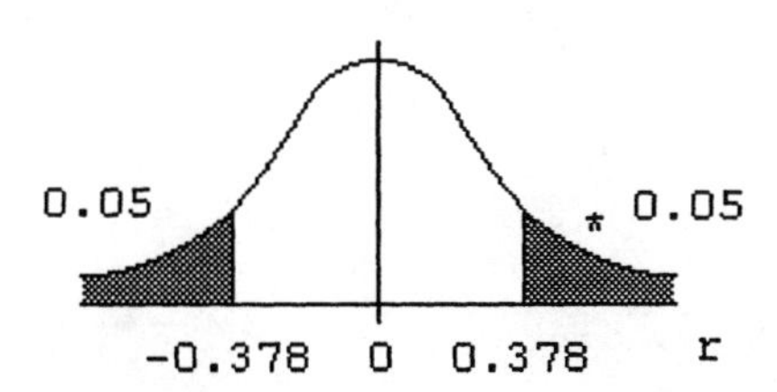

b. r^* falls in the critical region, see Step 4a.

Step 5: a. Reject H_o.

b. There is sufficient reason to reject the null hypothesis, at the 0.10 level of significance.

13.27 Step 1: a. The linear correlation coefficient between uplift rate and length of drainage basin, ρ.

b. $H_o: \rho = 0.0$

$H_a: \rho \neq 0.0$

Step 2: a. random sample, assume normality for y at each x

b. r, $df = n - 2 = 24 - 2 = 22$

c. $\alpha = 0.05$

Step 3: a. $n = 24$, $r = 0.16942$

b. $r^* = 0.16942$

Step 4: -- using p-value approach -------------------

a. $\mathbf{P} = P(r < -0.16942) + P(r > 0.16942)$

$= 2P(r > 0.16942)$;

Using Table 11, ES9-p727: $\mathbf{P} > 0.10$

b. $\mathbf{P} > \alpha$

-- using classical approach ------------------

a. critical region: $r \leq -0.423$ and $r \geq 0.423$

b. r^* falls in the noncritical region

Step 5: a. Fail to reject H_o.

b. There is not sufficient reason to reject the null hypothesis, at the 0.05 level of significance.

13.29 Summary of data: $n = 10$, $\Sigma x = 26.2$, $\Sigma y = 82.5$,
$\Sigma x^2 = 174.88$, $\Sigma xy = 256.41$, $\Sigma y^2 = 704.61$

$SS(x) = 174.88 - (26.2^2/10) = 106.236$
$SS(y) = 704.61 - (82.5^2/10) = 23.985$
$SS(xy) = 256.41 - [(26.2)(82.5)/10] = 40.26$

$r = SS(xy)/\sqrt{SS(x) \cdot SS(y)} = 40.26/\sqrt{(106.236)(23.985)} = \underline{0.798}$

Step 1: a. The linear correlation coefficient for size of a metropolitan area and its crime rate, ρ.
b. H_o: $\rho = 0.0$
H_a: $\rho \neq 0.0$

Step 2: a. random sample, assume normality for y at each x
b. r, df = n - 2 = 10 - 2 = 8
c. $\alpha = 0.05$

Step 3: a. n = 10, r = 0.798
b. $r^* = 0.798$

Step 4: -- using p-value approach --------------------
b. $\mathbf{P} = P(r < 0.798) + P(r > 0.798)$
$= 2P(r > 0.798)$;
Using Table 11, ES9-p727: $\mathbf{P} < 0.01$
b. $\mathbf{P} < \alpha$

-- using classical approach ------------------
a. critical region: $r \leq -0.632$ and $r \geq 0.632$
b. r^* falls in the critical region

Step 5: a Reject H_o
b. There is sufficient reason to conclude that the correlation coefficient is different than zero, at the 0.05 level of significance.

13.31 a. r = 0.613
b. Step 1: a. The linear correlation coefficient between personal income and the value of construction contracts, ρ.
b. H_o: $\rho = 0.0$
H_a: $\rho \neq 0.0$

Step 2: a. random sample,assume normality for y at each x
b. r, df = n - 2 = 7 - 2 = 5
c. $\alpha = 0.05$

Step 3: a. n = 7, r = 0.613
b. $r^* = 0.613$

Step 4: -- using p-value approach --------------------

a. **P** = P(r < -0.613) + P(r > 0.613)
= 2P(r > 0.613);
Using Table 11, ES9-p727: **P** > 0.10

b. **P** > α

-- using classical approach ------------------

a. critical region: r ≤ -0.754 and r ≥ 0.754
b. r* falls in the noncritical region

Step 5: a. Fail to reject H_o
b. At the 0.05 level of significance, we have failed to show that personal income and the value of new construction contracts are correlated.

SECTION 13.3 EXERCISES

Sample Regression Line: $\hat{y} = b_o + b_1 x$, where

$$b_1 = \frac{SS(xy)}{SS(x)} = \frac{\sum(xy) - [(\sum x)(\sum y) / n]}{\sum x^2 - (\sum x)^2 / n} \quad \text{and} \quad b_o = \frac{1}{n}\left(\sum y - b_1 \cdot \sum x\right)$$

Population Regression Line: $y = \beta_0 + \beta_1 x + \varepsilon$

Sample estimate of ε = e = $(y - \hat{y})$

Variance of y about the regression line = variance of the error e

$$s_e^2 = \frac{(\sum y^2) - [(b_0)(\sum y)] - [(b_1)(\sum xy)]}{n - 2} = \frac{SSE}{n - 2}$$

Standard Deviation of the error = $s_e = \sqrt{s_e^2}$

Explore the relationship between residuals and the line of best fit with Interactivity 13A.

Slight variations in sums of squares and further calculations can result from round-off errors.

13.33 Verify -- answers given in exercise.

13.35

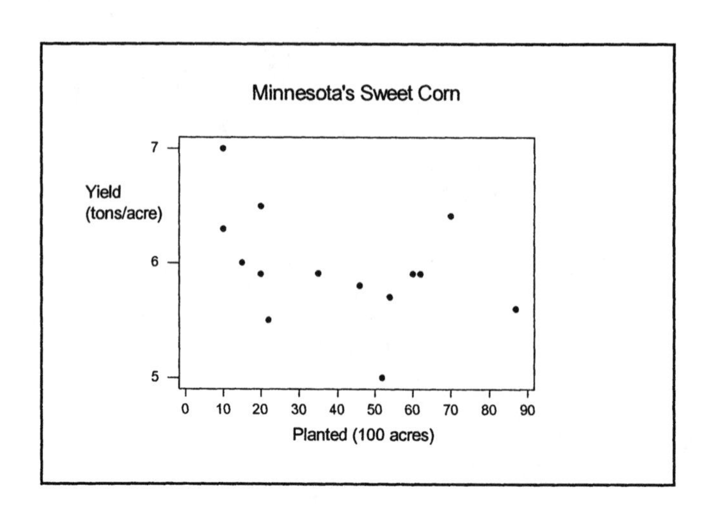

Summay of data: $n = 14$, $\Sigma x = 563$, $\Sigma y = 83.42$
$\Sigma x^2 = 30583$, $\Sigma xy = 3289.15$, $\Sigma y^2 = 500.126$

$SS(x) = 30583 - (563^2/14) = 7942.36$
$SS(y) = 500.126 - (83.42^2/14) = 3.06197$
$SS(xy) = 3289.15 - [(563)(83.42)/14] = -65.5257$

$r = SS(xy)/\sqrt{SS(x) \cdot SS(y)}$
$= -65.5257/\sqrt{(7942.36)(3.06197)} = \underline{-0.420}$

Using formula 3.6:
$b_1 = -65.5257/7942.36 = -0.00825$

Using formula 3.7:
$b_0 = [83.42 - (-0.00825)(563)]/14 = 6.29$

Best fit line: $\hat{y} = 6.29 - 0.00825x$

13.37

a. Scatter diagram and best fit line:

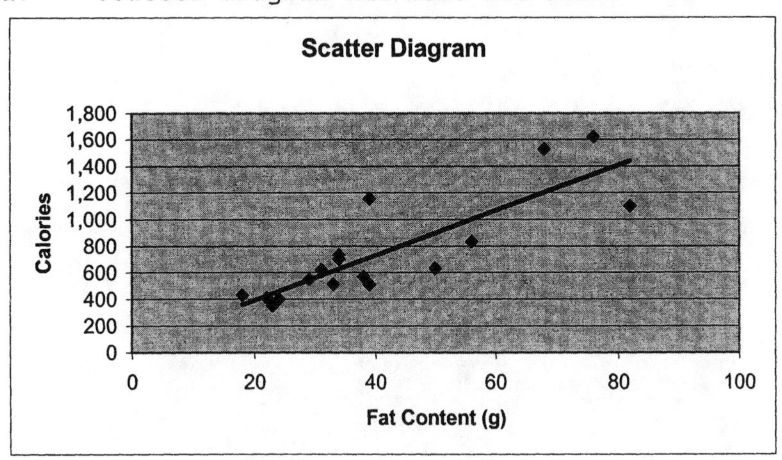

b. Summary of data: $n = 17$, $\Sigma x = 696$, $\Sigma y = 12{,}655$

$\Sigma x^2 = 34362$, $\Sigma xy = 617{,}415$, $\Sigma y^2 = 11{,}801{,}175$

$SS(x) = 34362 - (696^2/17) = 5866.941177$

$SS(xy) = 617{,}415 - [(696)(12655)/17] = 99304.41177$

Using formula 3.6:

$b_1 = 99304.41177/5866.941177 = 16.926$

Using formula 3.7:

$b_0 = [12655 - (16.926096)(696)]/17 = 51.4$

Best fit line: $\hat{y} = 16.926x + 51.4$

c. 16.926 (slope of regression line)

d.

$$s_e^2 = \frac{\sum y^2 - (b_0)\sum y - (b_1)\sum xy}{n-2} = \frac{11{,}801{,}175 - 51.4(12{,}655) - 16.93(617{,}415)}{15}$$

$$= \frac{11{,}801{,}175 - 650{,}467 - 10{,}452{,}836}{15} = 46{,}524.8$$

13.39 a. Summary of data: $n = 10$, $\Sigma x = 50$, $\Sigma y = 35$,
$\Sigma x^2 = 330$, $\Sigma xy = 215$, $\Sigma y^2 = 145$

$SS(x) = 330 - (50^2/10) = 80$
$SS(xy) = 215 - [(50)(35)/10] = 40$

Using formula 3.6:
$b_1 = 40/80 = 0.50$

Using formula 3.7:
$b_0 = [35 - (0.50)(50)]/10 = 1.0$

$\hat{y} = \underline{1.0 + 0.5x}$

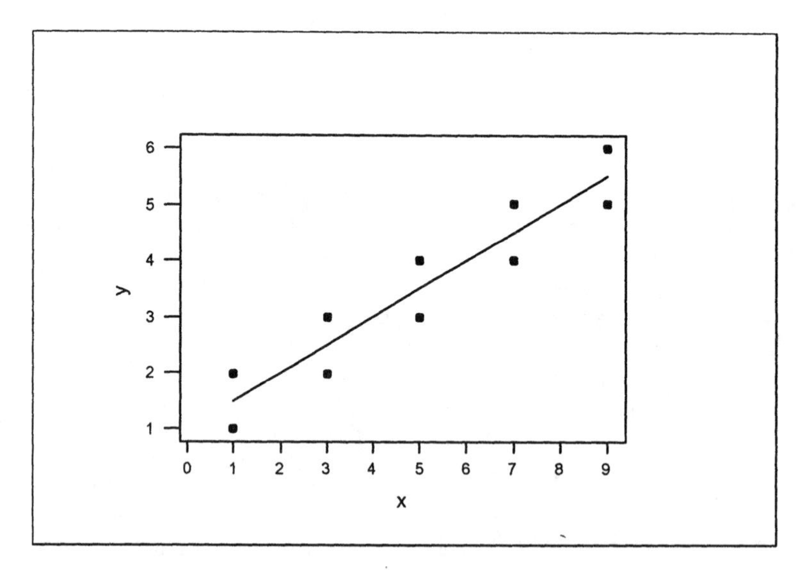

b. If $x = 1$, then $\hat{y} = 1.0 + 0.5(1) = \underline{1.5}$
If $x = 3$, then $\hat{y} = 1.0 + 0.5(3) = \underline{2.5}$
If $x = 5$, then $\hat{y} = 1.0 + 0.5(5) = \underline{3.5}$
If $x = 7$, then $\hat{y} = 1.0 + 0.5(7) = \underline{4.5}$
If $x = 9$, then $\hat{y} = 1.0 + 0.5(9) = \underline{5.5}$

c.

point	A	B	C	D	E	F	G	H	I	J
x	1	1	3	3	5	5	7	7	9	9
y	1	2	2	3	3	4	4	5	5	6
$\hat{y}$	1.5	1.5	2.5	2.5	3.5	3.5	4.5	4.5	5.5	5.5
$e=y-\hat{y}$	-.5	.5	-.5	.5	-.5	.5	-.5	.5	-.5	.5

d. $s_e^2 = 2.50/8 = \underline{0.3125}$

e. $s_e^2 = [145 - (1.0)(35) - (0.5)(215)]/8$
$= 2.50/8 = \underline{0.3125}$

SECTION 13.4 ANSWER NOW EXERCISES

13.41 From the scatter diagram, n = 72

a. $S_1 = -3953.85 + 3.13(12,600) = 35484.15 \approx 35,500$

b. From the scatter diagram, n = 72

$b_1 \pm t(df, \alpha/2) \cdot s_{b1}$
$3.130 \pm (1.99)(0.065) = 3.130 \pm 0.129$
$\underline{3.001 \text{ to } 3.259}$ (difference due to round-off)

SECTION 13.4 EXERCISES

To determine the p-value for the test of the slope of the regression line, the t-distribution is used. Review its use in: ES9-p422, ST-p276, if necessary. The only difference required is **df = n - 2**, since the data is bivariate. The test statistic is $t^* = (b_1 - \beta_1)/s_{b1}$.

13.43 a. **P** = $P(t>2.40|df=16) = \underline{0.0145}$

b. **P** = $2 \cdot P(t>2.00|df=13) = 2(0.0334) = \underline{0.0668}$

c. **P** = $P(t<-1.57|df=22) = \underline{0.0653}$

13.45 $s_{b1} = \sqrt{s_e^2/SS(x)} = \sqrt{21.088/587.8} = \underline{0.1894}$

The Confidence Interval Estimate of β_1

$$b_1 \pm t(df, \alpha/2) \cdot s_{b1} \quad \text{with } df = n - 2$$

13.47 a. $\hat{y} = -348 + 2.04x$

b. The 95% confidence interval for β_1 is 1.60 to 2.48.

$2.04 \pm (2.31)(0.1894)$

2.04 ± 0.44

1.60 to 2.48

c. With 95% confidence, it is believed that the slope of the line of best fit for the population is between 1.60 and 2.48.

Hypothesis tests will be completed using the same format as before. You may want to review: ES9-p370, ST-pp241&249. The only differences are:

1. **writing hypotheses**: (see box before ex. 13.42)
2. **using Table 6 or 7**: $df = n - 2$
3. **the calculated test statistic**: $t* = \dfrac{b_1 - \beta_1}{s_{b1}}$

Computer and/or calculator commands to determine the line of best fit and also perform a hypothesis test concerning the slope can be found in ES9-pp631&632. Excel also constructs the confidence interval for the slope.

Slight variations in sums of squares and further calculations can result from round-off errors.

13.49 a.

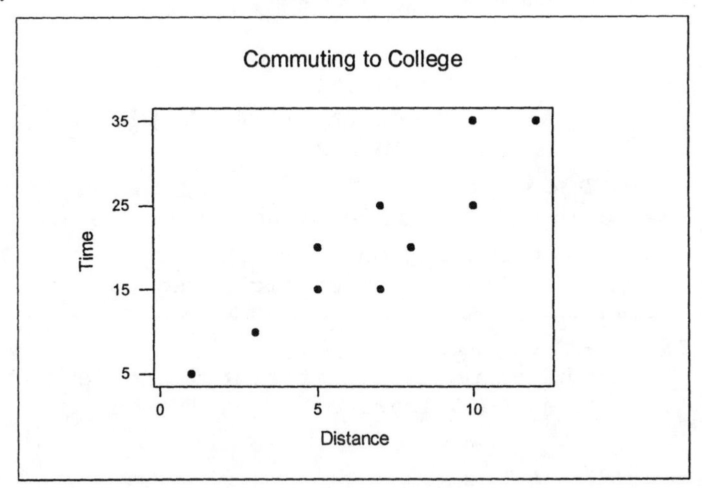

b. Summary of data: $n = 10$, $\Sigma x = 68$, $\Sigma y = 205$,
$\Sigma x^2 = 566$, $\Sigma xy = 1670$, $\Sigma y^2 = 5075$

$SS(x) = 566 - (68^2/10) = 103.6$
$SS(xy) = 1670 - [(68)(205)/10] = 276.0$
$b_1 = 276.0/103.6 = 2.664$
$b_0 = [205 - (2.664)(68)]/10 = 2.38$

$\hat{y} = \underline{2.38 + 2.664x}$

c.

Step 1: a. The slope β_1 of the line of best fit for the population of distances and their corresponding times required for students to commute to college.

b. H_o: $\beta_1 = 0$ (no value)
H_a: $\beta_1 > 0$

Step 2: a. random sample, assume normality for y at each x

b. t, df = 8 c. $\alpha = 0.05$

Step 3: a. $n = 10$, $b_1 = 2.664$,
$s_e^2 = [5075 - (2.38)(205) - (2.664)(1670)]/8$
$= 137.192/8 = 17.149$
$s_{b1} = \sqrt{s_e^2/SS(x)} = \sqrt{17.149/103.6} = 0.407$

b. $t = (b_1 - \beta_1)/s_{b1}$
$t^* = (2.664 - 0)/0.407 = 6.55$

Step 4: -- using p-value approach --------------------

a. $\mathbf{P} = P(t > 6.55 | df = 8)$;

Using Table 6, ES9-p717:
$\mathbf{P} < 0.005$
Using Table 7, ES9-p718:
$\mathbf{P} < 0.002$

b. $\mathbf{P} < \alpha$

-- using classical approach -----------------

a. critical region: $t \geq 1.86$
b. t^* falls in the critical region

Step 5: a. Reject H_o.
b. The slope is significantly greater than zero, at the 0.05 level of significance.

d. $b_1 \pm t(df, \alpha/2) \cdot s_{b1}$
$2.66 \pm (2.90)(0.407) = 2.66 \pm 1.18$
<u>1.48 to 3.84</u>, the 0.98 interval for β_1

13.51 a. Summary of data: $n = 10$, $\Sigma x = 1105$, $\Sigma y = 93327$,
$\Sigma x^2 = 126475$, $\Sigma xy = 10447010$, $\Sigma y^2 = 885420825$

$SS(x) = 126475 - (1105^2/10) = 4372.5$
$SS(xy) = 10447010 - [(1105)(93327)/10] = 134376.5$

$b_1 = 134376.5/4372.5 = 30.732$
$b_0 = [93327 - (30.732)(1105)]/10 = 5937$

$\hat{y} = \underline{5937 + 30.73x}$

b. $s_e^2 = [885420825-(5937)(93327)-(30.73)(10447010)]/8$
$= 1287726.088$
$s = s_e = \sqrt{1287726.088} = 1134.78 = 1135$

$s_{b1}^2 = 1287722.963/4372.5 = 294.50568$
standard error of slope $= s_{b1} = \sqrt{294.50568} = 17.16$

c. The observed standard deviation, $s = 1135$, measures the variability of the ordered pairs about the line of best fit.

The observed standard error of slope = 17.16 is the estimate for the standard deviation of the sampling distribution of the possible values of slope for all possible samples of size 10.

d.

Step 1: a. The slope β_1 of the line of best fit for the population of engine horsepowers and their corresponding base prices for various jet boats.

b. $H_o: \beta_1 = 0$ (no value)

$H_a: \beta_1 > 0$

Step 2: a. random sample, normality assumed for y at each x

b. t, df = 8 c. $\alpha = 0.05$

Step 3: a. $n = 10$, $b_1 = 30.73$, $s_{b1} = 17.16$

b. $t = (b_1 - \beta_1)/s_{b1}$

$t* = (30.73 - 0)/17.16 = 1.79$

Step 4: -- using p-value approach --------------------

a. $\mathbf{P} = P(t > 1.79 | df = 8)$;

Using Table 6, ES9-p717:

$0.05 < \mathbf{P} < 0.10$

Using Table 7, ES9-p718:

$0.055 < \mathbf{P} < 0.064$

b. $\mathbf{P} > \alpha$

-- using classical approach ------------------

a. critical region: $t \geq 1.86$

b. t* falls in the noncritical region

Step 5: a. Fail to reject H_o.

b. Horsepower does not play a significant role as a predictor of price, at the 0.05 level of significance.

SECTION 13.5 ANSWER NOW EXERCISES

13.53 Prediction intervals are wider than confidence intervals, therefore the prediction interval belts would be further apart on the graphs and contain all of the points.

SECTION 13.5 EXERCISES

Estimating $\mu_{y|x_o}$ and y_{x_o}

$\mu_{y|x_o}$ - the mean of the population y-values at a given x

y_{x_o} - the individual y-value selected at random for a given x

1. point estimate for $\mu_{y|x_o}$ and y_{x_o}: $\hat{y}$ $(\hat{y} = b_o + b_1x)$
2. confidence interval for $\mu_{y|x_o}$:
$$\hat{y} \pm t(n-2,\alpha/2)\cdot s_e\sqrt{(1/n)+[(x_o-\overline{x})^2/SS(x)]}$$
3. prediction interval for y_{x_o}:
$$\hat{y} \pm t(n-2,\alpha/2)\cdot s_e\sqrt{1+(1/n)+[(x_o-\overline{x})^2/SS(x)]}$$

Review the Five-Step Procedure in: ES9-p350, ST-p229, if necessary.

Slight variations in sums of squares and further calculations can result from round-off errors.

13.55 Step 1: $\mu_{y|x=70}$, the mean crutch length for individuals who say they are 70 in. tall.

Step 2: a. random sample, normality assumed for y at each x

b. t c. $1 - \alpha = 0.95$

Step 3: $n = 107$, $x_o = 70$, $\overline{x} = 68.84$, $s_e = \sqrt{0.50} = 0.707$

$SS(x) = (n-1)s^2 = 106(7.35^2) = 5726.385$

$\hat{y} = 4.8 + 0.68(70) = 52.4$

Step 4: a. $\alpha/2 = 0.05/2 = 0.025$; $df = 105$; $t(105,0.025) = 1.96$

b. $E = t(n-2,\alpha/2)\cdot s_e\sqrt{(1/n)+[(x_o-\overline{x})^2/SS(x)]}$

$E = (1.96)(0.707)\cdot\sqrt{(1/107)+[(70-68.84)^2/5726.385]}$

$E = (1.96)(0.707)\sqrt{0.0095808} = 0.14$

c. $\hat{y} \pm E = 52.4 \pm 0.14$

Step 5: <u>52.3 to 52.5</u>, the 0.95 interval for $\mu_{y|x=70}$

Computer and/or calculator commands for calculating the regressions line can be found in ES9-pp161&162.
MINITAB also provides confidence interval belts and prediction interval belts. See commands in ES9-p642.

13.57 Summary of data: $n = 10$, $\Sigma x = 16.25$, $\Sigma y = 152$,

13.57 Summary of data: $n = 10$, $\Sigma x = 16.25$, $\Sigma y = 152$,
$\Sigma x^2 = 31.5625$, $\Sigma xy = 275$, $\Sigma y^2 = 2504$

$SS(x) = 31.5625 - (16.25^2/10) = 5.15625$
$SS(xy) = 275 - [(16.25)(152)/10] = 28.0$
$b_1 = 28.0/5.15625 = 5.4303$
$b_0 = [152 - (5.4303)(16.25)]/10 = 6.3758$

$\hat{y} = 6.3758 + 5.4303x$

$s_e^2 = [2504 - (6.3758)(152) - (5.4303)(275)]/8$
$= 5.19324$

$s_e = \sqrt{5.19324} = 2.279$

When $x = 2.0$, then $\hat{y} = 6.3758 + 5.4303(2.0) = 17.24$
$t(8, 0.025) = 2.31$ and $\overline{x} = 16.25/10 = 1.625$

a. Step 1: $\mu_{y|x=2.00}$, the mean heart-rate reduction for a dose of 2.00 mg.
Step 2: a. random sample, normality assumed for y at each x
b. t   c. $1 - \alpha = 0.95$
Step 3: $n = 10$, $x_o = 2.00$, $\overline{x} = 1.625$, $s_e = 2.279$
$\hat{y} = 17.24$
Step 4: a. $\alpha/2 = 0.05/2 = 0.025$; $df = 8$; $t(8,0.025) = 2.31$
b. $E = t(n-2,\alpha/2) \cdot s_e \sqrt{(1/n) + [(x_o - \overline{x})^2/SS(x)]}$
$E = (2.31)(2.279) \cdot \sqrt{(1/10) + [(2.0-1.625)^2/5.15625]}$
$E = (2.31)(2.279)\sqrt{0.127273} = 1.88$
c. $\hat{y} \pm E = 17.24 \pm 1.88$
Step 5: 15.4 to 19.1, the 0.95 interval for $\mu_{y|x=2}$

b. Step 1: $y_{x=2.00}$, the heart-rate reduction expected for an individual receiving a dose of 2.00 mg.
Step 2: a. random sample, normality assumed for y at each x
b. t   c. $1 - \alpha = 0.95$
Step 3: $n = 10$, $x_o = 2.00$, $\overline{x} = 1.625$, $s_e = 2.279$
$\hat{y} = 17.24$

Step 4: a. $\alpha/2 = 0.05/2 = 0.025$; df = 8; $t(8,0.025) = 2.31$

b. $E = t(n-2,\alpha/2)\cdot s_e\sqrt{1+(1/n)+[(x_o-\overline{x})^2/SS(x)]}$

$E = (2.31)(2.279)\sqrt{1 + 0.127273} = 5.59$

c. $\hat{y} \pm E = 17.24 \pm 5.59$

Step 5: 11.6 to 22.8, the 0.95 interval for $y_{x=2}$

13.59 a. Summary of data: $n = 8$, $\Sigma x = 256593$, $\Sigma y = 454$,

$\Sigma x^2 = 9013638265$, $\Sigma xy = 15173202$, $\Sigma y^2 = 26316$

$SS(x) = 9013638265 - (256593^2/8) = 783642308.9$
$SS(xy) = 15173202 - [(256593)(454)/8] = 611549.25$

$b_1 = 611549.25/783642308.9 = 0.00078039$
$b_0 = [454 - (0.0007804)(256593)]/8 = 31.71935$

$\hat{y} = 31.72 + 0.0007804x$

b. Step 1: $\mu_{y|x=40,000}$, the mean percent with up-to-date immunizations for families with a median household income equal to \$40,000.

Step 2: a. random sample, normality assumed for y at each x

b. t c. $1 - \alpha = 0.95$

Step 3: $n = 8$, $x_o = 40,000$, $\overline{x} = 32074.13$,

$s_e = \sqrt{[26316-(31.72)(454)-(0.0007804)(15173202)]/6}$

$= \sqrt{12.3255} = 3.51$

$\hat{y} = 31.72 + 0.0007804(40000) = 62.936$

Step 4: a. $\alpha/2 = 0.05/2 = 0.025$; df = 6; $t(6,0.025) = 2.45$

b. $E = t(n-2,\alpha/2)\cdot s_e\sqrt{(1/n)+[(x_o-\overline{x})^2/SS(x)]}$

$E = (2.45)(3.51)\cdot\sqrt{(1/8)+[(40000-32074.13)^2/783642308.9]}$

$E = (2.45)(3.51)\sqrt{0.2051633912} = 3.895 = 3.90$

c. $\hat{y} \pm E = 62.94 \pm 3.90$

Step 5: 59.04% to 66.84%, the 0.95 interval for $\mu_{y|x=40,000}$

c. Step 1: $y_{x=40,000}$, the percent with up-to-date immunizations for families with a median household income equal to \$40,000.

Step 2: a. random sample, normality assumed for y at each x

b. t c. $1 - \alpha = 0.95$

Step 3: $n = 8$, $x_o = 40{,}000$, $\bar{x} = 32074.13$,

$s_e = \sqrt{[26316-(31.72)(454)-(0.0007804)(15173202)]/6}$

$= \sqrt{12.3255} = 3.51$

$\hat{y} = 31.72 + 0.0007804(40000) = 62.936$

Step 4: a. $\alpha/2 = 0.05/2 = 0.025$; df = 6; $t(6,0.025) = 2.45$

b. $E = t(n-2,\alpha/2)\cdot s_e\sqrt{1+(1/n)+[(x_o-\bar{x})^2/SS(x)]}$

$E = (2.45)(3.51)\sqrt{1 + 0.2051633912} = 9.44$

c. $\hat{y} \pm E = 62.94 \pm 9.44$

Step 5: 53.5% to 72.38%, the 0.95 interval for $y_{x=40,000}$

d. 'b' gives the interval estimate for the mean of all such percentages, whereas 'c' gives the interval estimate for an individual percentage.

13.61 The standard error for $\bar{x}$'s is much smaller than the standard deviation for individual x's (CLT). Thus the confidence interval will be narrower in accordance.

RETURN TO CHAPTER CASE STUDY

13.63 a.

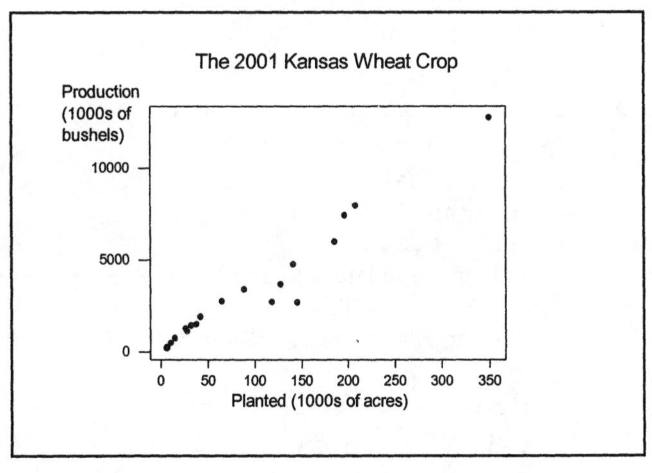

r = 0.974; $\hat{y}$ = 50 + 34.2x

b.

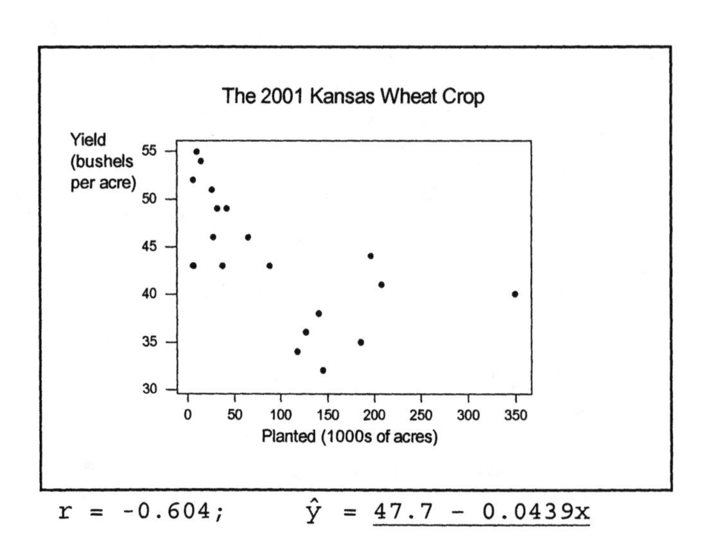

$r = -0.604$; $\hat{y} = \underline{47.7 - 0.0439x}$

13.65 - Answers will vary.

CHAPTER EXERCISES

13.67 Step 1: a. The linear correlation coefficient for the population, ρ.

b. $H_o: \rho = 0.0$

$H_a: \rho > 0.0$

Step 2: a. random sample, assume normality for y at each x

b. df = n - 2 = 45 - 2 = 43

c. $\alpha = 0.05$

Step 3: a. n = 45

b. $r* = 0.69$

Step 4: -- using p-value approach --------------------

a. **P** = P(r > 0.69)

Using Table 11, ES9-p727: **P** < 0.005

b. **P** < α

-- using classical approach ------------------

a. critical region: $r \geq 0.29$

b. r* falls in the critical region

Step 5: a. Reject H_o
b. There is sufficient reason to conclude that the correlation coefficient is positive.

13.69 a. (1) $r_{12} = 0.0186103 = 0.019$
(2) $r_{34} = 0.0961191 = 0.096$

b. There is a practically no correlation between street price and overall rating for either power desktops or budget PCs. Assuming overall rating is indicative of quality, the data indicates that there is no price-quality relationship for desktop PCs. Paying more does not mean that consumers will likely purchase a machine with a higher rating, and paying less does not mean that the machine will likely have a lower rating.

13.71 a. Step 1: a. The linear correlation coefficient for the population, ρ.

b. H_o: $\rho = 0.0$

c. H_a: $\rho \neq 0.0$

Step 2: a. random sample, assume normality for y at each x

b. $df = n - 2 = 17 - 2 = 15$

c. $\alpha = 0.05$

Step 3: a. $n = 17$

b. $r^* = 0.61$

Step 4: -- using p-value approach --------------------

a. $\mathbf{P} = P(r > 0.61)$

Using Table 11, ES9-p727: $\mathbf{P} < 0.01$

b. $\mathbf{P} < \alpha$

-- using classical approach ------------------

a. critical region: $r \leq -0.482$, $r \geq 0.482$

b. r^* falls in the critical region

Step 5: a. Reject H_o.

b. Yes, the correlation coefficient is significantly different from zero.

b. $\hat{y} = 1.8(50) + 28.7 = \underline{118.7}$

13.73 $4.23 \pm 2.23(0.13)$

$\underline{3.94 \text{ to } 4.52}$, the 0.95 interval for β_1

13.75 a.

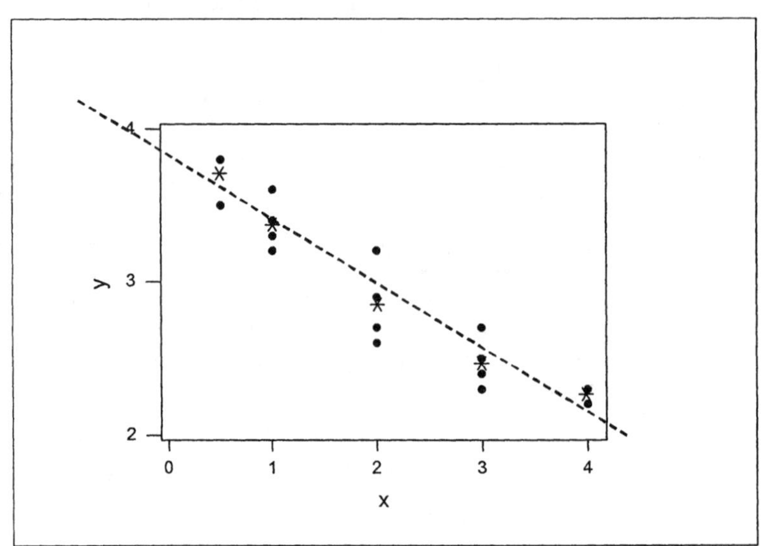

b. See dashed line on graph in (a).

c. See *'s on graph. The *'s seem to follow a curved path, not a straight line. The *'s are above the line at the ends and below the line in the middle.

d. Summary of data: $n = 18$ $\Sigma x = 37.5$, $\Sigma y = 52.7$,

$\Sigma x^2 = 104.75$, $\Sigma xy = 98.75$, $\Sigma y^2 = 159.49$

$SS(x) = 104.75 - (37.5^2/18) = 26.625$
$SS(xy) = 98.75 - [(37.5)(52.7)/18] = -11.0417$

$b_1 = -11.0417/26.625 = -0.4147$
$b_0 = [52.7 - (-0.4147)(37.5)]/18 = 3.792$

$\hat{y} = \underline{3.79 - 0.415x}$

e. $s_e^2 = [159.49 - (3.79)(52.7) - (-0.4147)(98.75)]/16$
$= 0.04429$

$s_e = \sqrt{0.04429} = \underline{0.21045}$

f. $s_{b1} = \sqrt{s_e^2/SS(x)} = \sqrt{0.04429/26.625} = \underline{0.041}$

$-0.415 \pm 2.12(0.041) = -0.415 \pm 0.087$

$\underline{-0.502 \text{ to } -0.328}$, the 0.95 interval for β_1

g. At x = 3.0:
Step 1: $\mu_{y|x=3.0}$, the mean value of y at x = 3.0.
Step 2: a. random sample, normality assumed for y at each x
b. t
c. $1 - \alpha = 0.95$
Step 3: a. $n = 18$, $x_o = 3.0$, $\overline{x} = 37.5/18 = 2.08$, $s_e = 0.21045$
b. $\hat{y} = 3.79 - 0.415(3.0) = 2.55$
Step 4: a. $\alpha/2 = 0.05/2 = 0.025$; df = 16; $t(16,0.025) = 2.12$
b. $E = t(n-2,\alpha/2)\cdot s_e\sqrt{(1/n)+[(x_o-\overline{x})^2/SS(x)]}\cdot$
$E = (2.12)(0.21045)\sqrt{(1/18) + [(3-2.08)^2/26.625]}$
$E = 0.13$
c. $\hat{y} \pm E = 2.55 \pm 0.13$
$\underline{2.42 \text{ to } 2.68}$, the 0.95 interval for $\mu_{y|x=3}$

At x = 3.5:
Step 1: $\mu_{y|x=3.5}$, the mean value of y at x = 3.5.
Step 2: a. random sample, normality assumed for y at each x
b. t
c. $1 - \alpha = 0.95$
Step 3: a. $n = 18$, $x_o = 3.5$, $\overline{x} = 2.08$, $s_e = 0.21045$
b. $\hat{y} = 3.79 - 0.415(3.5) = 2.34$
Step 4: a. $\alpha/2 = 0.05/2 = 0.025$; df = 6; $t(16,0.025) = 2.12$
b. $E = t(n-2,\alpha/2)\cdot s_e\sqrt{(1/n)+[(x_o-\overline{x})^2/SS(x)]}$
$E = (2.12)(0.21045)\sqrt{(1/18) + [(3.5-2.08)^2/26.625]}$
$E = 0.16$
c. $\hat{y} \pm E = 2.34 \pm 0.16$
$\underline{2.18 \text{ to } 2.50}$ the 0.95 interval for $\mu_{y|x=3.5}$

h. At $x = 3.0$:

Step 1: $y|_{x=3.0}$, the value of y at $x = 3.0$.

Step 2: a. random sample, normality assumed for y at each x

b. t

c. $1 - \alpha = 0.95$

Step 3: a. $n = 18$, $x_o = 3.0$, $\bar{x} = 37.5/18 = 2.08$, $s_e = 0.21045$

b. $\hat{y} = 3.79 - 0.415(3.0) = 2.55$

Step 4: a. $\alpha/2 = 0.05/2 = 0.025$; df = 16; $t(16,0.025) = 2.12$

b. $E = t(n-2,\alpha/2)\cdot s_e\sqrt{1+(1/n)+[(x_o-\bar{x})^2/SS(x)]}$

$E=(2.12)(0.21045)\sqrt{1 + (1/18) + [(3-2.08)^2/26.625]}$

$E = 0.47$

c. $\hat{y} \pm E = 2.55 \pm 0.47$

$\underline{2.08 \text{ to } 3.02}$, the 0.95 interval for $y_{x=3}$

At $x = 3.5$:

Step 1: $y|_{x=3.5}$, the value of y at $x = 3.5$.

Step 2: a. random sample, normality assumed for y at each x

b. t

c. $1 - \alpha = 0.95$

Step 3: a. $n = 18$, $x_o = 3.5$, $\bar{x} = 2.08$, $s_e = 0.\ 21045$

b. $\hat{y} = 3.79 - 0.415(3.5) = 2.34$

Step 4: a. $\alpha/2 = 0.05/2 = 0.025$; df = 16; $t(16,0.025) = 2.12$

b. $E = t(n-2,\alpha/2)\cdot s_e\sqrt{1+(1/n)+[(x_o-\bar{x})^2/SS(x)]}$

$E = (2.12)(0.21045)\sqrt{(1+1/18)+[(3.5-2.08)^2/26.625]}$

$E = 0.47$

c. $\hat{y} \pm E = 2.34 \pm 0.47$

$\underline{1.87 \text{ to } 2.81}$ the 0.95 interval for $y_{x=3.5}$

13.77 Summary of data: $n = 21$ $\Sigma x = 1177$, $\Sigma y = 567$,

$\Sigma x^2 = 70033$, $\Sigma xy = 32548$, $\Sigma y^2 = 15861$

$SS(x) = 70033 - (1177^2/21) = 4064.95$

$SS(xy) = 32548 - [(1177)(567)/21] = 769.0$

$SS(y) = 15861 - (567^2/21) = 552.0$

$r = 769.0/\sqrt{(4064.95)(552.0)} = 0.5133$

a. Step 1: a. The linear correlation coefficient for the number of stamens and the number of carpels in a particular species of flowers, ρ.

b. $H_o: \rho = 0.0$

c. $H_a: \rho \neq 0.0$

Step 2: a. random sample, assume normality for y at each x

b. $df = n - 2 = 21 - 2 = 19$

c. $\alpha = 0.05$

Step 3: a. $n = 21$

b. $r* = 0.513$

Step 4: -- using p-value approach --------------------

a. $\mathbf{P} = 2P(r > 0.513)$

Using Table 11, ES9-p727: $0.01 < \mathbf{P} < 0.02$

b. $\mathbf{P} < \alpha$

-- using classical approach ------------------

a. critical region: $r \leq -0.433$, $r \geq 0.433$

b. $r*$ is in the critical region

Step 5: a. Reject H_o.

b. There is sufficient reason to conclude that there is linear correlation.

b. $b_1 = 769.0/4064.95 = 0.1892$

$b_0 = [567 - (0.1892)(1177)]/21 = 16.3958$

$\hat{y} = \underline{16.40 + 0.189x}$

c. Step 1: a. The slope β_1 of the line of best fit for the population of number of stamens per flower and the corresponding number of carpels.

b. $H_o: \beta_1 = 0$ (no value)

c. $H_a: \beta_1 > 0$

Step 2: a. random sample, assume normality for y at each x

b. t, df = 19

c. $\alpha=0.05$

Step 3: a. $n = 21$, $b_1 = 0.1892$,
$s_e^2 = [15861 - (16.3958)(567) - (0.1892)(32548)]/19$
$s_e^2 = 21.3947$
$s_{b1} = \sqrt{s_e^2/SS(x)} = \sqrt{21.3947/4064.95} = 0.07255$

b. $t = (b_1 - \beta_1)/s_{b1}$
$t* = 0.189/0.07255 = 2.61$

Step 4: -- using p-value approach --------------------
a. $\mathbf{P} = P(t > 2.61 | df = 19)$;
Using Table 6, ES9-p717:
$0.005 < \mathbf{P} < 0.01$
Using Table 7, ES9-p718:
$0.007 < \mathbf{P} < 0.009$

b. $\mathbf{P} < \alpha$
-- using classical approach ------------------
a. critical region: $t \geq 1.73$
b. t* is in the critical region

Step 5: a. Reject H_o.
b. The slope is significantly greater than zero.

d. Step 1: $y|_{x=64}$, the number of carpels found in a mature flower if the number of stamens is 64.

Step 2: a. t
b. $1 - \alpha = 0.95$

Step 3: a. $n = 21$, $x_o = 64$, $\bar{x} = 1177/21 = 56.05$
$s_e = \sqrt{21.3947}$
b. $\hat{y} = 16.40 + 0.189(64) = 28.50$

Step 4: a. $\alpha/2 = 0.05/2 = 0.025$; $df = 19$;
$t(19, 0.025) = 2.09$

b. $E = t(n-2, \alpha/2) \cdot s_e \sqrt{1 + (1/n) + [(x_o - \bar{x})^2/SS(x)]}$
$E = (2.09)(\sqrt{21.3947})\sqrt{1 + (1/21) + [(64 - 56.05)^2/4064.95]}$
$E = 9.97$

c. $\hat{y} \pm E = 28.50 \pm 9.97$
18.53 to 38.47, the 0.95 interval for one $y_{x=64}$

13.79 Summary of data: $n = 5$, $\Sigma x = 16$, $\Sigma y = 38$,

$\Sigma x^2 = 66$, $\Sigma xy = 145$, $\Sigma y^2 = 326$

$SS(x) = 66 - (16^2/5) = 14.8$
$SS(xy) = 145 - [(16)(38)/5] = 23.4$
$SS(y) = 326 - (38^2/5) = 37.2$

$b_1 = 23.4/14.8 = 1.5811$

$r = 23.4/\sqrt{(14.8)(37.2)} = \underline{0.9973}$ [Formula 13.3]

$r = 1.5811\sqrt{14.8/37.2} = \underline{0.9973}$ [Formula in exercise]

CHAPTER 14 ∇ ELEMENTS OF NONPARAMETRIC STATISTICS

Chapter Preview

Chapter 14 introduces the concept of nonparametric statistics. Up to this point, especially in chapters 8, 9 and 10, the methods used were parametric methods. Parametric methods rely on the normality assumption through knowledge of the parent population or the central limit theorem. In nonparametric (distribution-free) methods, few assumptions about the parent population are required yet the methods are only slightly less efficient than their parametric counterparts. Chapter 14 will demonstrate nonparametric methods for hypothesis tests concerning one mean, two independent means, two dependent means, correlation and randomness.

A survey conducted by NFO Research, Inc. on the attitudes of teenagers toward social and moral values is used in this chapter's Case Study.

CHAPTER 14 CASE STUDY

14.1 a. There appears to be a "general" agreement; boys and girls agreed on the two most important, they agreed on the four least important, and they scrambled the order for the middle six.

b.

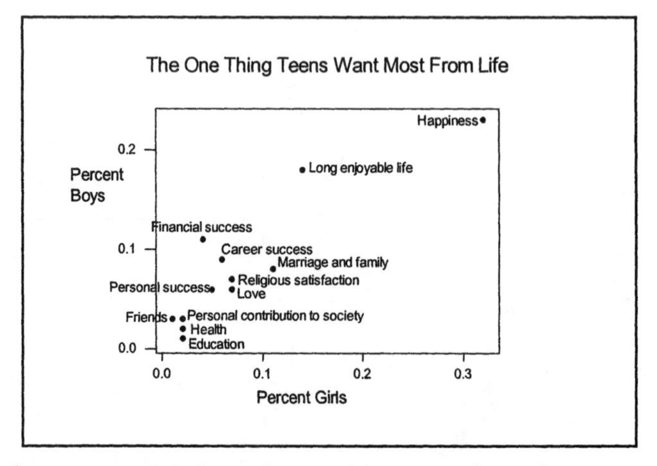

The importance placed on these twelve items appears to correlate as described in (a).

SECTION 14.3 ANSWER NOW EXERCISES

14.2 Ranked data:

66 [75] 80 82 84 88 90 91 [94] 110

For $n = 10$ and $1 - \alpha = 0.90$, the critical value from Table 12 is $k = 1$.

$x_{k+1} = x_2 = 75$ and $x_{n-k} = x_9 = 94$

75 to 94, the 0.90 interval for median M

SECTION 14.3 EXERCISES

Sign test $(1 - \alpha)$ Confidence Interval for M - the population median

1. Arrange data in ascending order (smallest to largest)
2. Assign the notation: x_1(smallest), x_2, x_3 ... x_n(largest) to the data.
3. Critical value = k (value from Table 12 using n and α)
4. $(1 - \alpha)$ confidence interval extends from the data values:

   x_{k+1} to x_{n-k}

14.3 Ranked data:

33 34 35 36 38 [39] 40 40 42 45

46 46 46 46 [47] 47 48 54 59 65

For $n = 20$ and $1 - \alpha = 0.95$, the critical value from Table 12 is $k = 5$.

$x_{k+1} = x_6 = 39$ and $x_{n-k} = x_{15} = 47$

39 to 47, the 0.95 interval for median M

14.5

-30	-16	-14	-14	-13	-12	-12	-11	-10	-10
-9	-9	-8	-8	-8	[-7]	-6	-6	-5	-4
-4	-4	-3	-3	-2	-1	-1	0	[1]	2
2	5	6	6	6	6	6	9	12	12
13	16	18	19						

For $n = 44$ and $1 - \alpha = 0.95$, the critical value from Table 12 is $k = 15$

$x_{k+1} = x_{16} = -7$ and $x_{n-k} = x_{29} = 1$

CI: −7 to 1 points of change in scores.

The Sign Test

The Sign Test is used to test one mean (median) or the median difference between paired data (dependent samples). Refer to the Five-Step Hypothesis Test Procedure in: ES9-pp370&386, ST-pp241&249, if necessary. The only changes are in:

1. **the hypotheses**
   a. null hypothesis
      H_o: M = # <u>or</u> H_o: p(preference) = 0.5 <u>or</u> H_o:p(sign)= 0.5
   b. possible alternative hypotheses:
      H_a: M ≠ # <u>or</u> H_o: p(preference) ≠ 0.5 <u>or</u> H_o:p(sign)≠ 0.5
      H_a: M < # <u>or</u> H_o: p(preference) < 0.5 <u>or</u> H_o:p(sign)< 0.5
      H_a: M > # <u>or</u> H_o: p(preference) > 0.5 <u>or</u> H_o:p(sign)> 0.5
2. **the critical value of the test statistic**,
   a. for sample sizes ≤ 100, x;Table 12(Appendix B,ES9-p728)
   b. for sample sizes > 100, z;Table 4(Appendix B,ES9-p715)
3. **the calculated test statistic**,
   a. for n ≤ 100; x = the number of the less frequent sign
   b. for n > 100; x = n(sign of preference)

$$z^* = (x' - (n/2))/(\tfrac{1}{2}\sqrt{n}) \quad \text{where}$$

$$x' = x - \tfrac{1}{2}, \text{ if } x > (n/2) \quad \text{or} \quad x' = x + \tfrac{1}{2}, \text{ if } x < (n/2)$$

14.7 a. H_o: Median = 32 vs. H_a: Median < 32

b. H_o: P(prefer new recipe) = 0.50 vs. P(prefer) < 0.50

c. H_o: P(+ gain) = 0.5 vs. H_a: P(+) ≠ 0.5
where (+) = weight gain

Table 12, Critical Values of the Sign Test, gives the maximum allowable number of the less frequent sign, k, that will cause rejection of H_o. k is based on n (the total number of signs, excluding zeros) and α.

Therefore, if x ≤ k, reject H_o and if x > k, fail to reject H_o.

14.9 Step 1: a. Median age of the population of all marrow transplantation patients.
b. H_o: Median = 30 years
H_a: Median ≠ 30 years

Step 2: a. Assume sample is random. Age is continuous.
b. x = n(least frequent sign)
c. α = 0.05

Step 3: a. + = over 30 years of age; n = 100
n(+) = 40, n(0) = 0, n(-) = 60
b. x = n(+) = 40

Step 4: -- using p-value approach --------------
a. **P** = 2P(x ≤ 40|n = 100);
Using Table 12, Appendix B, ES9-p728:
0.05 < **P** < 0.10
b. **P** > α
-- using classical approach ------------
a. critical region: n(least freq sign) ≤ 39
b. The test statistic is not in the critical region.

Step 5: a. Fail to reject H_o.
b. The evidence is not sufficient to show the median age is not equal to 30 years.

14.11 Step 1: a. Proportion who can solve the problem correctly.
b. H_o: P(+) = 0.5
H_a: P(+) ≠ 0.5

Step 2: a. Assume random sample. x is binomial and approximately normal.
b. x = n(least freq sign)
c. α is unspecified

Step 3: a. + = correct solution; n = 75

==

a.

b. x = n(+) = 20

Step 4: -- using p-value approach ---------------
a. $\mathbf{P} = 2P(x \le 20 | n = 75)$;
Using Table 12, Appendix B, ES9-p728:
$\mathbf{P} < 0.01$

----

b.

b. x = n(+) = 27

Step 4: -- using p-value approach ---------------
a. $\mathbf{P} = 2P(x \le 27 | n = 75)$;
Using Table 12, Appendix B, ES9-p728:
$0.01 < \mathbf{P} < 0.05$

----

c.

b. x = n(+) = 30

Step 4: -- using p-value approach ---------------
a. $\mathbf{P} = 2P(x \le 30 | n = 75)$;
Using Table 12, Appendix B, ES9-p728:
$0.10 < \mathbf{P} < 0.25$

----

d.

b. x = n(+) = 33

Step 4: -- using p-value approach ---------------
a. $\mathbf{P} = 2P(x \le 33 | n = 75)$;
Using Table 12, Appendix B, ES9-p728:
$\mathbf{P} > 0.25$

==

b. If $\mathbf{P} < \alpha$

Step 5: a. Reject H_o.
b. The evidence does show that the proportion is significantly different than one-half.

----

b. If $\mathbf{P} > \alpha$

Step 5: a. Fail to reject H_o.
b. The evidence does not show that the proportion is significantly different than one-half.

For testing the median difference between paired data, subtract corresponding pairs of data and use the signs of the differences. Hypotheses can be written in three forms.

null hypothesis:

H_o: No difference between the pairs <u>or</u> H_o: $M = 0$ <u>or</u> H_o: $p(+) = 0.5$

possible alternative hypotheses:

H_a: There is a difference between the pairs
H_a: $M \neq \#$ <u>or</u> H_a: $p(+) \neq 0.5$

H_a: One of the pairs is greater than the other
(subtract greater - smaller) <u>or</u>
H_a: $M > \#$ <u>or</u> H_a: $p(+) > 0.5$

H_a: One of the pairs is less than the other
(subtract smaller - greater) <u>or</u>
H_a: $M < \#$ <u>or</u> H_a: $p(+) < 0.5$

14.13 Step 1: a. Median salary of full professors in Colorado.
b. H_o: $M = 83{,}282$
H_a: $M < 83{,}282$

Step 2: a. Assume sample is random. Salaries are numerical.
b. x = n(least frequent sign)
c. $\alpha = 0.05$

Step 3: a. + = positive, - = negative; $n = 20$
$n(+) = 4$, $n(0) = 0$, $n(-) = 16$
b. $x = n(+) = 4$

Step 4: -- using p-value approach ---------------
a. $\mathbf{P} = P(x \leq 4 | n = 20)$;
Using Table 12, Appendix B, ES9-p728:
$0.005 < \mathbf{P} < 0.025$
Using a computer: $P = 0.0059$
b. $\mathbf{P} < \alpha$

-- using classical approach -------------
a. critical region: n(least freq sign)= $x \leq 5$
b. The test statistic is in the critical region.

Step 5: a. Reject H_o.
b. Sufficient evidence to support the claim that the median salary of full professors in Colorado is lower than the mean for the whole country, at the 0.05 level of significance.

14.15 Step 1: a. Preference for the taste of a new cola.

b. H_o: There is no preference; p = P(prefer) = 0.5

H_a: There is a preference for the new; p > 0.5

Step 2: a. x is binomial and approximately normal

b. z

c. $\alpha = 0.01$

Step 3: a. + = prefer new; n = 1228;

n(+) = 645, n(0) = 272, n(-) = 583

b. x = n(+) = 645; x' = 644.5

$z = (x' - (n/2))/(\frac{1}{2}\sqrt{n})$

$z^* = (644.5 - (1228/2))/(\frac{1}{2}\sqrt{1228})$

$= (644.5 - 614)/17.5214 = 1.74$

Step 4: -- using p-value approach ---------------

a. **P** = P(z > 1.74) = 0.5000 - 0.4591 = 0.0409

b. **P** > α

-- using classical approach -------------

a. critical region: $z \geq 2.33$

b. The test statistic is not in the critical region.

Step 5: a. Fail to reject H_o.

b. The evidence does not allow us to conclude that there is a significant preference for the new cola.

14.17 Step 1: a. Proportion of high school seniors that can solve problems involving fractions, decimals and percentages.

b. H_o: P(+) = 0.5

H_a: P(+) $\neq$ 0.5

Step 2: a. Assume random sample. x is binomial and approximately normal.

b. x = n(least frequent sign)

c. $\alpha = 0.05$

Step 3: a. n = 1500

Step 4: a. $\pm z(0.05) = 1.96$

b. $z = (x' - (n/2))/(\frac{1}{2}\sqrt{n})$

$-1.96 = (x' - (1500/2))/\frac{1}{2}\sqrt{1500}$

$-1.96 = (x' - 750)/19.3649$

x' = 712.04; critical value is <u>712</u>

SECTION 14.4 ANSWER NOW EXERCISES

14.18 The two sets of data seems to be spread out in a similar fashion, however the set making up sample B seems to be "slid" to the right by about ten points.

14.19 The null hypothesis for a Mann-Whitney U test is basically "the same" or "no change" therefore when the p-value is less than 0.01, the null hypothesis is rejected and there is significant evidence of change - the phrases, significant, drop dramatically, and increased steadily are associated with this. When the p-value is 0.20, the null hypothesis is not rejected and no change can be claimed.

SECTION 14.4 EXERCISES

The Mann-Whitney U Test

The Mann-Whitney U Test is used to test the difference between two independent means. Refer to the Five-Step Hypothesis Test Procedure in: ES9-pp370&386, ST-pp241&249, if necessary. The only changes are in:

1. **the hypotheses**
   a. null hypothesis
      H_o: The average value is the same for both groups.
   b. possible alternative hypotheses:
      H_a: The average value is not the same for both groups.
      H_a: The average value of one group is greater than that of the other group.
      H_a: The average value of one group is less than that of the other group.

2. **the critical value of the test statistic,**
   a. for sample sizes ≤ 20, U; Table 13 (Appendix B, ES9-p729)
   b. for sample sizes > 20, z; Table 4 (Appendix B, ES9-p715)

3. **the calculated test statistic,**
   a. for $n \leq 20$; U^* = smaller of U_a and U_b, where

$$U_a = n_a \cdot n_b + \frac{(n_a)(n_b + 1)}{2} - R_b \quad \text{and} \quad \ldots$$

$$U_b = n_a \cdot n_b + \frac{(n_a)(n_{a+1})}{2} - R_a$$

R_a = sum of ranks for sample A, R_b = sum of ranks for sample B

b. for n > 20; U^* = smaller of U_a and U_b

$z^* = (U - \mu_u) / (\sigma_u)$ where

$$\mu_u = \frac{n_a \cdot n_b}{2} \qquad \sigma_u = \sqrt{\frac{n_a n_b (n_a + n_b + 1)}{12}}$$

Table 13, Critical Values of U in the Mann-Whitney Test, gives only critical values for the left-hand tail. $U(n_1, n_2, \alpha)$ is based on the two sample sizes and the amount of α for a one or two-tailed test.

If $U^* \le U(n_1,n_2,\alpha)$, reject H_o and

if $U^* > U(n_1,n_2,\alpha)$, fail to reject H_o.

14.21 a. Critical region: $U \le 88$

b. Critical region: $z \le -1.65$

14.23 MINITAB verify -- answers given in exercise.

14.25 Step 1: a. Story response scores for two groups of children.

H_o: Group 1 scores are not higher than Group 2 scores.

H_a: Group 1 scores are higher

Step 2: a. Independent samples and cholesterol values are numerical.

b. U

c. $\alpha = 0.05$

Step 3: a. ranked data: (underlined = group 2)

<u>25</u> <u>27</u> <u>30</u> 30 <u>32</u> 35 36 <u>39</u> 40 42 45

ranks:

1 2 3 4 5 6 7 8 9 10 11

(3 and 4: 3.5 each)

R_1 = 3.5 + 6 + 7 + 9 + 10 + 11 = 46.5

R_2 = 1 + 2 + 3.5 + 5 + 8 = 19.5

b. $U_1 = n_1 \cdot n_2 + [(n_2)(n_2+1)/2] - R_2$

$U_1 = (6)(5) + [(5)(5+1)/2] - 19.5 = 25.5$

$U_2 = n_2 \cdot n_1 + [(n_1)(n_1+1)/2] - R_1$

$U_2 = (5)(6) + [(6)(6+1)/2] - 46.5 = 4.5$; $U^* = 4.5$

Step 4: -- using p-value approach ---------------

a. **P** = P(U < 4.5)

Using Table 13, Appendix B, ES9-p729

0.025 < **P** < 0.05

b. **P** < α

-- using classical approach -------------

a. critical values: U(7,5,0.05) = 6; U $\leq$ 6

b. The test statistic is in the critical region.

Step 5: a. Reject H_o

b. Group 1 scores are significantly higher than Group 2 scores.

14.27 Step 1: a. Rainfall amounts based on cloud unseeding and cloud seeding.

b. H_o: Rainfall amount is the same for the two methods.

H_a: Rainfall amount is higher with cloud seeding.

Step 2: a. Independent samples and rainfall amounts are numerical.

b. U

c. $\alpha = 0.05$

Step 3: a. $n_U = n_S = 25$; $R_U = 503$, $R_S = 772$,

$U_U = 178$, $U_S = 447$

$\mu_u = (25 \cdot 25)/2 = 312.5$,

$\sigma_u = \sqrt{[(25)(25)(25+25+1)]/12} = 51.54$

b. $U^* = 178$

c. $z^* = (U - \mu_u)/\sigma_u$

$z^* = (178 - 312.5)/51.54 = -2.61$

Step 4: -- using p-value approach --------------

a. **P** = $P(z > 2.61) = 0.5000-0.4955 = 0.0045$

b. **P** $< \alpha$

-- using classical approach ------------

a. critical values: $z(0.05) = 1.65$

b. The test statistic is in the critical region.

Step 5: a. Reject H_o.

b. The evidence does allow us to conclude that there is a significant increase in the average amount of rainfall with cloud seeding, at the 0.05 level of significance.

14.26 Step 1: a. Amount of change in writing and reading scores.

b. H_o: Equal improvement in writing and reading scores.

H_a: Improvement was not equal in writing and reading scores.

Step 2: a. Independent samples and score values are numerical.

b. U

c. $\alpha = 0.05$

Step 3: a. $n_W = n_R = 45$; $R_W = 1798.5$, $R_R = 2296.5$,

$\mu_u = (45 \cdot 45)/2 = 1012.5$,

$\sigma_u = \sqrt{(45 \cdot 45 \cdot (45+45+1))/12} = 123.92$

b. $U^* = 763.5$

c. $z^* = (U - \mu_u)/\sigma_u$

$z^* = (763.5 - 1012.5)/123.92 = -2.01$

Step 4: -- using p-value approach --------------

a. **P** = $2P(z > 2.01) = 2(0.5000-0.4478) = 0.0444$

b. **P** $< \alpha$

-- using classical approach ------------

a. critical values: $\pm z(0.025) = \pm 1.96$

b. The test statistic is in the critical region.

Step 5: a. Reject H_o.

b. The evidence does allow us to conclude that there is a significant difference in the improvement in fourth-grade writing and reading scores, at the 0.05 level of significance.

SECTION 14.5 ANSWER NOW EXERCISES

14.29a. Answers will vary.
One set generated: 9, 1, 6, 7, 6, 8, 4, 3, 4, 9

b.

Step 1: a. Randomness of generated integers.
b. H_o: The sequence is random above and below the median value.
H_a: The sequence is not of random order

Step 2: a. Each data fits one of two categories.
b. V
c. $\alpha = 0.05$

Step 3: a. n(above) = 6, n(below) = 4
b. V* = 5

Step 4: -- using p-value approach ---------------
a. Using Table 14, Appendix B, ES9-p730
P > 0.05
b. **P** > α
-- using classical approach -------------
a. Critical regions: $V \leq 2$ or $V \geq 9$
b. The test statistic is not in the critical region.

Step 5: a. Fail to reject H_o.
b. The evidence is not significant, we can not conclude
that this sequence is not random, at the 0.05 level of significance.

c. Answers will vary.
One set generated: 5, 6, 3, 9, 9, 2, 6, 4, 1, 1

d.

Step 1: a. Randomness of generated integers.
b. H_o: The sequence is random above and below the median value.
H_a: The sequence is not of random order

Step 2: a. Each data fits one of two categories.
b. V
c. $\alpha = 0.05$

Step 3: a. n(above) = 4, n(below) = 5
b. V* = 6

Step 4: -- using p-value approach --------------

a. Using Table 14, Appendix B, ES9-p730
   P > 0.05

b. **P** > α

-- using classical approach -------------

a. Critical regions: $V \leq 2$ or $V \geq 9$

b. The test statistic is not in the critical region.

Step 5: a. Fail to reject H_o.

b. The evidence is not significant, we can not conclude
   that this sequence is not random, at the 0.05 level of significance.

Same results as part b.

14.30 a. By comparing the number of actual occurrences to the expected number of occurrences using a multinomial or contingency table test the relative frequency of occurrences can be tested.

b. The runs test will test the order, or sequence, of occurrence for the numbers generated.

c. The correlation will test the independence of side-by-side outcomes to be sure there is no influence of one part of a game with another part of the same game.

d. When testing for randomness, it is the null hypothesis that states random, thereby making the "fail to reject" decision the desired outcome. The probability associated with that result is $1 - \alpha$, not the level of significance, and $1 - \alpha$ is known as the level of confidence.

SECTION 14.5 EXERCISES

The Runs Test

The Runs Test is used to test the randomness of a set of data. Refer to the Five-Step Hypothesis Test Procedure in: ES9-pp370&386, ST-pp241&249, if necessary. The only changes are in:

1. **the hypotheses**
   H_o: The data occurred in a random order
   H_a: The data is not in random order

2. **the critical value of the test statistic**,
   a. for sample sizes ≤ 20, V;Table 14 (Appendix B,ES9-p730)

   b. for sample sizes > 20, z;Table 4 (Appendix B,ES9-p715)

3. **the calculated test statistic**,
   a. for n ≤ 20; V = the number of runs
   b. for n > 20; V = the number of runs

$z^* = (V^* - \mu_V) / (\sigma_V)$ where

$$\mu_v = \frac{2n_1 \cdot n_2}{n_1 + n_2} + 1 \qquad \sigma_v = \sqrt{\frac{(2n_1 \cdot n_2)(2n_1 \cdot n_2 - n_1 - n_2)}{(n_1 + n_2)^2(n_1 + n_2 - 1)}}$$

14.31 a. H_o: The data did occur in a random order
H_a: The data did not occur in a random order

b. H_o: Sequence of odd/even is in random order
H_a: Not in random order.

c. H_o: The order of entry by gender was random
H_a: The order of entry was not random

14.33 Step 1: a. Randomness; P(women) and P(men).
b. H_o: The hiring sequence is random.
H_a: The hiring sequence is not of random order

Step 2: a. Each data fits one of two categories.
b. V
c. $\alpha = 0.05$

Step 3: a. n(M) = 15, n(F) = 5
b. V* = 9

Step 4: -- using p-value approach ---------------
a. Using Table 14, Appendix B, ES9-p730
P > 0.05
b. **P** > α
-- using classical approach -------------
a. Critical regions: $V \leq 4$ or $V \geq 12$
b. The test statistic is not in the critical region.

Step 5: a. Fail to reject H_o.
b. The evidence is not significant, we can not conclude
that this sequence is not random.

14.35 Step 1: a. Randomness; P(late bus).
b. H_o: Random order of increase and decrease in value from previous value.
H_a: Lack of randomness (a trend, an increase in wait time)
Step 2: a. Each data fits one of two categories.
b. V
c. $\alpha = 0.05$
Step 3: a. n(decreases) = 4, n(increases) = 13
b. $V^* = 8$
Step 4: -- using p-value approach ---------------
a. Using Table 14, Appendix B, ES9-p730
P > 0.05
b. **P** > α
-- using classical approach -------------
a. Critical regions: $V \leq 3$ or $V \geq 10$
b. The test statistic is not in the critical region.

Step 5: a. Fail to reject H_o.
b. The evidence is not significant, we can not conclude that there is an increase in wait time.

14.37 a. Median = <u>22.5</u>; V = <u>8</u>
b. Step 1: a. Randomness; P(occurrence).
b. H_o: Random ages above and below median.
H_a: The data did not occur randomly.

Step 2: a. Each data fits one of two categories.
b. V
c. $\alpha = 0.05$

Step 3: a. n(a) = 7, n(b) = 7
b. V* = 8

Step 4: -- using p-value approach ---------------
a. Using Table 14, Appendix B, ES9-p730
P > 0.05
b. **P** > α
-- using classical approach ------------
a. Critical regions: $V \leq 3$ or $V \geq 13$
b. The test statistic is not in the critical region.

Step 5: a. Fail to reject H_o.
b. The evidence is not significant, we are unable to conclude that this sequence lacks randomness.

14.39 a. MINITAB verify -- answers given in exercise.

b. z* = <u>-3.76</u>
P = 2P(z < -3.76) = 2(0.0001) = <u>0.0002</u>

c. Yes, reject the hypothesis of random runs above and below the median.

d.

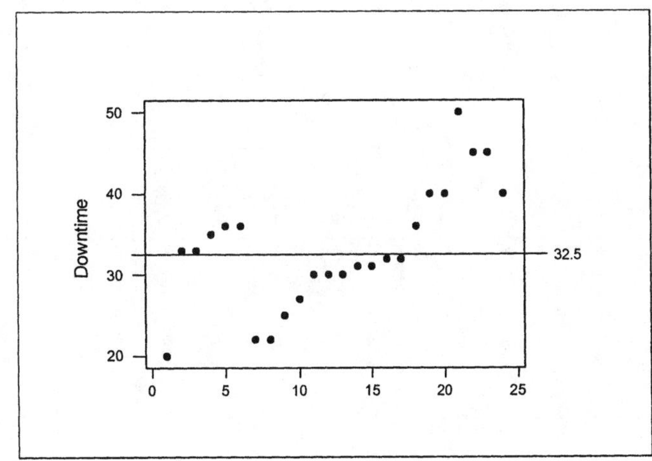

14.41 a. Step 1: a. Randomness of data above and below the median.

b. H_o: Randomness in number of absences.
H_a: The data did not occur randomly.

Step 2: a. Each data fits one of two categories.

b. V

c. $\alpha = 0.05$

Step 3: a. $\tilde{x} = 10.5$; n(above) = 13, n(below) = 13

b. V* = 9

Step 4: -- using p-value approach ---------------

a. Using Table 14, Appendix B, ES9-p730
P > 0.05

b. **P** > α

-- using classical approach -------------

a. Critical regions: $V \leq 8$ or $V \geq 20$

b. The test statistic is not in the critical region.

Step 5: a. Fail to reject H_o.

b. The evidence is not significant, we can not conclude that this sequence is not random.

b. Step 1: a. Randomness of data above and below the median.

b. H_o: Randomness in number of absences.
H_a: The data did not occur randomly.

Step 2: a. Each data fits one of two categories.

b. V

c. $\alpha = 0.05$

Step 3: a. $\tilde{x} = 10.5$; n(above) = 13, n(below) = 13

b. V* = 9

c. n(1) = 125, n(2) = 125

$\mu = [(2n_1n_2)/(n_1 + n_2)] + 1$

$\mu_V = [(2)(13)(13)/(13+13)] + 1 = 14$

$$\sigma_V = \sqrt{\frac{(2n_1n_2)(2n_1n_2 - n_1 - n_2)}{(n_1 + n_2)^2(n_1 + n_2 - 1)}}$$

$$= \sqrt{[(2)(13)(13)][2(13)(13)-13-13]/(13+13)^2(13+13-1)}$$

$= 2.498$

$z = (V - \mu_V)/\sigma_V$

$z^* = (9 - 14)/2.498 = -2.00$

Step 4: -- using p-value approach --------------

a. $\mathbf{P} = 2P(z > 2.00)$

Using Table 3, Appendix B, ES9-p714

$\mathbf{P} = 2(0.5000 - 0.4772) = 0.0456$

b. $\mathbf{P} < \alpha$

-- using classical approach -------------

a. Critical values: $z(0.025) = \pm 1.96$

b. The test statistic is in the critical region.

Step 5: a. Reject H_o.

b. The evidence is significant, we can conclude that this sequence is not random.

SECTION 14.6 ANSWER NOW EXERCISE

14.43 a. The null hypothesis, " the mean VMI score from Study 1 is the same as the mean VMI score from Study 2" is rejected for all significance levels larger than 0.0001. The t-test for the comparison of 2 dependent means might have been used since the two studies used a common set of 58 subjects.

b. The null hypothesis, "$\rho = 0.00$, there is no linear correlation between the two sets of VMI scores" is rejected for all significance levels larger than 0.0001.

c. The null hypothesis, "$\rho_s = 0.00$, there is no rank correlation between the two sets of VMI scores" is rejected for all significance levels larger than 0.0001.

d. Basically they are testing the same thing, the difference being that Pearson's correlation uses the VMI score, while Spearman's correlation uses the ranks of the VMI scores.

SECTION 14.6 EXERCISES

The Spearman Rank Correlation Test

The Rank Correlation Test is used to test for the correlation or relationship between two variables. Refer to the Five-Step Hypothesis Test Procedure in: ES9pp370&386, ST-pp421&429, if necessary. The only changes are in:

1. **the hypotheses**
   a. null hypothesis
      H_o: There is no correlation or relationship between the two variables <u>or</u> $\rho_s = 0$
   b. possible alternative hypotheses:
      H_a: There is a correlation or relationship between the two variables <u>or</u> $\rho_s \neq 0$.
      H_a: There is a positive correlation <u>or</u> $\rho_s > 0$.
      H_a: There is a negative correlation <u>or</u> $\rho_s < 0$.

2. **the critical value of the test statistic,**
   a. two-tailed test, $\pm r_s$; Table 15 (Appendix B, ES9-p731)

   b. one-tailed test, $+r_s$ or $-r_s$; Table 15 (Appendix B, ES9-p731)

3. **the calculated test statistic,**

$$r_s^* = 1 - \frac{6[\sum(d_i)^2]}{n(n^2 - 1)}$$

Table 15, Critical Values of Spearman's Rank Correlation Coefficient, gives positive critical values based on sample size and the level of significance. For a two-tailed test, add a plus and minus sign to the table value. For a one-tailed test, double the level of significance, then apply a plus or minus sign, whichever is appropriate.

14.45 a.

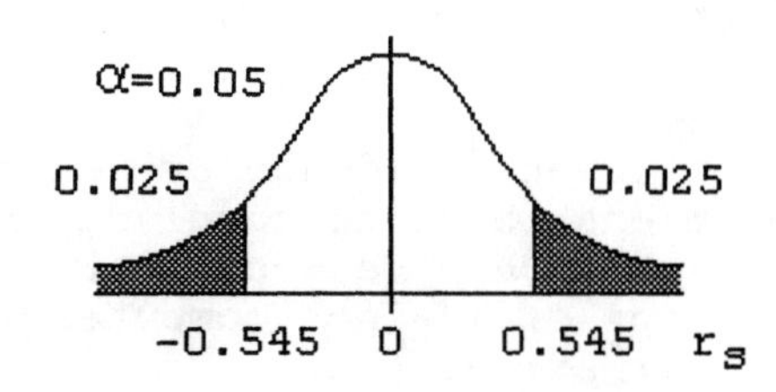

b.

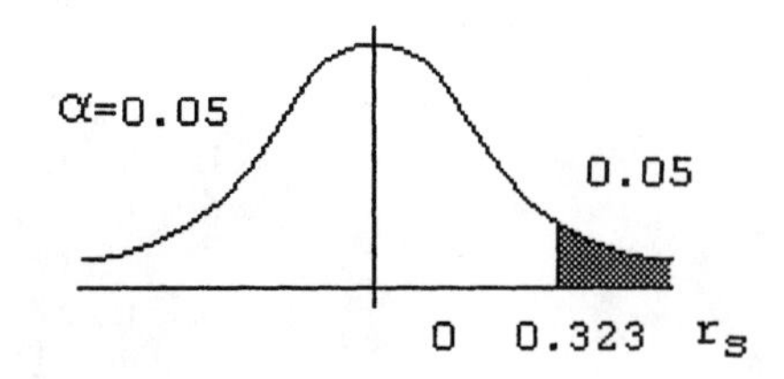

c.

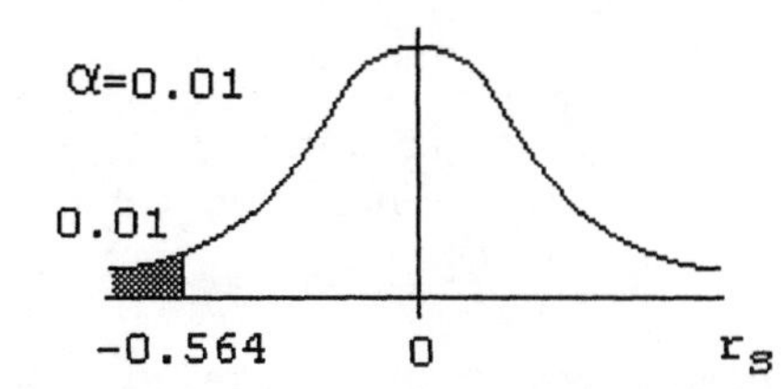

14.47 a. The formula for the rank correlation coefficient: 0.133

$$r_s = 1 - \frac{6\sum(d_i)^2}{n(n^2-1)}: \qquad r_s = 1 - \frac{6(143)}{10(99)} = 1 - .867 = 0.133$$

b. Step 1: a. Correlation between the overall rating and the street price.

b. H_o: $\rho_s = 0$
H_a: $\rho_s > 0$

Step 2: a. Assume random sample of ordered pairs, one ordinal variable and one numerical variable

b. r_s

c. $\alpha = 0.05$

Step 3: a. $n = 10$, $\Sigma d^2 = 143$

b. $r_s* = 0.133$

Step 4: -- using p-value approach --------------

a. Using Table 15, Appendix B, ES9-p731
P > 0.10
Using computer: **P** = 0.3652

b. **P** > α

-- using classical approach ------------
a. critical region: $r_s \geq 0.564$
b. r_s* is not in the critical region

Step 5: a. Fail to reject H_o.
b. There is not sufficient evidence presented by these data to enable us to conclude that there is any relationship between overall performance ratings of 17-inch computer monitors and their street price.

14.49 Summary of data: $n = 12$, $\Sigma d^2 = 70.5$

$r_s = 1 - [(6)(70.5)/(12)(12^2-1)] = \underline{0.753}$

14.51 Step 1: a. Correlation between undergraduate GPA and GPA at graduation from a graduate nursing program.
b. H_o: $\rho_s = 0$
H_a: $\rho_s > 0$

Step 2: a. Assume random sample of ordered pairs, both variables are numerical
b. r_s
c. $\alpha = 0.05$

Step 3: a. $n = 10$, $\Sigma d^2 = 43.5$
b. $r_s = 1 - [(6)(\Sigma d^2)/(n)(n^2-1)]$
$r_s^* = 1 - [(6)(43.5)/(10)(99)] = 0.736$
$r_s^* = 0.732$ (using MINITAB)

Step 4: -- using p-value approach ---------------
a. Using Table 15, Appendix B, ES9-p731
$0.01 < \mathbf{P} < 0.025$
Using computer: $\mathbf{P} = 0.016$
b. $\mathbf{P} < \alpha$

-- using classical approach ------------
a. critical region: $r_s \geq 0.564$
b. r_s* is in the critical region

Step 5: a. Reject H_o.
b. There is sufficient reason to conclude there is a positive relationship.

14.53 a.

Lee Co	Florida	U.S.	Lee Rank	FL Rank	US Rank
0.02	0.02	0.026	1	1	1
0.08	0.05	0.065	6	3	2.5
0.05	0.08	0.160	3.5	6	6
0.05	0.05	0.071	3.5	3	4.5
0.25	0.21	0.168	8	8	7
0.03	0.05	0.071	2	3	4.5
0.06	0.06	0.065	5	5	2.5
0.30	0.34	0.359	9	9	8
0.16	0.15	*	7	7	*

b. Step 1: a. Correlation between Lee County job classification rates and all of Florida.

b. H_o: $\rho_s = 0$

H_a: $\rho_s \neq 0$

Step 2: a. Assume random sample of ordered pairs, both variables are numerical

b. r_s

c. $\alpha = 0.05$

Step 3: a. $n = 9$, $\Sigma d^2 = 16.5$

b. $r_s = 1 - [(6)(\Sigma d^2)/(n)(n^2-1)]$

$r_s* = 1 - [(6)(16.5)/(9)(80)] = 0.8625$

Step 4: -- using p-value approach ---------------

a. Using Table 15, Appendix B, ES9-p731

P < 0.01

b. **P** < α

-- using classical approach -------------

a. critical region: $r_s \leq -0.700$ and $r_s \geq 0.700$

b. r_s* is in the critical region

Step 5: a. Reject H_o.

b. There is sufficient reason to conclude there is a correlation, at the 0.05 level of significance.

c. Step 1: a. Correlation between Lee County job classification rates and all of the US.

b. H_o: $\rho_s = 0$

H_a: $\rho_s \neq 0$

Step 2: a. Assume random sample of ordered pairs, both variables are numerical

b. r_s

c. $\alpha = 0.05$

Step 3: a. $n = 8$, $\Sigma d^2 = 32$

b. $r_s = 1 - [(6)(\Sigma d^2)/(n)(n^2-1)]$

$r_s* = 1 - [(6)(32)/(8)(63)] = 0.619$

Step 4: -- using p-value approach ---------------

a. Using Table 15, Appendix B, ES9-p731

P > 0.10

b. **P** > α

-- using classical approach -------------

a. critical region: $r_s \leq -0.738$ and $r_s \geq 0.738$

b. r_s* is in the critical region

Step 5: a. Fail to reject H_o.

b. There is sufficient reason to conclude there is no correlation, at the 0.05 level of significance.

d. Step 1: a. Correlation between all of Florida's job classification rates and all of the US.

b. H_o: $\rho_s = 0$

H_a: $\rho_s \neq 0$

Step 2: a. Assume random sample of ordered pairs, both variables are numerical

b. r_s

c. $\alpha = 0.05$

Step 3: a. $n = 8$, $\Sigma d^2 = 11$

b. $r_s = 1 - [(6)(\Sigma d^2)/(n)(n^2-1)]$

$r_s* = 1 - [(6)(11)/(8)(63)] = 0.869$

Step 4: -- using p-value approach ---------------

a. Using Table 15, Appendix B, ES9-p731

0.01 < **P** < 0.02

b. **P** < α

-- using classical approach -------------

a. critical region: $r_s \leq -0.738$ and $r_s \geq 0.738$

b. r_s* is in the critical region

Step 5: a. Reject H_o.

b. There is sufficient reason to conclude there is a correlation, at the 0.05 level of significance.

e. Lee Co. and Florida share a significant correlation, Florida and the US share a significant correlation, however Lee Co. and the US are not correlated. {No transitive property!!}

14.55 a.

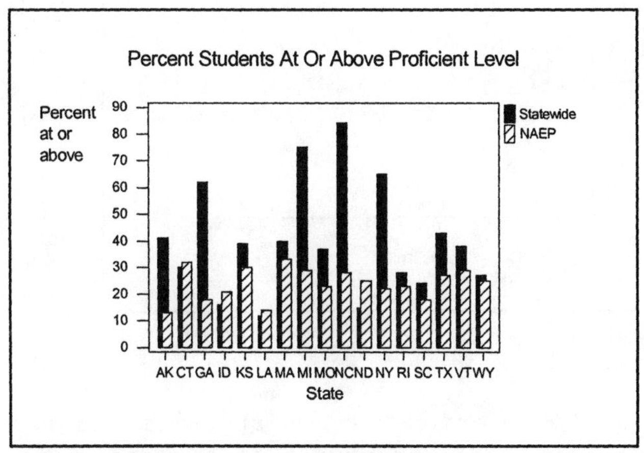

There appears to be very little relationship between the two sets of percentages. Some of the largest state percentages are paired with some of the lower NAEP percentages, while some of the lowest state percentages are also paired with some of the lowest NAEP percentages.

b.

	State	NAEP	Rstate	R NAEP	d	dsq	sumdsq
Arkansas	41	13	12	1	-11	121	594
Connecticut	30	32	7	16	9	81	
Georgia	62	18	14	3.5	-11	110.25	
Idaho	16	21	3	5	2	4	
Kansas	39	30	10	15	5	25	
Louisiana	12	14	1	2	1	1	
Massachusetts	40	33	11	17	6	36	
Michigan	75	29	16	13.5	-2.5	6.25	
Missouri	37	23	8	7.5	-0.5	0.25	
New York	65	22	15	6	-9	81	
North Carolina	84	28	17	12	-5	25	
North Dakota	15	25	2	9.5	7.5	56.25	
Rhode Island	28	23	6	7.5	1.5	2.25	
South Carolina	24	18	4	3.5	-0.5	0.25	
Texas	43	27	13	11	-2	4	
Vermont	38	29	9	13.5	4.5	20.25	
Wyoming	27	25	5	9.5	4.5	20.25	

c.

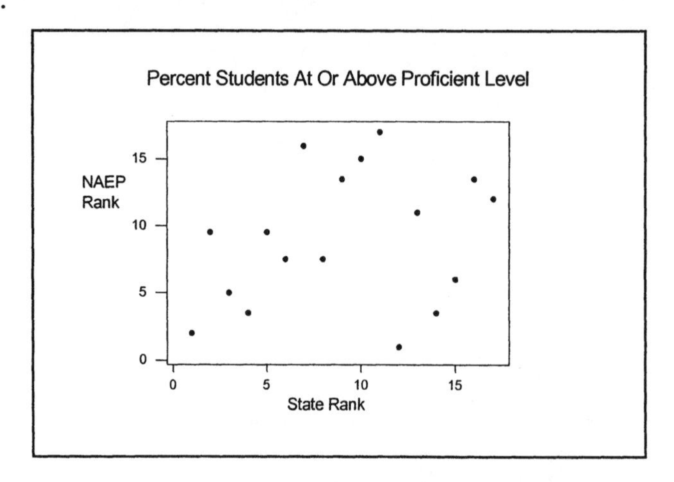

d.

Step 1: a. Correlation between statewide assessment and national assessment of education.

b. H_o: $\rho_s = 0$
H_a: $\rho_s \neq 0$

Step 2: a. Assume random sample of ordered pairs, both variables are numerical

b. r_s

c. $\alpha = 0.05$

Step 3: a. $n = 17$, $\Sigma d^2 = 594$

b. $r_s = 1 - [(6)(\Sigma d^2)/(n)(n^2-1)]$
$r_s* = 1 - [(6)(594)/(17)(288)] = 0.272$

Step 4: -- using p-value approach ---------------

a. Using Table 15, Appendix B, ES9-p731
P > 0.10

b. **P** > α

-- using classical approach -------------

a. critical region: $r_s \leq -0.490$ and $r_s \geq 0.490$

b. r_s* is not in the critical region

Step 5: a. Fail to reject H_o.

b. There is not sufficient reason to conclude there is a correlation between the two sets of percentages, at the 0.05 level of significance.

14.57 a.

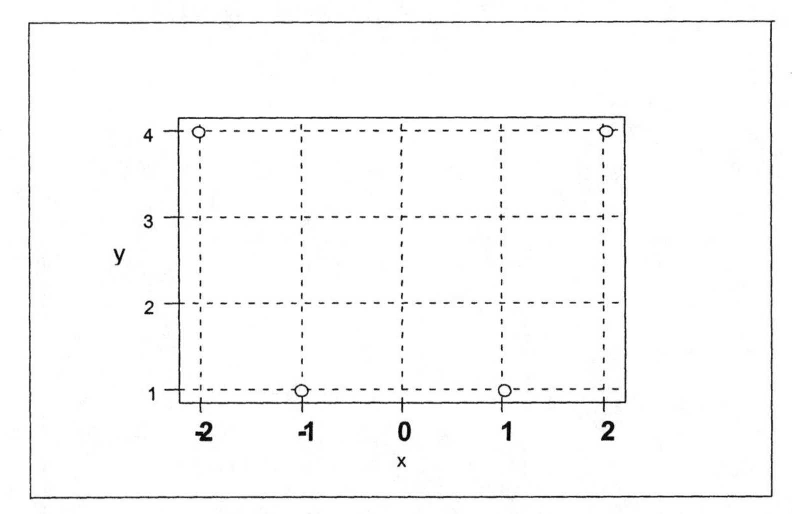

b.

x	y	rank(x)	rank(y)	d	d^2
-2	4	1	3.5	-2.5	6.25
-1	1	2	1.5	0.5	0.25
1	1	3	1.5	1.5	2.25
2	4	4	3.5	0.5	0.25
				Σ	9.00

$r_s = 1 - [6(9.0)/4(16-1)] = \underline{0.10}$

c.

x	y	x^2	xy	y^2
-2	4	4	-8	16
-1	1	1	-1	1
1	1	1	1	1
2	4	4	8	16
0	10	10	0	34

$r = \{0 - [(0)(10)/4]\}/\sqrt{\{10-[0^2/4]\}\{34-[10^2/4]\}} = \underline{0.00}$

d. The two results are not identical. Both values are near zero. The rank correlation measures the correlation between the rank numbers, while the Pearson coefficient uses the numerical values.

RETURN TO CHAPTER CASE STUDY

14.58 a. They do not respond with exactly the same rate for each choice, but they are very similar.

b.

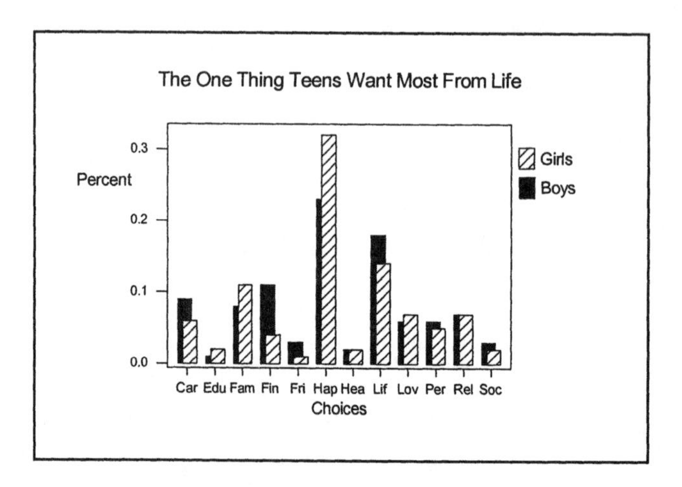

Notice how the two sets of bars mirror each other. That is; shortest with shortest, the tallest with the tallest and so on.

c. The information given is percentages (relative frequencies) and the chi-square tests studied in Chapter 11 require the use of frequencies (counts).

14.59 a.

Ranks:

Choices	Boys	Girls
Happiness	1	1
Long enjoyable life	2	2
Financial success	3	8
Career success	4	6
Marriage and family	5	3
Religious satisfaction	6	4.5
Love	7.5	4.5
Personal success	7.5	7
Personal contribution to society	9.5	10
Friends	9.5	12
Health	11	10
Education	12	10

b.

Step 1: a. Boys and girls responses to what they want most from life.

b. H_o: Boys and girls responses have the same distribution.
H_a: Boys and girls responses have different distributions.

Step 2: a. Independent samples and yields are numerical.

b. U

c. $\alpha = 0.05$

Step 3: a. ranked data: (see part a)

$R_B = 1.5 + 4.5 + 7.5 + 7.5 + 12 + 12 + 15 + 17 + 18 + 19.5 + 22 + 23 = 159.5$

$R_G = 1.5 + 4.5 + 4.5 + 4.5 + 9 + 10 + 12 + 15 + 15 + 19.5 + 21 + 24 = 140.5$

$U_B = n_B \times n_G + [(n_G)(n_G + 1)] \div 2 - R_G$
$= 12 \times 12 + [12(13) \div 2] - 140.5$
$= 144 + 78 - 140.5 = 81.5$

$U_G = n_B \times n_G + [(n_B)(n_B + 1)] \div 2 - R_B$
$= 12 \times 12 + [12(13) \div 2] - 159.5$
$= 144 + 78 - 159.5 = 62.5$

Check: $81.5 + 62.5 = 144 = 12 \times 12$
$U^* = 62.5$

Step 4: -- using p-value approach --------------

a. $\mathbf{P} = P(U < 62.5)$
Using Table 13, Appendix B, ES9-p729
$\mathbf{P} > 0.10$

b. $\mathbf{P} > \alpha$

-- using classical approach ------------

a. critical region: $U \leq 37$

b. The test statistic is not in the critical region.

Step 5: a. Fail to reject H_o

b. There is no significant difference in the choices of boys and girls, at the 0.05 level of significance.

c.

Step 1: a. Correlation between boys preferences and girls preferences

b. H_o: $\rho_s = 0$
H_a: $\rho_s \neq 0$

Step 2: a. Assume random sample of ordered pairs, both variables are numerical

b. r_s

c. $\alpha = 0.05$

Step 3: a. $n = 12$, $\Sigma d^2 = 56$

b. $r_s = 1 - [(6)(\Sigma d^2)/(n)(n^2-1)]$

$r_s* = 1 - [(6)(56)/(12)(143)] = 0.804$

Step 4: -- using p-value approach ---------------

a. Using Table 15, Appendix B, ES9-p731

P < 0.01

Using computer: **P** = 0.002

b. **P** < α

-- using classical approach -------------

a. critical region: $r_s \leq -0.591$ and $r_s \geq 0.591$

b. r_s* is in the critical region

Step 5: a. Reject H_o.

b. There is sufficient reason to conclude there is a correlation, at the 0.05 level of significance.

d. Both results suggest agreement between the boys and girls on the "one thing teens want most from life." The null hypothesis for the Mann-Whitney U test states agreement between the two distributions and the decision was "fail to reject." The null hypothesis for the rank correlation test states that there is no relationship between the two sets of choices and it was rejected.

CHAPTER EXERCISES

14.61 a.

```
Stem-and-leaf of Hydrogen  N  = 52
Leaf Unit = 0.10
    3    2 078
   10    3 1356688
   18    4 13355689
  (12)   5 001112444589
   22    6 11137
   17    7 2388
   13    8 3378
    9    9 4479
    5   10 56
    3   11 1
    2   12 4
    1   13
    1   14
    1   15 2
```

b. skewed right

c. For $n = 52$ and $1 - \alpha = 0.95$, the critical value from Table 12 is $k = 18$.
$x_{k+1} = x_{19} = 5.0$ and $x_{n-k} = x_{34} = 6.3$

5.0 to 6.3, the 0.95 interval for median M

14.63 a. Using the Sign Test (one median):

Step 1: a. Median score on exam.
b. H_o: Median = 50
H_a: Median ≠ 50

Step 2: a. Assume sample is random. Exam score is continuous.
b. x = n(least frequent sign)
c. $\alpha = 0.05$

Step 3: a. + = above 50, - = below 50, 0 = 50; n = 30
n(+) = 10, n(0) = 2, n(-) = 20
b. x = n(+) = 10

Step 4: -- using p-value approach ---------------
a. **P** = $2P(x \leq 10 | n = 30)$;
Using Table 12, Appendix B, ES9-p728:
P ≈ 0.10
b. **P** > α
-- using classical approach -------------
a. critical region: n(least freq sign)= $x \leq 9$
b. The test statistic is not in the critical region.

Step 5: a. Fail to reject H_o.
b. The sample evidence is not sufficient to justify the claim that median is different than 50, at the 0.05 level of significance.

b. Step 1: a. Median score on exam.
b. H_o: Median = 50
H_a: Median < 50

Step 2: a. Assume sample is random. Exam score is continuous.
b. x = n(least frequent sign)
c. $\alpha = 0.05$

Step 3: a. + = above 50, - = below 50;
n = 30, n(+) = 10, n(-) = 20
b. x = n(+) = 10

Step 4: -- using p-value approach ---------------
a. $\mathbf{P} = P(x \le 10 | n = 30)$;
Using Table 12, Appendix B, ES9-p728:
$\mathbf{P} \approx 0.05$
b. $\mathbf{P} \le \alpha$

-- using classical approach -------------
a. critical region: n(least freq sign) = $x \le 10$
b. The test statistic is in the critical region.

Step 5: a. Reject H_o.
b. The sample evidence is sufficient to justify the claim that median is less than 50, at the 0.05 level of significance.

14.65 Using the Sign Test (dependent samples):
Step 1: a. Time required to run 220 yd sprint on two tracks.
b. H_o: No difference between average times (no faster)
H_a: Average time on B is less than on A (B is faster)
Step 2: a. Assume sample is random. Time is continuous.
b. x = n(least frequent sign)
c. $\alpha = 0.05$
Step 3: a. + = A time is greater; n = 10
n(+) = 8, n(0) = 0, n(-) = 2
b. x = n(-) = 2
Step 4: -- using p-value approach ---------------
a. $\mathbf{P} = P(x \le 2 | n = 10)$;
Using Table 12, Appendix B, ES9-p728:
$\mathbf{P} > 0.10$
b. $\mathbf{P} > \alpha$

-- using classical approach -------------
a. critical region: n(least freq sign) = $x \le 1$
b. The test statistic is not in the critical region.

Step 5: a. Fail to reject H_o.
b. The evidence is not sufficient to justify the claim that track B is faster, at the 0.05 level of significance.

14.67 Reject for $U \leq 127$

14.69 Using Mann-Whitney U Test (independent samples):

Step 1: a. Line width.

b. H_o: No difference in line width.

H_a: There is a difference in line width.

Step 2: a. Independent samples and line widths are numerical.

b. U

c. $\alpha = 0.05$

Step 3: a. ranked data and ranks: (underlined = normal group)

27.5	28.0	28.5	29.5	<u>30.5</u>	<u>30.6</u>	30.7	<u>30.9</u>	<u>32.9</u>	<u>35.1</u>
1	2	3	4	5	6	7	8	9	10

$R_n = 38, \quad R_m = 17$

b. $U_n = n_n \cdot n_m + [(n_m)(n_m+1)/2] - R_m$

$U_n = (5)(5) + [(5)(5+1)/2] - 17 = 23$

$U_m = n_m \cdot n_n + [(n_n)(n_n+1)/2] - R_n$

$U_m = (5)(5) + [(5)(5+1)/2] - 38 = 2; \; U^* = 2$

Step 4: -- using p-value approach --------------

a. $\mathbf{P} = 2P(U \leq 2 | \; n_n=5, \; n_m=5)$;

Using Table 13, Appendix B, ES9-p729:

$\mathbf{P} \approx 0.05$

b. $\mathbf{P} \leq \alpha$

-- using classical approach ------------

a. $U(5,5,0.05) = 2; \; U \leq 2$

b. The test statistic is in the critical region.

Step 5: a. Reject H_o.

b. There is a significant difference in line width.

14.71 a.

Batting Avg.	***League***	***Rank***	***ERA***	***League***	***Rank***
0.282	A	1.0	5.29	A	1.0
0.277	A	2.0	5.21	A	2.0
0.275	A	3.5	5.20	N	3.0
0.275	A	3.5	5.15	A	4.0
0.274	N	5.0	4.92	A	5.0
0.272	A	6.0	4.91	A	6.0
0.270	A	7.0	4.80	A	7.0
0.269	A	8.0	4.72	N	8.0
0.268	N	9.0	4.62	N	9.0
0.267	N	10.5	4.55	A	10.0
0.267	N	10.5	4.46	A	11.0
0.263	N	12.0	4.36	N	12.0
0.262	N	13.0	4.29	N	13.0
0.261	N	15.5	4.27	N	14.0
0.261	N	15.5	4.23	N	15.0
0.261	A	15.5	4.17	N	16.0
0.261	A	15.5	4.12	A	17.0
0.260	N	18.0	4.07	A	18.0
0.259	N	19.0	4.00	N	19.0
0.256	N	20.5	3.97	N	20.0
0.256	A	20.5	3.92	N	21.0
0.253	N	23.5	3.89	N	22.0
0.253	N	23.5	3.87	A	23.0
0.253	N	23.5	3.75	A	24.0
0.253	A	23.5	3.70	N	25.0
0.249	A	26.0	3.69	N	26.5
0.248	A	27.0	3.69	A	26.5
0.246	N	28.5	3.68	A	28.0
0.246	A	28.5	3.54	N	29.0
0.244	N	30.0	3.13	N	30.0

b. Batting Averages:

Step 1: a. Batting averages.

b. H_o: Batting averages in AL are not higher. (≤)
H_a: Batting averages in AL are higher. (>)

Step 2: a. Independent samples and batting averages are numerical.

b. U

c. $\alpha = 0.05$

Step 3: a.

$R_a = 1 + 2 + 3.5 + 3.5 + 6 + 7 + 8 + 15.5 + 15.5 + 20.5 + 23.5 + 26 + 27 + 28.5 = 187.5$

$R_n = 5 + 9 + 10.5 + 10.5 + 12 + 13 + 15.5 + 15.5 + 18 + 19 + 20.5 + 23.5 + 23.5 + 23.5 + 28.5 + 30 = 277.5$

b.

$U_a = n_a \times n_n + [(n_n)(n_n + 1)] \div 2 - R_n = 14 \times 16 + 16(17) \div 2 - 187.5 = 224 + 136 - 277.5 = 82.5$

$U_n = n_a \times n_n + [(n_a)(n_a + 1)] \div 2 - R_a = 14 \times 16 + 14(15) \div 2 - 277.5 = 224 + 105 - 187.5 = 141.5$

Check: $82.5 + 141.5 = 224 = 14 \times 16$

$U^* = 82.5$

Step 4: -- using p-value approach ---------------

a. **P** = $P(U < 82.5)$

Using Table 13, Appendix B, ES9, p729

P > 0.05

b. **P** $> \alpha$

-- using classical approach -------------

a. critical region: $U \leq 71$

b. The test statistic is not in the critical region.

Step 5: a. Fail to reject H_o

b. The American League batting average in 2002 was not higher than the National League at the 0.05 level of significance.

Earned Run Averages:

Step 1: a. Earned run averages

b. H_o: Earned run average for NL is not lower. ($\geq$)

H_a: Earned run average for NL is lower. ($<$)

Step 2: a. Independent samples and earned run averages are numerical.

b. U

c. $\alpha = 0.05$

Step 3: a.

R_a = 1 + 2 + 4 + 5 + 6 + 7 + 10 + 11 + 17 + 18 + 23 + 24 + 26.5 + 28 = 182.5

R_n = 3 + 8 + 9 + 12 + 13 + 14 + 15 + 16 + 19 + 20 + 21 + 22 + 25 + 26.5 + 29 + 30 = 282.5

b.

$U_a = n_a \times n_n + [(n_n)(n_n + 1)] \div 2 - R_n = 14 \times 16 + 16(17) \div 2 - 282.5 = 224 + 136 - 282.5 = 77.5$

$U_n = n_a \times n_n + [(n_a)(n_a + 1)] \div 2 - R_a = 14 \times 16 + 14(15) \div 2 - 182.5.5 = 224 + 105 - 182.5 = 146.5$

Check: $77.5 + 146.5 = 224 = 14 \times 16$

$U^* = 77.5$

Step 4: -- using p-value approach ---------------
- a. **P** = $P(U < 77.5)$
  Using Table 13, Appendix B, ES9-p729
  P > 0.05
- b. **P** > α

-- using classical approach -------------
- a. critical region: $U \leq 71$
- b. The test statistic is not in the critical region.

Step 5: a. Fail to reject H_o
b. The National League earned run average in 2002 is not lower than the American League at the 0.05 level of significance.

14.73 Using the Runs Test:

Step 1: a. Randomness in sequence of occurrence of defective and nondefective parts.
b. H_o: Random order.
H_a: Lack of randomness.

Step 2: a. Each data fits into one of two categories
b. V
c. $\alpha = 0.05$

Step 3: a. n(n) = 20, n(d) = 4
b. V* = 9

Step 4: -- using p-value approach ---------------

a. $\mathbf{P} = P(V \leq 9)$

$\mathbf{P} > 0.05$

b. $\mathbf{P} > \alpha$

-- using classical approach -------------

a. critical region: $V \leq 4$ and $V \geq 10$

b. the test statistic is not in the critical region

Step 5: a. Fail to reject H_o.

b. The sample results do not show a significant lack of randomness.

14.75 a. Median = 22.5

Company		Job Growth	Company		Job Growth
1	b	26	11	a	23
2	a	54	12	b	13
3	a	34	13	b	17
4	b	10	14	a	23
5	a	31	15	b	9
6	a	48	16	b	3
7	a	26	17	b	15
8	b	22	18	b	11
9	a	24	19	b	1
10	b	10	20	a	122

Runs above: 6 Runs below: 6

b.

Step 1: a. Randomness of job growth rate percentages.

b. H_o: The job growth rate percentages are listed in a random sequence

H_a: Lack of randomness.

Step 2: a. Each data fits into one of two categories

b. V

c. $\alpha = 0.05$

Step 3: a. $n_a = 9$ $n_b = 11$

b. $V^* = 12$

Step 4: -- using p-value approach --------------

a. $\mathbf{P} = P(V \leq 12)$

$\mathbf{P} > 0.05$

b. $\mathbf{P} > \alpha$

-- using classical approach ------------
a. critical region: $V \le 6$ and $V \ge 16$
b. The test statistic is not in the critical region

Step 5: a. Fail to reject H_o.

c. Conclusion: There is not sufficient evidence to reject the null hypothesis that the job growth rate percentages are listed in a random sequence at the 0.05 level of significance. Based on this sample evidence, a higher job growth rate does not imply a higher rank in attractiveness.

14.77 Using Spearman's Rank Correlation:

Step 1: a. Correlation between two daily high temperatures.
b. H_o: $\rho_s = 0$ (Independence)
H_a: $\rho_s > 0$ (Positive correlation)

Step 2: a. Assume random sample of ordered pairs, numerical variables
b. r_s
c. $\alpha = 0.05$

Step 3: a. $n = 18$, $\Sigma d^2 = 116.5$
b. $r_s = 1 - [(6)(\Sigma d^2)/(n)(n^2-1)]$
$r_s* = 1 - [(6)(116.5)/(18)(18^2-1)] = 0.880$

Step 4: -- using p-value approach --------------
a. Using Table 15, Appendix B, ES9-p731
P < 0.01
Using computer: **P** $= 0.000$
b. **P** $< \alpha$

-- using classical approach ------------
a. critical regions: $r_s \ge 0.399$
b. r_s* is in the critical region

Step 5: a. Reject H_o
b. There is a significant amount of correlation shown between the two sets of temperatures.

14.79 a. $r_{12} = 1 - \frac{6(1766)}{25(624)} = 1 - 0.679 = 0.321 = r_{12}$

$r_{13} = 1 - \frac{6(1824)}{25(624)} = 1 - 0.702 = 0.298 = r_{13}$

$r_{23} = 1 - \frac{6(30)}{25(624)} = 1 - 0.012 = 0.988 = r_{23}$

b. The three tests are identical for Steps 1-3:

Step 1: a. Correlation between the rankings.
b. H_o: $\rho_s = 0$
H_a: $\rho_s \neq 0$

Step 2: a. Assume random sample of ordered pairs, ordinal variables
b. r_s
c. $\alpha = 0.05$

Step 3: a. n = 25
b. r_s* for each are listed in (a)
For (1) vs. (2), $r_s^* = 0.321$

Step 4: -- using p-value approach ---------------
a. Using Table 15, Appendix B, ES9-p731
P > 0.10
b. **P** > α

-- using classical approach -------------
a. critical regions: $r_s \leq -0.400$ and $r_s \geq 0.400$
b. r_s* is not in the critical region

Step 5: a. Fail to reject H_o.
b. There is not sufficient evidence presented by these data to enable us to conclude that the SI preseason poll has any relationship to the USA Today/ESPN poll, at the 0.05 level of significance.

For (1) vs. (3), $r_s^* = 0.298$

Step 4: -- using p-value approach ---------------
a. Using Table 15, Appendix B, ES9-p731
P > 0.10
b. **P** > α

-- using classical approach -------------
a. critical regions: $r_s \leq -0.400$ and $r_s \geq 0.400$
b. r_s* is not in the critical region

Step 5: a. Fail to reject H_o.

b. There is not sufficient evidence presented by these data to enable us to conclude that the SI preseason poll has any relationship to the AP Top 25 poll, at the 0.05 level of significance.

For (2) vs. (3), $r_s^* = 0.988$

Step 4: -- using p-value approach ---------------

a. Using Table 15, Appendix B, ES9-p731

P < 0.01

b. **P** $< \alpha$

-- using classical approach -------------

a. critical regions: $r_s \leq -0.400$ and $r_s \geq 0.400$

b. r_s^* is in the critical region

Step 5: a. Reject H_o.

b. The USA Today/ESPN poll and the AP Top 25 poll have a strong positive relationship, at the 0.05 level of significance.

INTRODUCTORY CONCEPTS

SUMMATION NOTATION

The Greek capital letter sigma (Σ) is used in mathematics to indicate the summation of a set of addends. Each of these addends must be of the form of the variable following Σ. For example:

1. $\sum x$ means sum the variable x.
2. $\sum (x-5)$ means sum the set of addends that are each 5 less than the values of each x.

When large quantities of data are collected, it is usually convenient to index the response variable so that at a future time its source will be known. This indexing is shown on the notation by using i (or j or k) and affixing the index of the first and last addend at the bottom and top of the Σ. For example,

$$\sum_{i=1}^{3} x_i$$

means to add all the consecutive values of x's starting with source number 1 and proceeding to source number 3.

∇ ILLUSTRATION 1

Consider the inventory in the following table concerning the number of defective stereo tapes per lot of 100.

Lot Number (I)	1	2	3	4	5	6	7	8	9	10
Number of Defective Tapes per Lot (x)	2	3	2	4	5	6	4	3	3	2

a. Find $\sum_{i=1}^{10} x_i$. b. Find $\sum_{i=4}^{8} x_i$.

Solution

a. $$\sum_{i=1}^{10} x_i = x_1 + x_2 + x_3 + x_4 + \cdots + x_{10}$$

$$= 2 + 3 + 2 + 4 + 5 + 6 + 4 + 3 + 3 + 2 = 34$$

b.

$$\sum_{i=4}^{8} x_i = x_4 + x_5 + x_6 + x_7 + x_8 = 4 + 5 + 6 + 4 + 3 = 22$$

ΔΔ

The index system must be used whenever only part of the available information is to be used. In statistics, however, we will usually use all the available information, and to simplify the formulas we will make an adjustment. This adjustment is actually an agreement that allows us to do away with the index system in situations where all values are used. Thus in our previous illustration, $\sum_{i=1}^{10} x_i$ could have been written simply as $\sum x$.

NOTE The lack of the index indicates that all data are being used.

∇ ILLUSTRATION A-2

Given the following six values for x, 1, 3, 7, 2, 4, 5, find $\sum x$.

Solution

$$\sum x = 1 + 3 + 7 + 2 + 4 + 5 = 22$$

ΔΔ

Throughout the study and use of statistics you will find many formulas that use the $\sum$ symbol. Care must be taken so that the formulas are not misread. Symbols like $\sum x^2$ and $(\sum x)^2$ are quite different. $\sum x^2$ means "square each *x* value and then add up the squares," while $(\sum x)^2$ means "sum the *x* values and then square the sum."

∇ ILLUSTRATION A-3

Find (a) $\sum x^2$ and (b) $(\sum x)^2$ for the sample in Illustration A-2.

Solution

a.

x	1	3	7	2	4	5
x^2	1	9	49	4	16	25

$$\sum x^2 = 1 + 9 + 49 + 4 + 16 + 25 = 104$$

b. $\sum x = 22$, as found in Illustration A-2.

Thus,

$$\left(\sum x\right)^2 = (22)^2 = 484$$

As you can see, there is quite a difference between $\sum x^2$ and $(\sum x)^2$.

ΔΔ

Likewise, $\sum xy$ and $\sum x \sum y$ are different. These forms will appear only when there are paired data, as shown in the following illustration.

∇ ILLUSTRATION A-4

Given the five pairs of data shown in the following table, find (a) $\sum xy$ and (b) $\sum x \sum y$.

x	1	6	9	3	4
y	7	8	2	5	10

Solution

a. $\sum xy$ means to sum the products of the corresponding *x* and *y* values. Therefore, we have

x	1	6	9	3	4
y	7	8	2	5	10
xy	7	48	18	15	40

$$\sum xy = 7 + 48 + 18 + 15 + 40 = 128$$

b. $\sum x \sum y$ means the product of the two summations, $\sum x$ and $\sum y$. Therefore, we have

$$\sum x = 1 + 6 + 9 + 3 + 4 = 23$$
$$\sum y = 7 + 8 + 2 + 5 + 10 = 32$$
$$\sum x \sum y = (23)(32) = 736$$

ΔΔ

There are three basic rules for algebraic manipulation of the $\sum$ notation.

NOTE *c* represents any constant value.

RULE 1: $\sum_{i=1}^{n} c = nc$

To prove this rule, we need only write down the meaning of $\sum_{i=1}^{n} c$:

$$\sum_{i=1}^{n} c = \underbrace{c + c + c + \ldots + c}_{n \text{ addends}}$$

Therefore, $\sum_{i=1}^{n} c = n \cdot c$

∇ ILLUSTRATION A-5

Show that $\sum_{i=1}^{5} 4 = (5)(4) = 20$.

Solution

$$\sum_{i=1}^{5} 4 = \underbrace{4(\text{when } i=1) + 4(\text{when } i=2) + 4(i=3) + 4(i=4) + 4(i=5)}_{\text{five 4s added together}}$$

$= (5)(4) = 20$

ΔΔ

RULE 2: $\sum_{i=1}^{n} cx_i = c \cdot \sum_{i=1}^{n} x_i$

To demonstrate the truth of Rule 2, we will need to expand the term $\sum_{i=1}^{n} cx_i$, and then factor our the common term *c*.

$$\sum_{i=1}^{n} cx_i = cx_1 + cx_2 + cx_3 + \cdots + cx_n$$

$$= c(x_1 + x_2 + x_3 + \cdots + x_n)$$

Therefore,

$$\sum_{i=1}^{n} cx_i = c \cdot \sum_{i=1}^{n} x_i$$

RULE 3: $$\sum_{i=1}^{n} (x_i + y_i) = \sum_{i=1}^{n} x_i + \sum_{i=1}^{n} y_i$$

The expansion and regrouping of $\sum_{i=1}^{n} (x_i + y_i)$ is all that is needed to show this rule.

$$\sum_{i=1}^{n} (x_i + y_i) = (x_1 + y_1) + (x_2 + y_2) + \cdots + (x_n + y_n)$$

$$= (x_1 + x_2 + \cdots + x_n) + (y_1 + y_2 + \cdots + y_n)$$

Therefore,

$$\sum_{i=1}^{n} (x_i + y_i) = \sum_{i=1}^{n} x_i + \sum_{i=1}^{n} y_i$$

∇ ILLUSTRATION A-6

Show that $\sum_{i=1}^{3} (2x_i + 6) = 2 \cdot \sum_{i=1}^{3} x_i + 18$.

Solution

$$\sum_{i=1}^{3}(2x_i + 6) = (2x_1 + 6) + (2x_2 + 6) + (2x_3 + 6)$$

$$= (2x_1 + 2x_2 + 2x_3) + (6 + 6 + 6)$$
$$= (2)(x_1 + x_2 + x_3) + (3)(6)$$
$$= 2\sum_{i=1}^{3} x_i + 18$$

ΔΔ

∇ ILLUSTRATION A-7

Let $x_1 = 2$, $x_2 = 4$, $x_3 = 6$, $f_1 = 3$, $f_2 = 4$, and $f_3 = 2$. Find $\sum_{i=1}^{3} x_i \cdot \sum_{i=1}^{3} f_i$.

Solution

$$\sum_{i=1}^{3} x_i \cdot \sum_{i=1}^{3} f_i = (x_1 + x_2 + x_3) \cdot (f_1 + f_2 + f_3)$$
$$= (2 + 4 + 6) \cdot (3 + 4 + 2)$$
$$= (12)(9) = 108$$

ΔΔ

∇ ILLUSTRATION A-8

Using the same values for the x's and f's as in Illustration A-7, find $\sum(xf)$.

Solution Recall that the use of no index numbers means "use all data."

$$\sum(xf) = \sum_{i=1}^{3}(x_i f_i) = (x_1 f_1) + (x_2 f_2) + (x_3 f_3)$$

$$= (2 \cdot 3) + (4 \cdot 4) + (6 \cdot 2) = 6 + 16 + 12 = 34$$

ΔΔ

∇Δ EXERCISES

A.1 Write each of the following in expanded form (without the summation sign):

a. $\sum_{i=1}^{4} x_i$ b. $\sum_{i=1}^{3} (x_i)^2$

c. $\sum_{i=1}^{5} (x_i + y_i)$ d. $\sum_{i=1}^{5} (x_i + 4)$

e. $\sum_{i=1}^{8} x_i y_i$ f. $\sum_{i=1}^{4} x_i^2 f_i$

A.2 Write each of the following expressions as summations, showing the subscripts and the limits of summation:

a. $x_1 + x_2 + x_3 + x_4 + x_5 + x_6$

b. $x_1 y_1 + x_2 y_2 + x_3 y_3 + \cdots x_7 y_7$

c. $x_1^2 + x_2^2 + \cdots x_9^2$

d. $(x_1 - 3) + (x_2 - 3) \cdots + (x_n - 3)$

A.3 Show each of the following to be true:

a. $\sum_{i=1}^{4}(5x_i + 6) = 5 \cdot \sum_{i=1}^{4} x_i + 24$

b. $\sum_{i=1}^{n}(x_i - y_i) = \sum_{i=1}^{n} x_i - \sum_{i=1}^{n} y_i$

A.4 Given $x_1 = 2$, $x_2 = 7$, $x_3 = -3$, $x_4 = 2$, $x_5 = -1$, and $x_6 = 1$, find each of the following:

a. $\sum_{i=1}^{6} x_i$ b. $\sum_{i=1}^{6} x_i^2$

c. $\left(\sum_{i=1}^{6} x_i\right)^2$

A.5 Given $x_1 = 4$, $x_2 = -1$, $x_3 = 5$, $f_1 = 4$, $f_2 = 6$, $f_3 = 2$, $y_1 = -3$, $y_2 = 5$, and $y_3 = 2$, find each of the following:

a. $\sum x$ b. $\sum y$ c. $\sum f$

d. $\sum (x - y)$

e. $\sum x^2$ f. $\left(\sum x\right)^2$ g. $\sum xy$

h. $\sum x \cdot \sum y$

i. $\sum xf$ j. $\sum x^2 f$ k. $\left(\sum xf\right)^2$

A.6 Suppose that you take out a $12,000 small-business loan. The terms of the loan are that each month for 10 years (120 months) you will pay back $100 plus accrued interest. The accrued interest is calculated by multiplying 0.005 (6 percent/12) times the amount of the loan still outstanding.

That is, the first month you pay $\$12,000 \times 0.005$ in accrued interest, the second month $(\$12{,}000 - 100) \times 0.005$ in interest, the third month $[\$12{,}000 - (2)(100)] \times 0.005$, and so forth. Express the total amount of interest paid over the life of the loan by using summation notation.

The answers to these exercises can be found in the back of the manual.

USING THE RANDOM NUMBER TABLE

The random number table is a collection of random digits. The term *random* means that each of the 10 digits (0,1,2,3,...,9) has an equal chance of occurrence. The digits, in Table 1 (Appendix B, JES2-p515) can be thought of as single-digit numbers (0-9), as two-digit numbers (00-99), as three-digit numbers (000-999), or as numbers of any desired size. The digits presented in Table 1 are arranged in pairs and grouped into blocks of five rows and five columns. This format is used for convenience. Tables in other books may be arranged differently.

Random numbers are used primarily for one of two reasons: (1) to identify the source element of a population (the source of data) or (2) to simulate an experiment.

∇ ILLUSTRATION 1

A simple random sample of 10 people is to be drawn from a population of 7564 people. Each person will be assigned a number, using the numbers from 0001 to 7564. We will view Table 1 as a collection of four-digit numbers (two columns used together), where the numbers 0001, 0002, 0003, ..., 7564 identify the 7564 people. The numbers 0000, 7565, 7566, ..., 9999 represent no one in our population; that is, they will be discarded if selected.

Now we are ready to select our 10 people. Turn to Table 1 (Appendix B, ES8-p719). We need to select a starting point and a path to be followed. Perhaps the most common way to locate a starting point is to look away and arbitrarily point to a starting point. The number we located this way was 3909 on page 516. (It is located in the upper left corner of the block that is in the fourth large block from the left and the second large block down.) From here we will proceed down the column, then go to the top of the next set of columns, if necessary. The person identified by number 3909 is the first source of data selected. Proceeding down the column, we find

8869 next. This number is discarded. The number 2501 is next. Therefore, the person identified by 2501 is the second source of data to be selected. Continuing down this column, our sample will be obtained from those people identified by the numbers 3909, 2501, 7485, 0545, 5252, 5612, 0997, 3230, 1051, 2712. (The numbers 8869, 8338, and 9187 were discarded.)

ΔΔ

∇ ILLUSTRATION 2

Let's use the random number table and simulate 100 tosses of a coin. The simulation is accomplished by assigning numbers to each of the possible outcomes of a particular experiment. The assignment must be done in such a way as to preserve the probabilities. Perhaps the simplest way to make the assignment for the coin toss is to let the even digits (0,2,4,6,8) represent heads and the odd digits (1,3,5,7,9) represent tails. The correct probabilities are maintained: $P(H) = P(0,2,4,6,8) = \frac{5}{10} = 0.5$ and $P(T) = P(1,3,5,7,9) = \frac{5}{10} = 0.5$. Once this assignment is complete, we are ready to obtain our sample.

Since the question asked for 100 tosses and there are 50 digits to a "block" in Table 1 (Appendix B, ES8-p719), let's select two blocks as our sample of random one-digit numbers (instead of a column 100 lines long). Let's look away and point to one block on p. 719 and then do the same to select one block from p. 720. We picked the second block down in the first column of blocks on p. 719 (24 even and 26 odd numbers) and the second block down in the third column of blocks on p. 720 (23 even and 27 odd numbers). Thus we obtain a sample of 47 heads and 53 tails for our 100 simulated tosses.

ΔΔ

There are, of course, many ways to use the random number table. You must use your good sense in assigning the numbers to be used and in choosing the path to be followed through the table. One bit of advice is to make the assignments in as simple and easy a method as possible to avoid errors.

∇Δ EXERCISES

1. A random sample of size 8 is to be selected from a population that contains 75 elements. Describe how the random sample of the 8 objects could be made with the aid of the random number table.

2. A coin-tossing experiment is to be simulated. Two coins are to be tossed simultaneously and the number of heads appearing is to be recorded for each toss. Ten such tosses are to be observed. Describe two ways to use the random number table to simulate this experiment.

3. Simulate five rolls of three dice by using the random number table.

The answers to these exercises can be found in the back of the manual.

ROUND-OFF PROCEDURE

When rounding off a number, we use the following procedure.

STEP 1 Identify the position where the round-off is to occur. This is shown by using a vertical line that separates the part of the number to be kept from the part to be discarded. For example,

125.267 to the nearest tenth is written as 125.2|67
7.8890 to the nearest hundredth is written as 7.88|90

STEP 2 Step 1 has separated all numbers into one of four cases. (*X*'s will be used as placeholders for number values in front of the vertical line. These *X*'s can represent any number value.)

Case I: *XXXX*|000...
Case II: *XXXX*|---(any value from 000...1 to 499 ...9)
Case III: *XXXX*|5000...0
Case IV: *XXXX*|---(any value from 5000...1 to 999...9)

STEP 3 Perform the rounding off.

Case I requires no round-off. It's exactly *XXXX*.

∇ ILLUSTRATION 1

Round 3.5000 to the nearest tenth.
3.5|000 becomes 3.5

ΔΔ

Case II requires rounding. We will round down for this case. That is, just drop the part of the number that is behind the vertical line.

∇ ILLUSTRATION 2

Round 37.6124 to the nearest hundredth.
37.61|24 becomes 37.61

ΔΔ

Case III requires rounding. This is the case that requires special attention. **When a 5 (exactly a 5) is to be rounded off, round to the even digit.** In the long run, half of the time the 5 will be preceded by an even digit (0,2,4,6,8) and you will round down, while the other half of the time the 5 will be preceded by an odd digit (1,3,5,7,9) and you will round up.

∇ ILLUSTRATION 3

Round 87.35 to the nearest tenth.
87.3|5 becomes 87.4

Round 93.445 to the nearest hundredth.
93.44|5 becomes 93.44

(**Note:** 87.35 is 87.35000... and 93.445 is 93.445000...)

ΔΔ

Case IV requires rounding. We will round up for this case. That is, we will drop the part of the number that is behind the vertical line and we will increase the last digit in front of the vertical line by one.

∇ ILLUSTRATION 4

Round 7.889 to the nearest tenth.

7.8|89 becomes 7.9

ΔΔ

NOTE **Case I, II, and IV describe what is commonly done. Our guidelines for Case III are the only ones that are different from typical procedure.**

If the typical round-off rule (0,1,2,3,4 are dropped; 5,6,7,8,9 are rounded up) is followed, then $(n + 1)/(2n + 1)$ of the situations are rounded up. (n is the number of different sequences of digits that fall into each of Case II and Case IV.) That is more than half. You (as many others have) may say, "So what?" In today's world that tiny, seemingly insignificant amount becomes very significant when applied repeatedly to large numbers.

∇Δ EXERCISES

1. Round each of the following to the nearest integer:

   a. 12.94 b. 8.762
   c. 9.05 d. 156.49
   e. 45.5 f. 42.5
   g. 102.51 h. 16.5001

2. Round each of the following to the nearest tenth:
   a. 8.67 b. 42.333
c. 49.666 d. 10.25
   e. 10.35 f. 8.4501
   g. 27.35001 h. 5.65
   i. 3.05 j. $\frac{1}{4}$

3. Round each of the following to the nearest hundredth:

a. 17.6666 b. 4.444
c. 54.5454 d. 102.055
e. 93.225 f. 18.005
g. 18.015 h. 5.555
i. 44.7450 j. $\frac{2}{3}$

The answers to these exercises can be found in the back of the manual.

REVIEW LESSONS

THE COORDINATE-AXIS SYSTEM AND THE EQUATION OF A STRAIGHT LINE

The rectangular coordinate-axis system is a graphic representation of points. Each point represents an ordered pair of values. (Ordered means that when values are paired, one value is always listed first, the other second.) The pair of values represents a horizontal location (the x-value, called the abscissa) and a vertical location (the y-value, the ordinate) in a fixed reference system. This reference system is a pair of perpendicular real number lines whose point of intersection is the 0 of each line (Figure 1-1, below).

Any point (x, y) is located by finding the point that satisfies both positional values. For example, the point $P(2,3)$ is exactly 2 units to the right of 0 along the horizontal axis and 3 units above 0 along the vertical axis. Figure 1-2 (next page) shows two lines, A and B. Line A represents all the points that are 2 units to the right of 0 along the x- (horizontal) axis. Line B represents all the points that are 3 units above the 0 along the y-(vertical) axis. Point P is the one point that satisfies both conditions. Typically we think of locating point P by moving along the x-axis 2 units in the positive direction and then moving parallel to the y-axis 3 units in the positive direction (Figure 1-3, next page).

If either value is negative, we just move in a negative direction a distance equal to the number value.

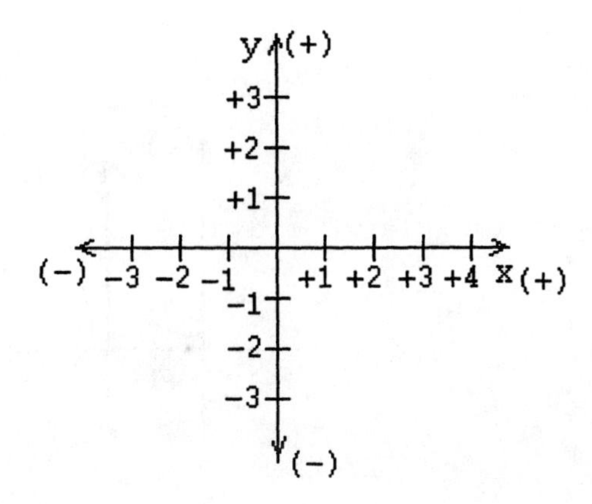

Figure 1-1

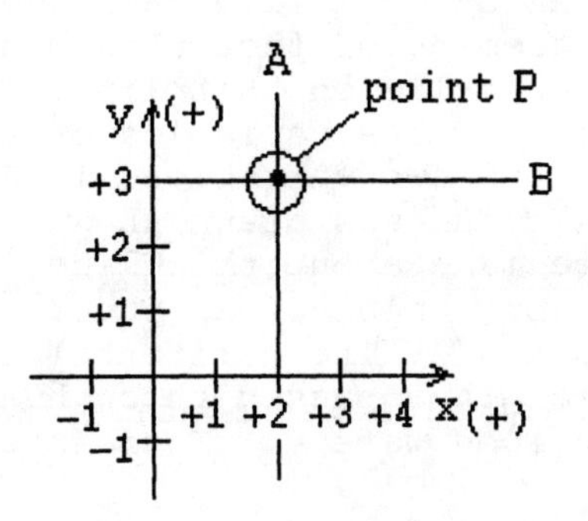

Figure 1-2

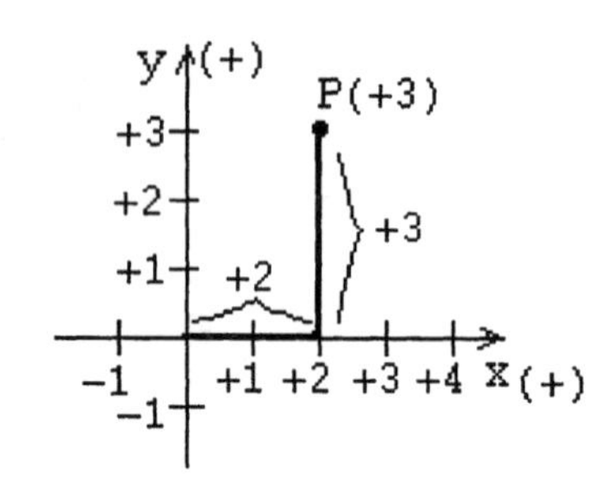

Figure 1-3

∇Δ EXERCISES

1. On a rectangular coordinate axis, drawn on graph paper, locate the following points.

   A(5,2) B(-5,2) C(-3,-2)
   D(3,0) E(0,-2) F(-2,5)

The equation of a line on a coordinate-axis system is a statement of fact about the coordinates of all points that lie on that line. This statement may be about one of the variables or about the relationship between the two variables. In Figure 1-2 above, a vertical line was drawn at $x = +2$. A statement that could be made about this line is that every point on it has an x-value of 2. Thus, the equation of this line is $x = 2$. The horizontal line that was drawn on the same graph passed through all the points where the y-values were +3. Therefore the equation of this line is $y = +3$.

All vertical lines will have an equation of $x = a$, where a is the value of the abscissa of every point on that line. Likewise, all horizontal lines will have an equation of $y = b$, where b is the ordinate of every point on that line.

In statistics, straight lines that are neither vertical nor horizontal are of greater interest. Such a line will have an equation that expresses the relationship between the two variables x and y. For example, it might be that y is always one less than the double of x; this would be expressed equationally as $y = 2x - 1$. There are an unlimited number of ordered pairs that fit this relationship; to name a few: (0,-1), (1,1), (2,3), (-2,-5), (1.5,2), (2.13,3.26) (see Figure 1-4).

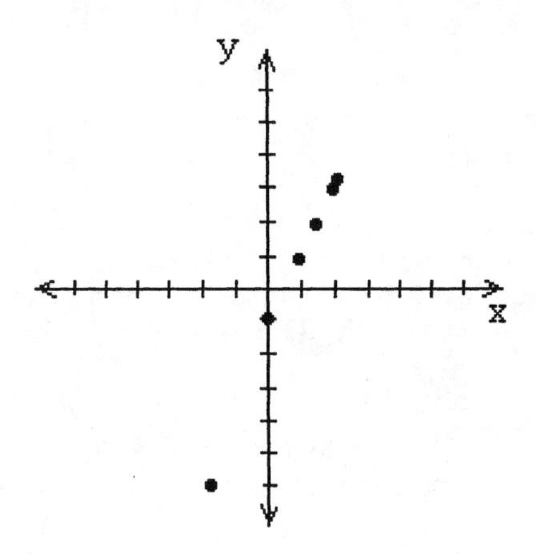

Figure 1-4

Notice that these points fall on a straight line. Many more points could be named that also fall on this same straight line; in fact, all the pairs of values that satisfy $y = 2x - 1$ lie somewhere along it. The converse is also true: all the points that lie on this straight line have coordinates that make the equation $y = 2x - 1$ a true statement. When drawing the line that represents the equational relationship between the coordinates of points, one needs to find only two points of the straight line; however, it is often useful to locate three of four to ensure accuracy.

2. a. Find the missing values in the accompanying chart of ordered pairs, where $x + y = 5$ is the relationship.

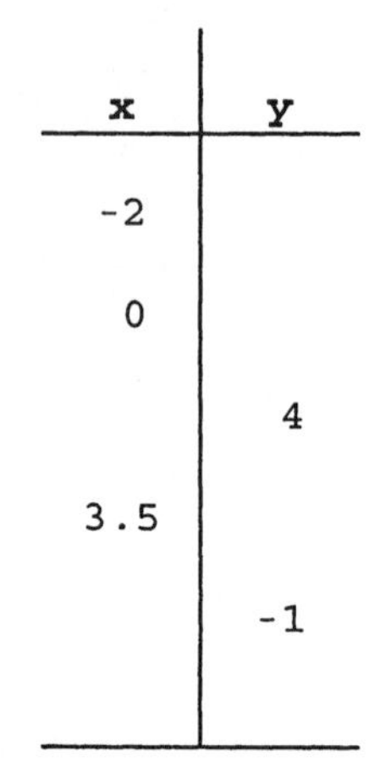

x	y
-2	
0	
	4
3.5	
	-1

b. On a coordinate axis, locate the same five points and then draw a straight line that passes through all of them.

3. Find five points that belong to the relationship expressed by $y = (3x/2) - 4$; then locate them on the axis system and draw the line representing $y = (3x/2) - 4$.

The form of the equation of a straight line that we are interested in is called the slope-intercept form. Typically in mathematics this slope-intercept form is expressed by $y = mx + b$, where m represents the concept of *slope* and b is the *y-intercept*.

Let's look at the y-intercept first. If you will look back at Exercise 3, you will see that the y-intercept for $y = (3x/2) - 4[y = (3x/2) + (-4)]$ is -4. This value is simply the value of y at the point where the graph of the line intersects the y-axis, and all nonvertical lines will have this property. The value of the y-intercept may be found on the graph or from the equation. From the graph it is as simple as identifying it, but we will need to have the equation solved for y in order to identify it from the equation. For example, in Exercise 2 we had the equation $x + y = 5$; if we solve for y, we have $y = -x + 5$, and the y-intercept is 5 (the same as is found on the graph for Exercise 2).

The *slope* of a straight line is a measure of its inclination. This measure of inclination can be defined as the amount of vertical change that takes place as the value of x increases by exactly one unit. This amount of change may be found anywhere on the line since this value is the same everywhere on a given straight line. If we inspect the graph drawn for Exercise 2(Figure 1-5), we will see that the slope is −1, meaning that each x increase of 1 unit results in a decrease of 1 unit in the y-value.
Figure 1-5

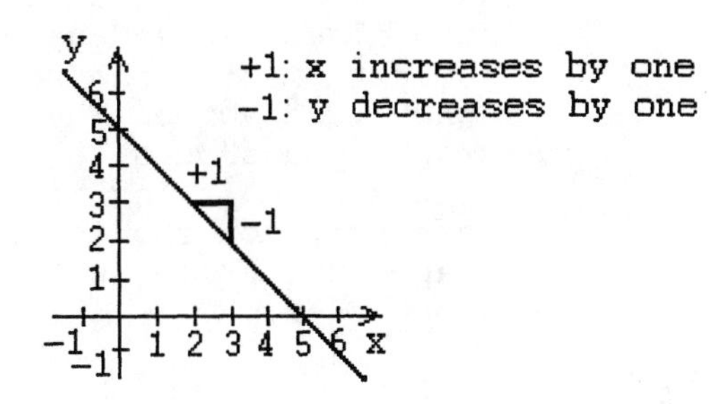

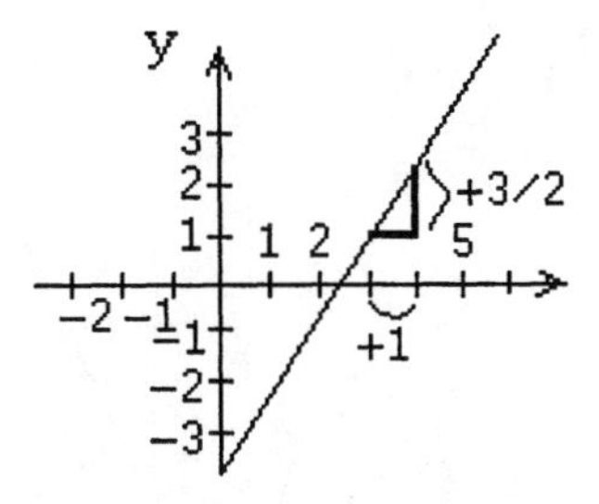

Figure 1-6

An inspection of the graph drawn for Exercise 3 (Figure 1-6) will reveal a slope of +3/2. This means that y increased by 3/2 for every increase of one unit in x.

Illustration: Find the slope of the straight line that passes through the points $(-1,1)$ and $(4,11)$.

Solution: $m = \dfrac{\Delta y}{\Delta x} = \dfrac{11 - 1}{4 - (-1)} = \dfrac{10}{5} = 2$

NOTES:

1. $\Delta y = y_2 - y_1$ and $\Delta x = x_2 - x_1$ where (x_1, y_1) and (x_2, y_2) are the two points that the line passes through.
2. As stated before, these properties are both algebraic and graphic. If you know about them from one source, then the other must agree. Thus, if you have the graph, you should be able to read these values from it, or if you have the equation, you should be able to draw the graph of the line with these given properties.

ILLUSTRATION: Graph $y = 2x + 1$.

SOLUTION: $m = 2$ and $b = 1$. Locate the y-intercept $(y = +1)$ on the y-axis. Then draw a line that has a slope of 2 (Figure 1-7).

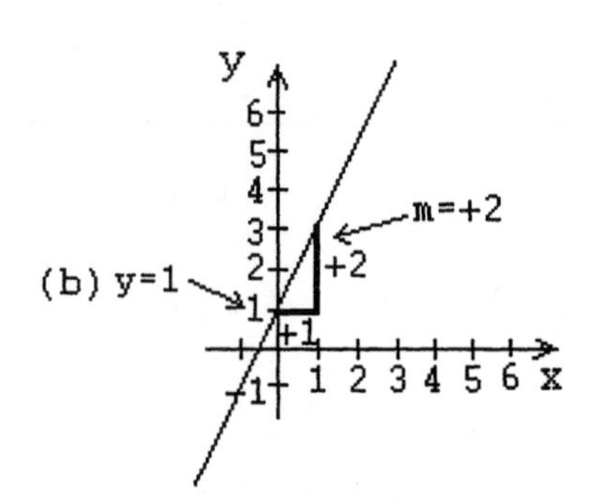

Figure 1-7

Illustration: Find the equation of the line that passes through $(-1,1)$ and $(2,7)$.

Solution: (Algebraically)

$$m = \frac{\Delta y}{\Delta x} = \frac{7-1}{2-(-1)} = \frac{6}{3} = 2$$

$y = 2x + b$, and the line passes through $(2,7)$. Therefore $x = 2$ and $y = 7$ must satisfy (make the statement true) $y = 2x + b$. In order for that to happen, b must be equal to $3[7 = 2(2) + b]$. Therefore the equation of such a line is $y = 2x + 3$.

(Graphically) Draw a graph of a straight line that passes through $(-1,1)$ and $(2,7)$; then read m and b from it (Figure 1-8).

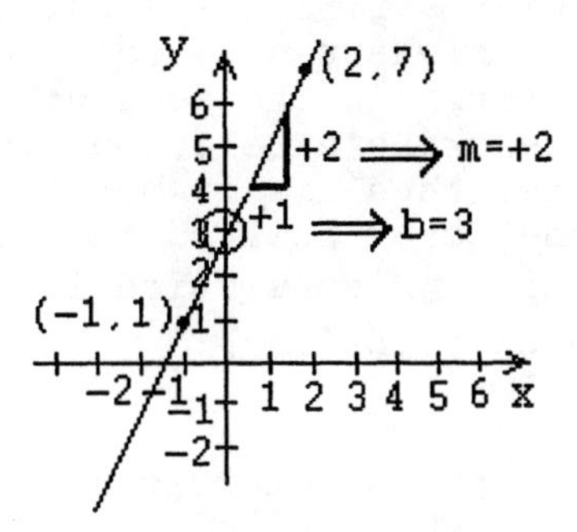

Figure 1-8

Therefore $y = mx + b$ becomes $y = 2x + 3$.

4. Write the equation of a straight line whose slope is 10 and whose y-intercept is -3.

5. Draw the graph of each of the following equations (use graph paper).
   a. $y = x + 2$
   b. $y = -2x + 10$
   c. $y = (1/3)x - 2$

6. Find the equation of the straight line that passes through each of the following pairs of points (a) algebraically and (b) graphically (use graph paper).
   I. $(3,1)$ and $(9,5)$
   II. $(-2,3)$ and $(6,-1)$

The preceding discussion about the equation of a straight line is presented from a mathematical point of view. Mathematicians and statisticians often approach concepts differently. For instance, the statistician typically places the terms of a linear equation in exactly the opposite order from the mathematician's equation.
To the statistician, for example, the equation of the straight line is $y = b + mx$, while to the mathematician it is $y = mx + b$. m and b represent exactly the same concepts in each case — the different order is a matter of emphasis. The mathematician's first interest in an equation is usually the highest-powered term; thus he or she places it first in the sequence. The statistician tends to describe a relationship in as simple a form as possible; thus his or her first interest is usually the lower-powered terms. The equation of the straight line in statistics is $y = b_0 + b_1x$, where b_0 is the y-intercept and b_1 represents the slope.

The answers to the exercises can be found in the back of the manual.

TREE DIAGRAMS

The purpose of this lesson is to learn how to construct and read a tree diagram. A tree diagram is a drawing that schematically represents the various possible outcomes of an experiment. It is called a tree diagram because of the branch concept that it demonstrates.

Let's consider the experiment of tossing one coin one time. We will start the experiment by tossing the coin and will finish it by observing a result (heads or tails)(see Figure 2-1).

Start **Observation**

Heads

Tails

Figure 2-1

This information is expressed by the tree shown in Figure 2-2.

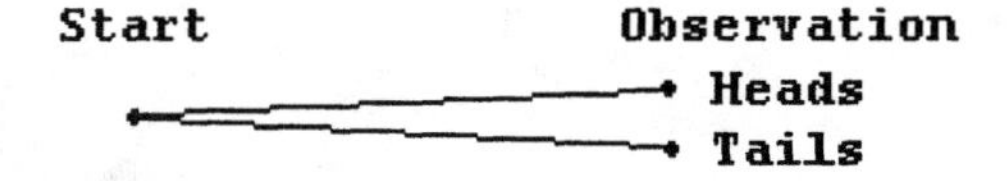

Figure 2-2

In reading a diagram like this, the single point at the left is simply interpreted as we are ready to start and do not yet know the outcome. The branches starting from this point must represent all of the different possibilities. With one coin there are only two possible outcomes, thus two branches.

∇Δ EXERCISES

1. Draw a tree diagram that shows the possible results from rolling a single die once.

2. Draw a tree diagram showing the possible methods of transportation that could be used to travel to a resort area. The possible choices are car, bus, train, and airplane.

Now let's consider the experiment of tossing a coin and single die at the same. What are the various possibilities? The coin can result in a head (H) or a tail (T) and the die could show a 1, 2, 3, 4, 5, or 6. Thus to show all the possible pairs of outcomes, we must decide which to observe first. This is an arbitrary decision, as the order observation does not affect the possible pairs of results. To construct the tree to represent this experiment we list the above-mentioned possibilities in columns, as shown in Figure 2-3.

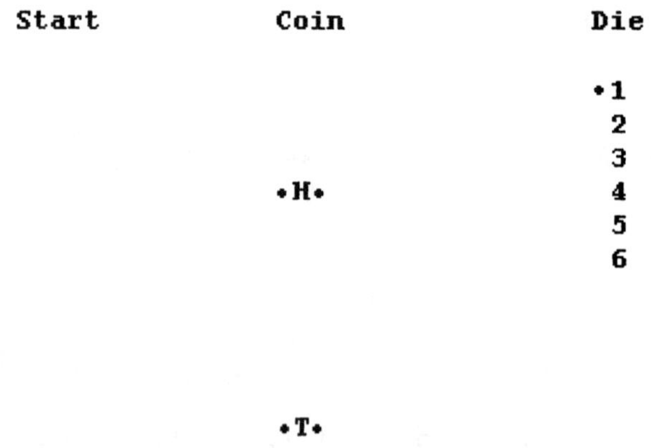

Start	Coin	Die
		•1
		2
		3
	•H•	4
		5
		6
	•T•	

Figure 2-3

From "start" we draw two line segments that represent the possibility of *H* or *T* (Figure 2-4). The top branch means that we might observe a head on the coin. Paired with it is the result of the die, which could be any of the six numbers. We see this represented in Figure 2-5.

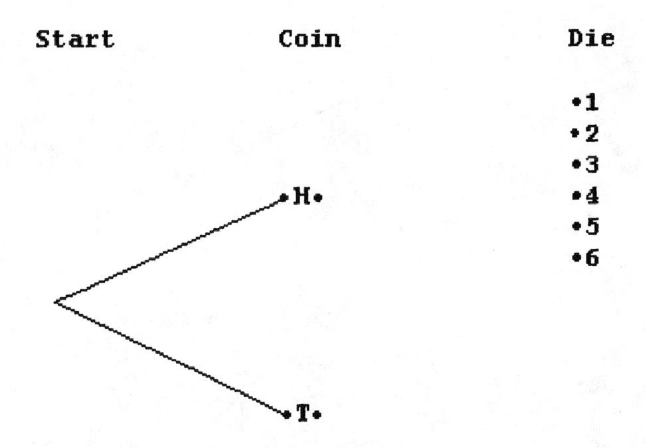

Figure 2-4

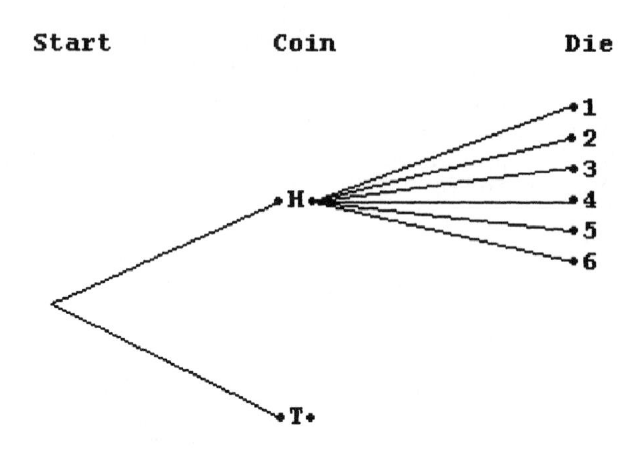

Figure 2-5

However, the outcome could have been T, so T must have branches going to each of the numbers 1 through 6. To make the diagram easier to read we list these outcomes again and draw another set of branches (Figure 2-6).

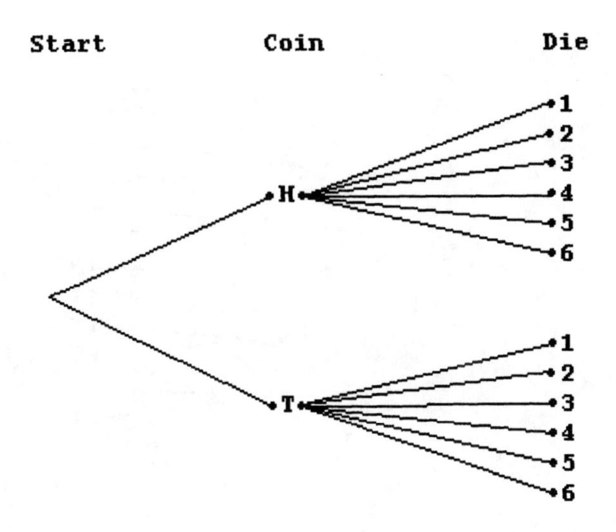

Figure 2-6

Figure 2-6 shows 12 different possible pairs of results — each one of the 12 branches on the tree represents one of these pairs. (A complete branch is a path from the start to an end.) The 12 branches in Figure 2-6, from top to bottom, are H1, H2, H3, H4, H5, H6, T1, T2, T3, T4, T5, T6.

The ordering could have been reversed: the tree diagram in Figure 2-7 shows the die result first and the coin result second.

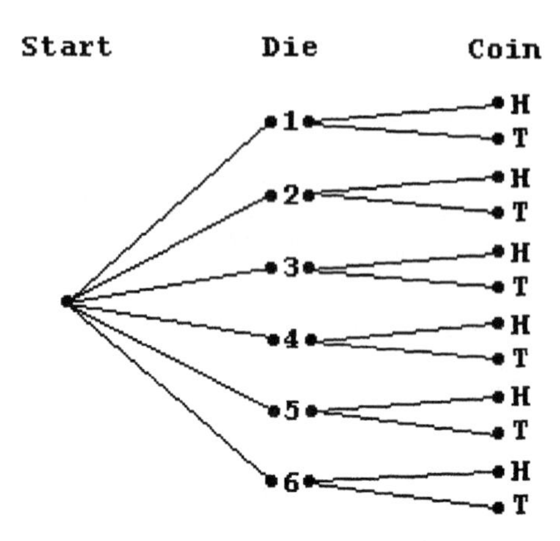

Figure 2-7

Is there any difference in the listing of the 12 possible results? Only the order of observation and the vertical order of the possibilities as shown on the diagram have changed. They are still the same 12 pairs of possibilities. If the experiment contains more than two stages of possible events we may expand this tree as far as needed.

EXERCISES

3. a. Draw a tree diagram that represents the possible results from tossing two coins.
   b. Repeat part (a) considering one coin a nickel and the other a penny.
   c. Is the list of possibilities for part (b) any different than it was for part (a)?

4. Draw a tree diagram representing the tossing of the coins.

5. a. Draw a tree diagram representing the possible results that could be obtained when two dice are rolled.
   b. How many branch ends does your tree have?

On occasion the stages of an experiment will be ordered; when this is the case the tree diagram must show ordered sets of branches, as in Figure 2-8, next page. The experiment consists of rolling a die. Then the result of the die will dictate your next trial. If an odd number results, you will toss coin. If a two or a six occurs, you stop. If any other number (a four) occurs, you roll the die again. Notice that the tree diagram becomes a very convenient "road map" showing all the various possibilities that may occur in an experiment of this nature. Remember that an event is represented by a complete branch (a broken line from the start to an end), and the number of ends of branches is the same as the number of possibilities for the experiment. There are 14 branches in the tree diagram in Figure 2-8. Do you agree?

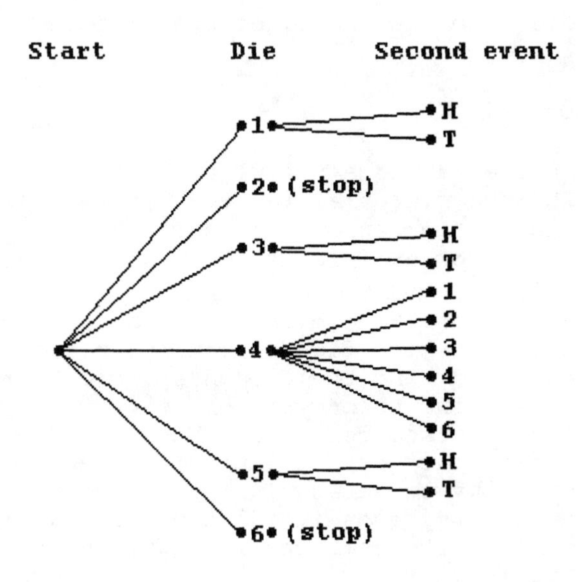

Figure 2-8

6. Students at our college are to be classified as male or female, graduates of public or private high schools, and by the type of curriculum they are enrolled in, liberal arts or career. Draw a tree diagram which shows all of the various possible classifications.

7. There are two scenic routes (A and B) as well as one business route (C) by which you may travel from your home to a nearby city. You are planning to drive to that city by way of one route and come home by a different route.
   a. Draw a tree diagram representing all of your possible choices for going and returning.
   b. How many different trips could you plan?
   c. How many of these trips are scenic in both directions?

The answers to the exercises can be found in the back of the manual.

VENN DIAGRAMS

The Venn diagram is a useful tool for representing sets. It is a pictorial representation that uses geometric configurations to represent *set containers*. For example, a set might be represented by a circle — the circle acts like a "fence" and encloses all of the elements that belong to that particular set. The figure drawn to represent a set must be closed, and the elements are either inside the boundary and belong to that set or they are outside and do not belong to that set. The universal set (sample space or population) is generally represented by a rectangular area, and its subsets are generally circles inside the rectangle. Complements, intersections, and unions of sets then become regions of various shapes as prescribed by the situation. The Venn diagram in Figure 3-1 shows a universal set and a subset *P*.

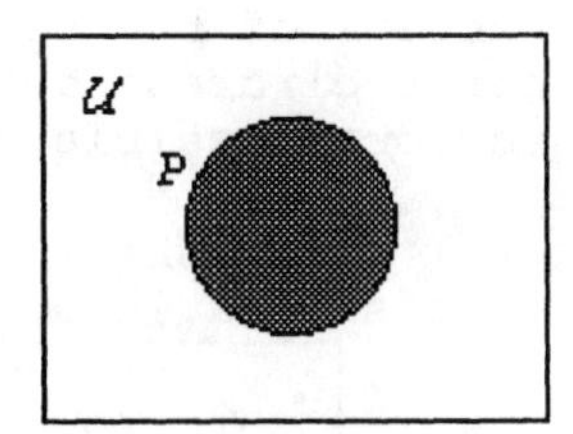

Figure 3-1

Any element that is represented by a point inside the rectangle is an element of the universal set. Likewise, any element represented by a point inside the circle, P(the shaded area), is a member of set P. The unshaded area of the rectangle then represents $\overline{P}$.

The Venn diagrams in Figure 3-2 show the regions representing $A \cap B$, $A \cup B$, $\left(\overline{A \cap B}\right)$, and $\left(\overline{A \cup B}\right)$. The shaded regions represent the identified sets.

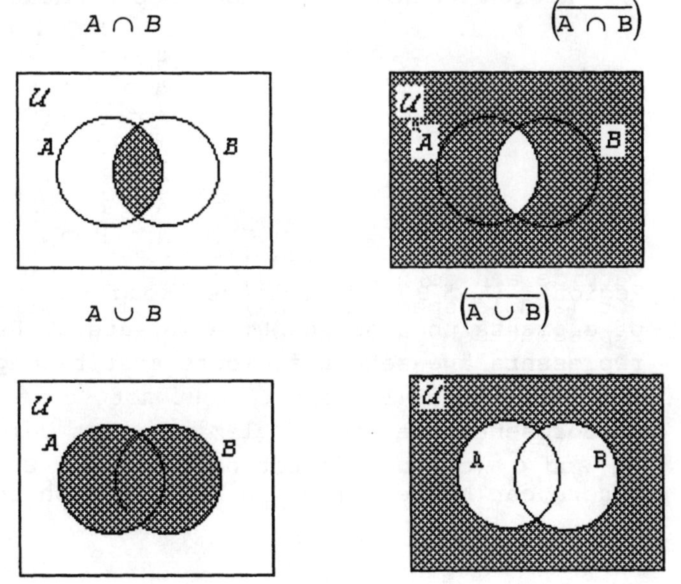

Figure 3-2

When three subsets of the same population are being discussed, three circles can be used to represent all of the various possible situations.

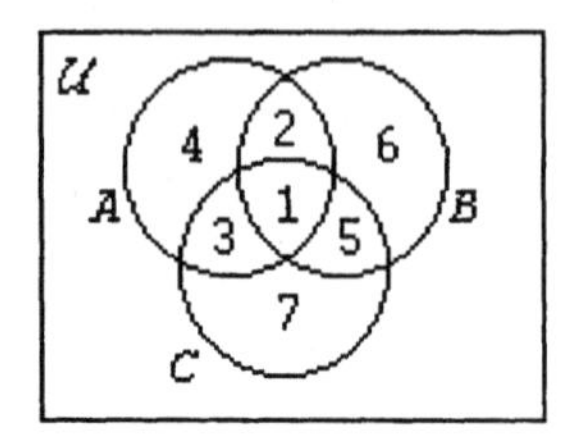

Figure 3-3

In Figure 3-3 the eight regions that are formed by intersecting the three sets have been numbered for convenience. Each of these regions represents the intersection of three sets (sets and/or complements of sets), as shown below.

Region Number	Set Representation
1	$A \cap B \cap C$
2	$A \cap B \cap \overline{C}$
3	$A \cap \overline{B} \cap C$
4	$A \cap \overline{B} \cap \overline{C}$
5	$\overline{A} \cap B \cap C$
6	$\overline{A} \cap B \cap \overline{C}$
7	$\overline{A} \cap \overline{B} \cap C$
8	$\overline{A} \cap \overline{B} \cap \overline{C}$

Region 1 $(A \cap B \cap C)$ might be thought of as the set of elements that belong to A, B, and C. Region 4 represents the set of elements that belong to A, $\overline{B}$ (but not to set B), and $\overline{C}$ (but not to set C). Region 8 represents the set of elements that belong to $\overline{A}$, $\overline{B}$, and $\overline{C}$ (or that do not belong to A, B, or C). The others can be described in similar fashion.

Figure 3-4 shows the union as sets B and C in the shaded areas of all three sets. Notice that $B \cup C$ is composed of regions 1, 2, 3, 5, 6, and 7. (You might note that three of these regions are inside A and three are outside).

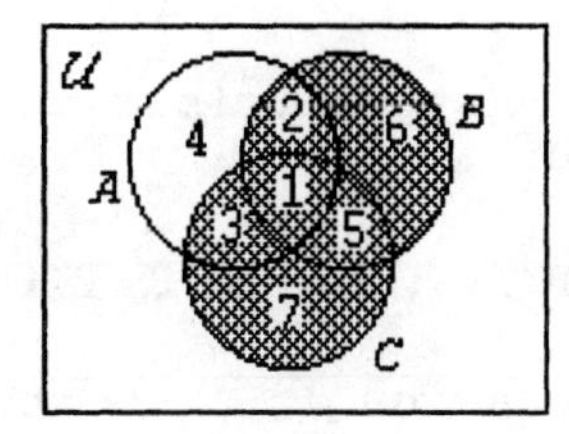

Figure 3-4

∇Δ EXERCISES

1. Shade the regions that represent each of the following sets on a Venn diagram as shown in Figure 3-5.

   a. A b. B c. $A \cap B$

   d. $A \cup B$ e. $\overline{A} \cup B$ f. $\overline{A} \cup \overline{B}$

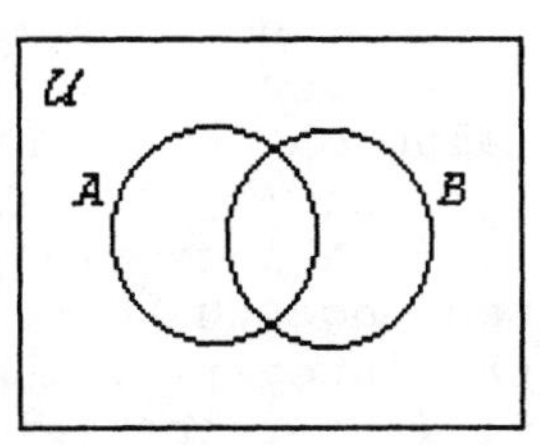

Figure 3-5

2. On a diagram showing three sets, P, Q, and R, shade the regions that represent the following sets.

   a. P b. $P \cap Q$ c. $P \cup R$

   d. $\overline{P}$ e. $P \cap Q \cap R$ f. $P \cup \overline{Q}$

   g. $P \cup Q \cup R$ h. $P \cup Q \cup \overline{R}$

The answers to these exercises can be found in the back of the manual.

THE USE OF FACTORIAL NOTATION

The factorial notation is a shorthand way to identify the product of a particular set of integers. 5! (five factorial) stands for the product of all positive integers starting with the integer 5 and proceeding downward (in value) until the integer 1 is reached. That is, $5! = 5 \times 4 \times 3 \times 2 \times 1$, which is 120. Likewise, $n!$ symbolizes the product of the integer n multiplied by the next smaller integer $(n - 1)$ multiplied by the next smaller integer $(n - 2)$ and so on, until the last integer, the number 1, is reached.

NOTES:

1. The number in front of the factorial symbol (!) will always be a positive integer or 0.
2. The last integer in the sequence is always the integer 1, with one exception: 0! (zero factorial). The value of zero factorial is defined to be 1, that is, $0! = 1$.
   1! (one factorial) is the product of a sequence that starts and ends with the integer 1, thus $1! = 1$.
   2! (two factorial) is the product of 2 and 1. That is, $2! = (2)(1) = 2$.
   $3! = (3)(2)(1) = 6$
   $n! = (n)(n - 1)(n - 2)(n - 3)\ldots(2)(1)$
   $(n - 2)! = (n - 2)(n - 3)(n - 4)\ldots(2)(1)$

$$(4!)(6!) = (4 \cdot 3 \cdot 2 \cdot 1)(6 \cdot 5 \cdot 4 \cdot 3 \cdot 2 \cdot 1) = (24)(720) = 17280$$

$$4(6!) = (4)(6!) = 4(6 \cdot 5 \cdot 4 \cdot 3 \cdot 2 \cdot 1) = (4)(720) = 2880$$

$$\frac{6!}{4!} = \frac{(6)(5)(4)(3)(2)(1)}{(4)(3)(2)(1)} = (6)(5) = 30$$

∇Δ Exercises

Evaluate each of the following factorials.

1. $4!$
2. $6!$
3. $8!$
4. $(6!)(8!)$
5. $\frac{8!}{6!}$
6. $\frac{8!}{4!4!}$
7. $\frac{8!}{6!2!}$
8. $2\frac{8!}{[5!]}$

The answers to these exercises can be found in the back of the manual.

ANSWERS TO INTRODUCTORY CONCEPTS AND REVIEW LESSONS EXERCISES

Summation Notation Exercises

1. (a) $x_1 + x_2 + x_3 + x_4$

   (b) $x_1^2 + x_2^2 + x_3^2$

   (c) $(x_1 + y_1) + (x_2 + y_2) + (x_3 + y_3) + (x_4 + y_4) + (x_5 + y_5)$

   (d) $(x_1 + 4) + (x_2 + 4) + (x_3 + 4) + (x_4 + 4) + (x_5 + 4)$

   (e) $x_1y_1 + x_2y_2 + x_3y_3 + x_4y_4 + x_5y_5 + x_6y_6 + x_7y_7 + x_8y_8$

   (f) $x_1^2f_1 + x_2^2f_2 + x_3^2f_3 + x_4^2f_4$

2. (a) $\sum_{i=1}^{6} x_i$ (b) $\sum_{i=1}^{7} x_iy_i$ (c) $\sum_{i=1}^{9} (x_i)^2$

   (d) $\sum_{i=1}^{n} (x_i - 3)$

4. (a) 8 (b) 68 (c) 64

5. (a) 8 (b) 4 (c) 12 (d) 4
   (e) 42 (f) 64 (g) -7 (h) 32
   (i) 20 (j) 120 (k) 400

6. $\sum_{i=1}^{120} [0.005(12,000 - (i - 1)100)]$

Using the Random Number Table Exercises

2. (a) Use a two-digit number to represent the results obtained. Let the first digit represent one of the coins and the second digit represent the other coin. Let an even digit indicate heads and an odd digit tails. Observe 10 two-digit numbers from the table. If a 16 is observed, it represents tails and heads on two coins. One head was therefore observed. The probabilities have been preserved.

(b) A second way to simulate this experiment is to find the probabilities associated with the various possible results. The number of heads that can be seen on two coins is 0, 1, or 2. (HH,HT,TH,TT is the sample space.) P(no heads) = 1/4; P(one head) = 1/2; P(two heads) = 1/4. Using two-digit numbers, let the numbers 00 to 24 stand for no head appeared, 25 to 74 stand for one head appeared, and 75 to 99 stand for two heads appeared. The probabilities have again been preserved. Observe 10 two-digit numbers.

Round-Off Procedure Exercises

1. (a) 13 (b) 9 (c) 9
   (d) 156 (e) 46 (f) 42
   (g) 103 (h) 17

2. (a) 8.7 (b) 42.3 (c) 49.7
   (d) 10.2 (e) 10.4 (f) 8.5
   (g) 27.4 (h) 5.6 (i) 3.0
   (j) 0.2

3. (a) 17.67 (b) 4.44 (c) 54.55
   (d) 102.06 (e) 93.22 (f) 18.00
   (g) 18.02 (h) 5.56 (i) 44.74
   (j) 0.67

Review Lessons

The Coordinate-Axis System and the Equation of a Straight Line

1.

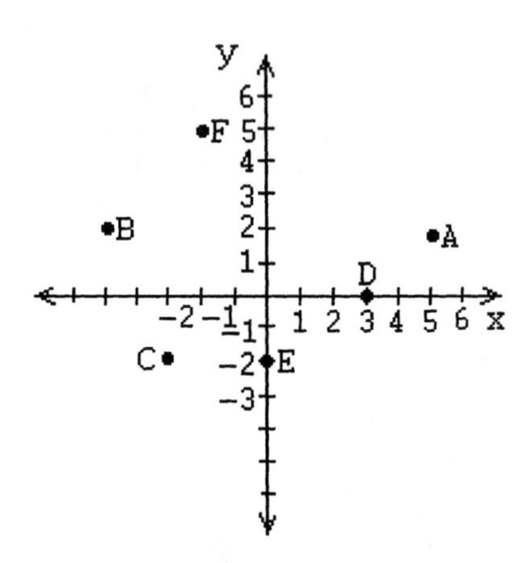

2. a.

Point	*x*	*y*
A	-2	7
B	0	5
C	1	4
D	3.5	1.5
E	6	-1

b.

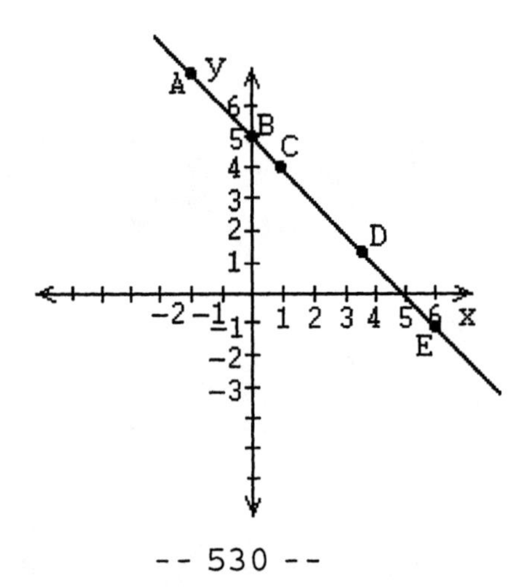

3. Pick any values of x you wish; $x = -2, 0, 1, 2, 4$ will be convenient

$x = -2$	0	1	2	4
$y = -7$	-4	-2.5	-1	2

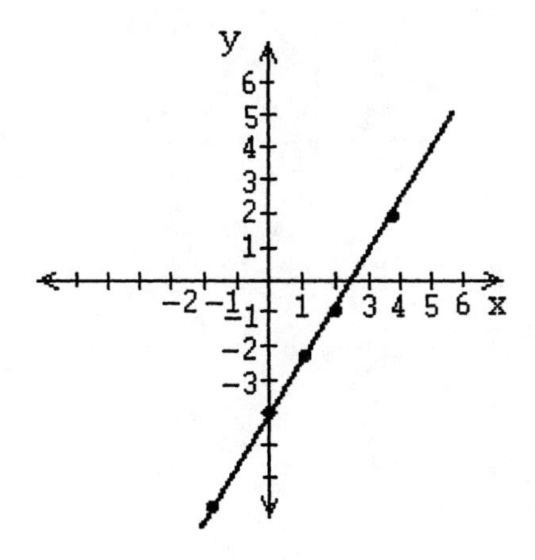

4. $m = 10$, $b = -3$, $y = 10x - 3$

5. a.

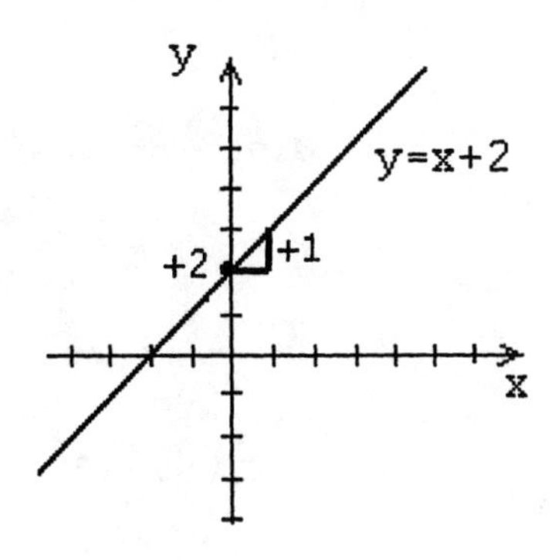

b.

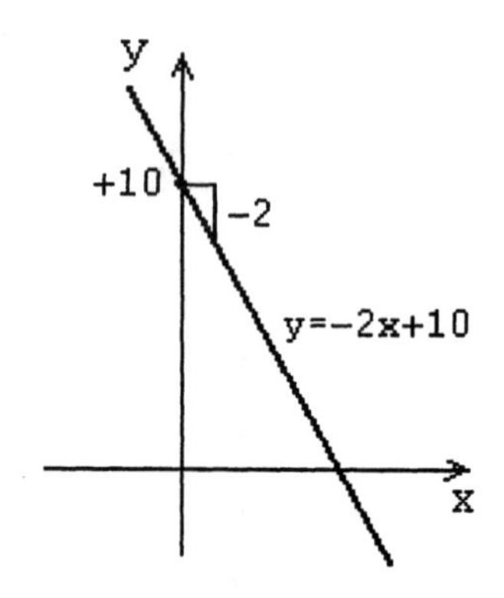

c.

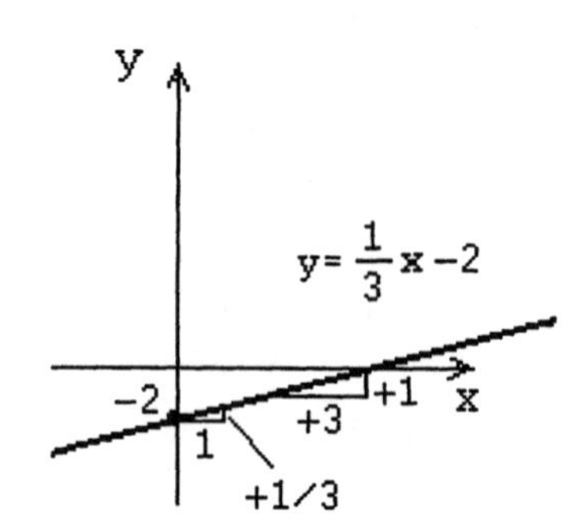

6. I. (a) Given points $(3,1)$ and $(9,5)$.

$$m=\frac{\Delta y}{\Delta x}=\frac{5-1}{9-3}=\frac{4}{6}=\frac{2}{3}$$

$y=\frac{2}{3}x+b$ and passes through $(3,1)$

$1=\frac{2}{3}(3)+b$ implies that $b=-1$

Thus $y=\frac{2}{3}x-1$

I. (b)

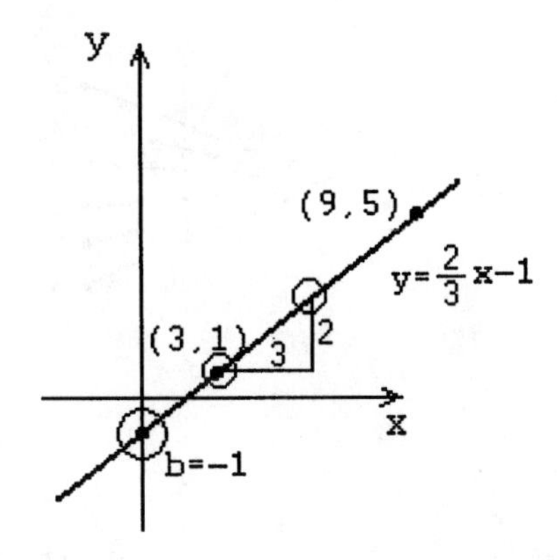

II. (a) Given points $(-2,3)$ and $(6,-1)$.

$$m=\frac{\Delta y}{\Delta x}=\frac{(-1)-(-3)}{6-(-2)}=\frac{-4}{8}=\frac{-1}{2}$$

$y=\frac{-1}{2}x+b$ and passes through $(-2,3)$

$3=\frac{-1}{2}(-2)+b$ implies that $b=2$

Thus $y=\frac{-1}{2}x+2$

(b)

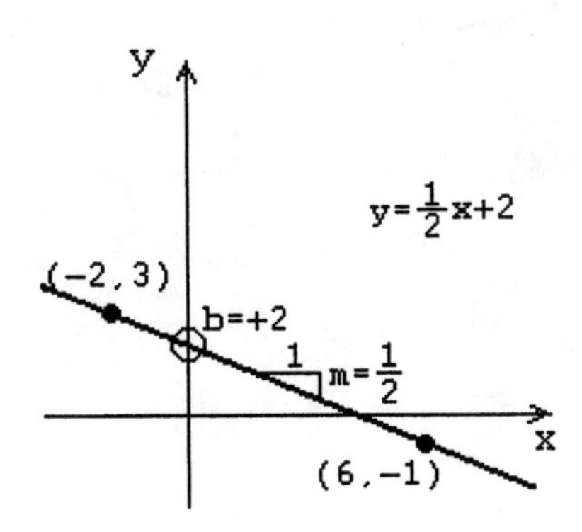

Tree Diagrams

1.

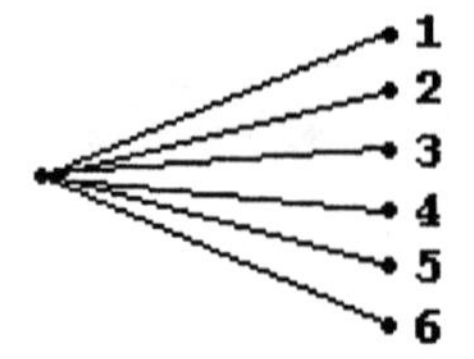

2.

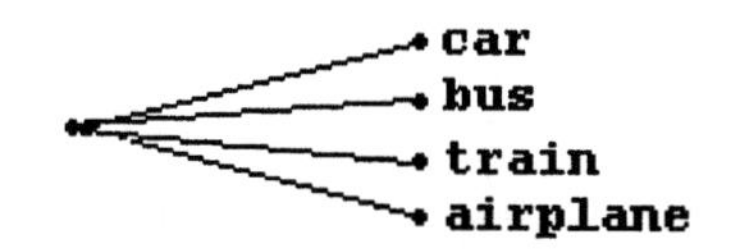

3. a.

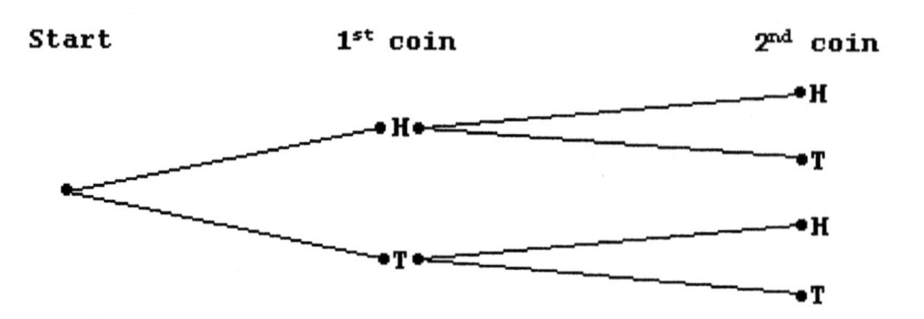

b.

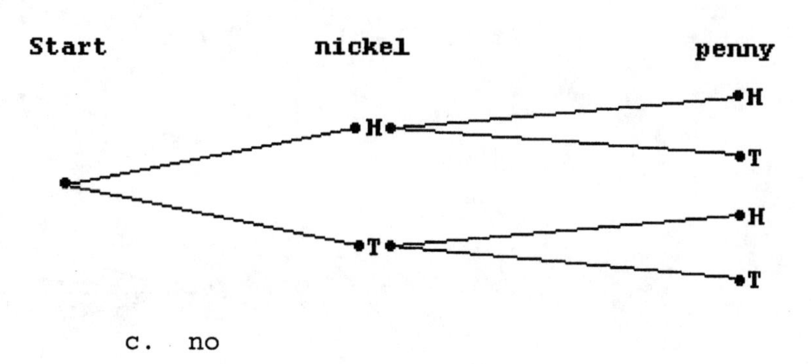

c. no

4.

Start 1st coin 2nd coin 3rd coin

H H H
T
T H
T
T H H
T
T H
T

5. a.

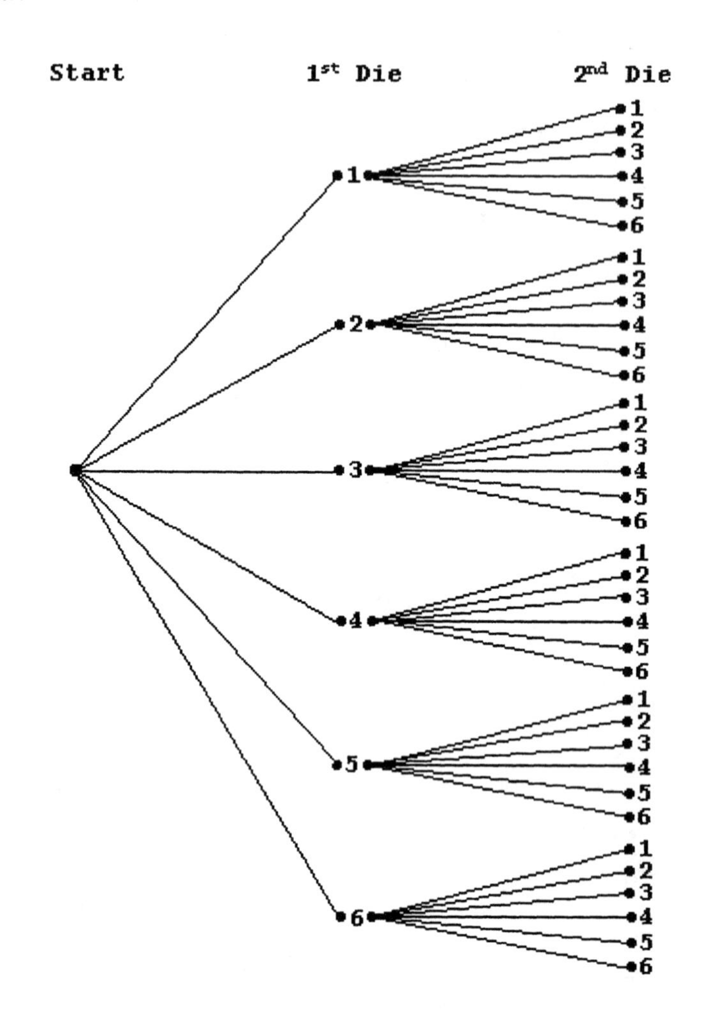

b. 36 branch ends

6.

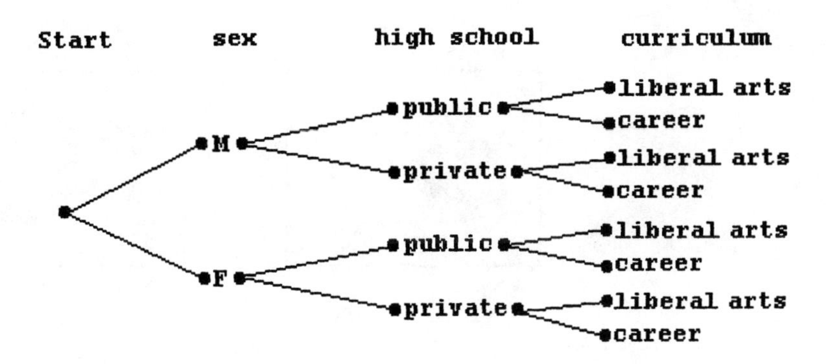
Start
sex
high school
curriculum
M
F
public
private
public
private
liberal arts
career
liberal arts
career
liberal arts
career
liberal arts
career

7. a.

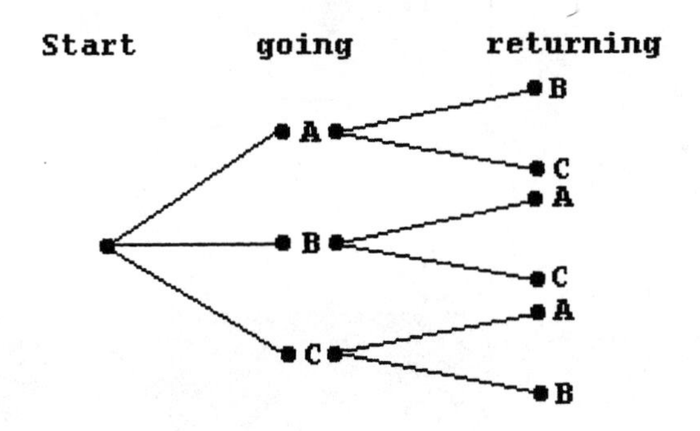
Start
going
returning
A
B
C
B
C
A
C
A
B

b. 6
c. 2

Venn Diagrams

1. a.

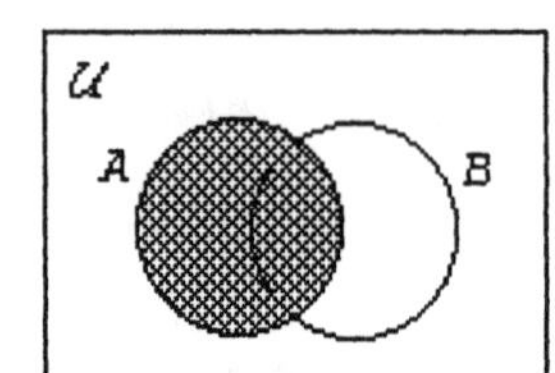

b.

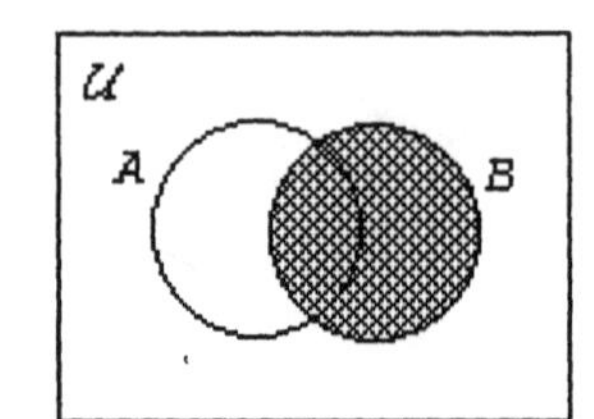

c.

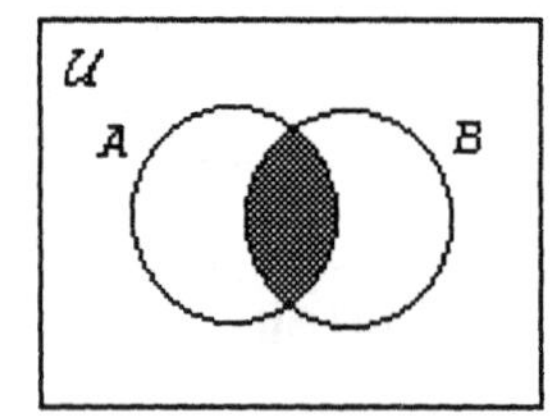

d.

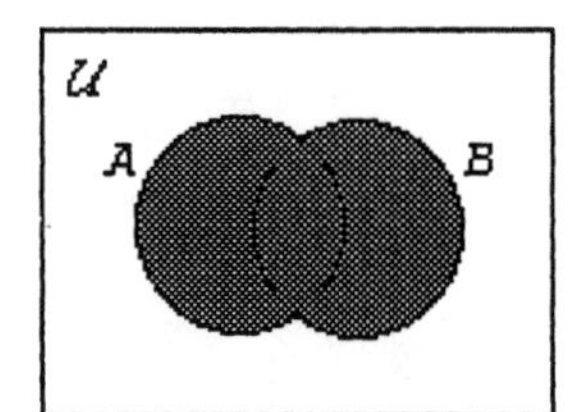

e.

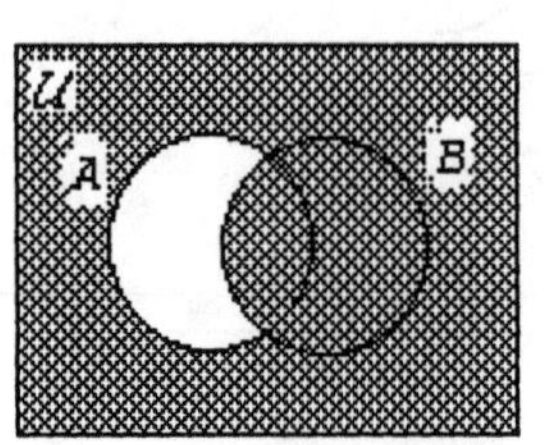

f.

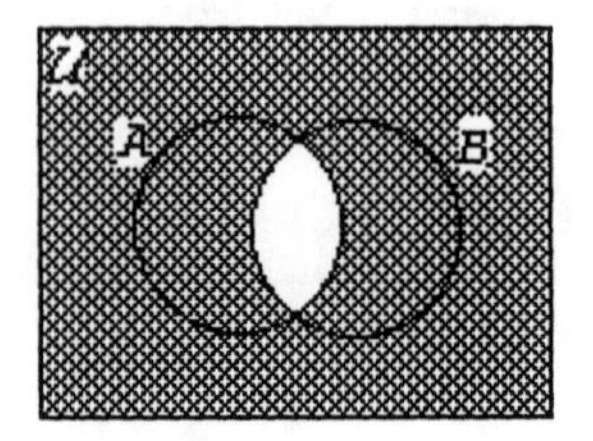

2. a.

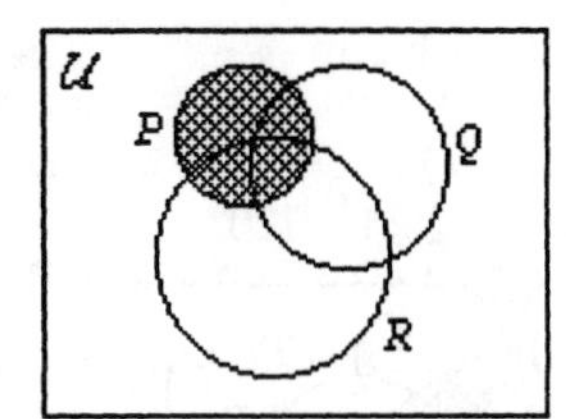

b.

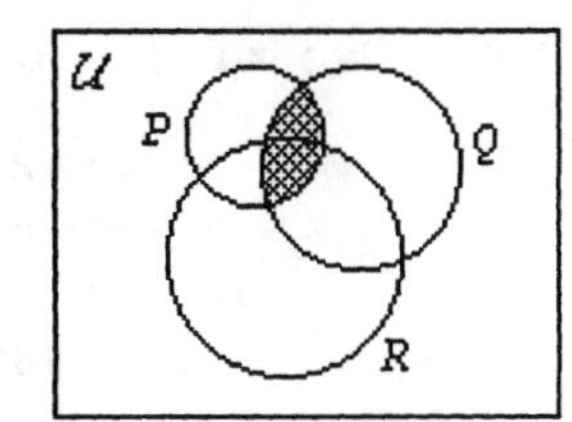

c.

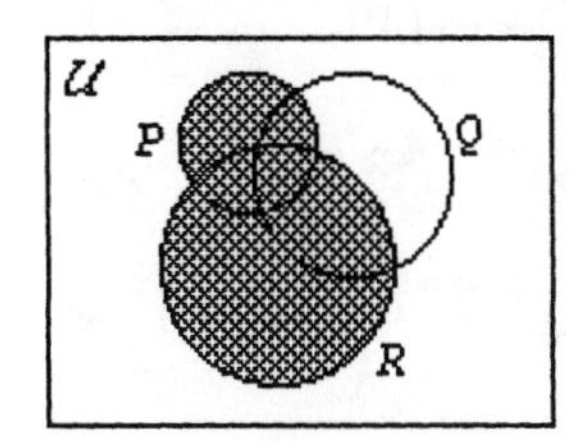

d.

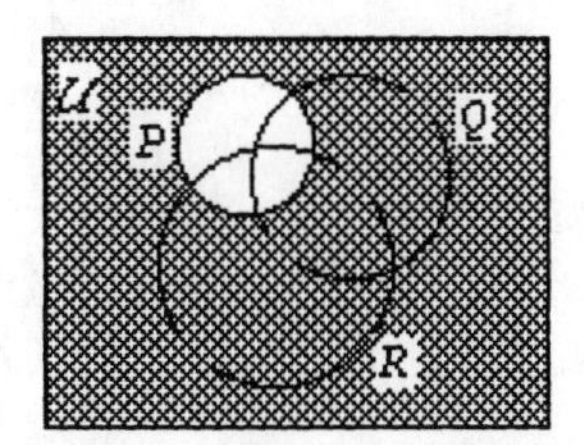

e.

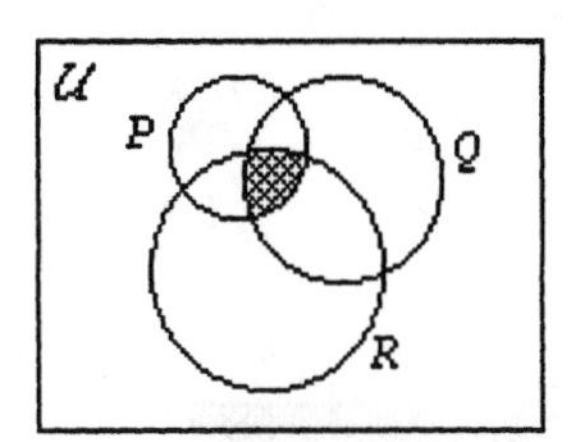

f.

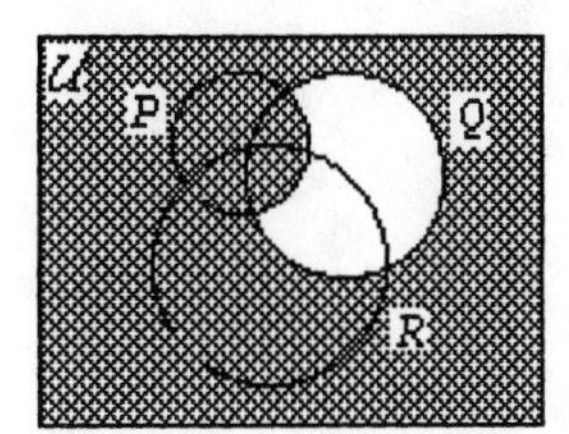

g.

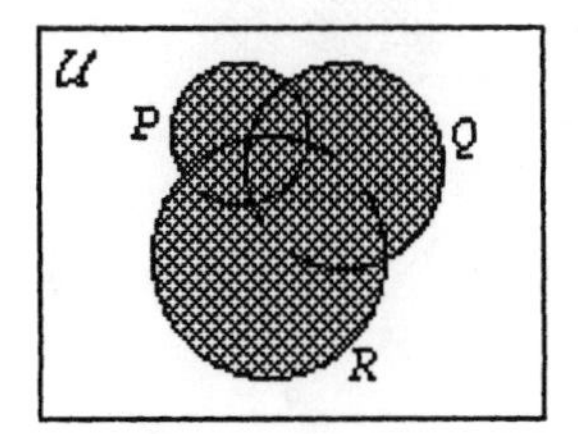

h.

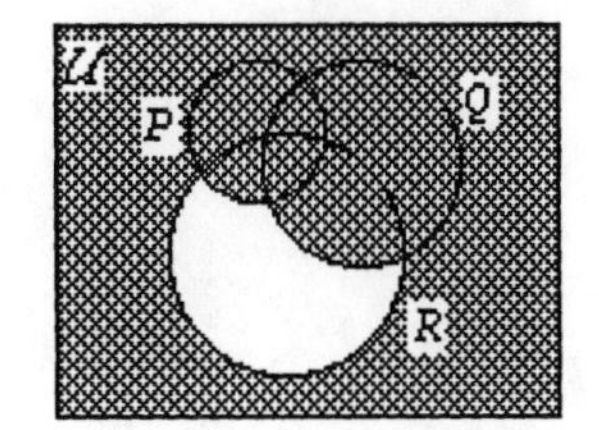

The Use of Factorial Notation

1. $4! = 4 \times 3 \times 2 \times 1 = 24$
2. $6! = 6 \times 5 \times 4 \times 3 \times 2 \times 1 = 720$
3. $8! = 8 \times 7 \times 6 \times 5 \times 4 \times 3 \times 2 \times 1 = 40320$
4. $(6!)(8!) = (720)(40320) = 29{,}030{,}400$
5. $\dfrac{8!}{6!} = \dfrac{8 \times 7 \times (6!)}{6!} = 8 \times 7 = 56$
6. $\dfrac{8!}{4!4!} = \dfrac{8 \times 7 \times 6 \times 5 \times (4!)}{4 \times 3 \times 2 \times 1 \times (4!)} = 2 \times 7 \times 5 = 70$
7. $\dfrac{8!}{6!2!} = \dfrac{8 \times 7 \times (6!)}{(6!) \times 2 \times 1} = 4 \times 7 = 28$
8. $2\left(\dfrac{8!}{5!}\right) = 2\left(\dfrac{8 \times 7 \times 6 \times (5!)}{5!}\right) = 2 \times 8 \times 7 \times 6 = 672$